What every aircraft owner needs to know about the design, operation, condition monitoring, maintenance and troubleshooting of piston aircraft engines

Mike Busch on Engines

Mike Busch A&P/IA

Copyright © 2018 by Michael D. Busch
All Rights Reserved.

ISBN-13: 978-1718608955
ISBN-10: 1718608950

Published in the United States by Savvy Aviator, Inc.
www.SavvyAviator.com

No part of this publication can be reproduced, stored in a retrieval system, or transmitted in any form or by any means, electronic, mechanical, photocopying, recording, scanning, or otherwise, except as permitted under the United States Copyright Act of 1976. Requests for permission should be addressed to Savvy Aviator, Inc. via email: permission@SavvyAviator.com.

Cover and interior design by Lynn Stuart.

This book is dedicated to the three wise men who taught me most of what I know about piston aircraft engines: George Braly, Jimmy Tubbs, and the late Bob Moseley. If you look closely, you'll find fingerprints of their generous mentorship on nearly every page.

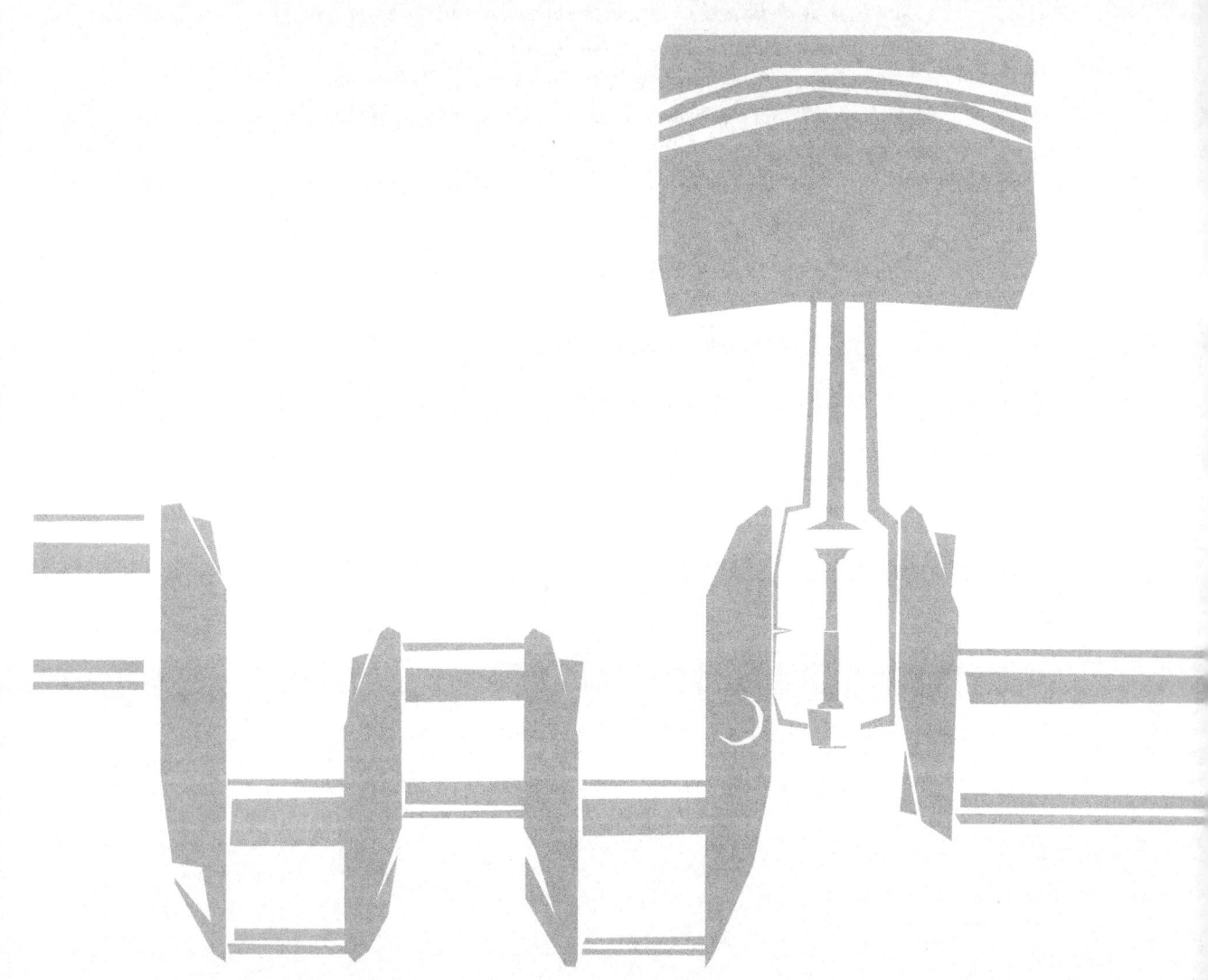

The creation of this book would not have been possible without the generous support of these patrons:

Century Club
Michael Hong
Davis Hunt
Michael Lepore
Michael Silvernagle
Daniel B. Wilson

Grand Benefactors
Garrison C. Cavell
Jim Elliott
Don H. La Ferney Jr.
Robert Lawrie
Joseph P. Soldo

Benefactors
Dane Coyle
James Gamble
Ronald Goro
Jeff Iskierka
Michael McCain
William McGlynn
Gary Mullen

Patrons
Luther Tuck Anderson
Robert Brahm
Darrell Chandler
Haydn Chandler
Jonathan Fisher
Michael Fishman
Ronald Gatewood
Sven Girsperger
Joe Goebel
Wiggy Greacen
Kevin Hanson
John Holman
David Huber
Martin Jansen
Mark Jensen
Brian Kindergan
William M Black
Peter Misunas
Rodolfo Paiz
Tim Rasmusson
Paul Reavis
James Riggs
Margaret Schindel
William Todd Self
Adam Smith
Phil Steklenski
Leslie James Story
Eric Westphal
Thomas Wiley
Ronald Williams
Richard Woodward
George C. Yendrey
Tommy Zamberlan

Sponsors
Ray Ackley
John Amdor
Michael Baraz
David Barnett
Melvyn Becerra
Richard Blain
Tom Bohannon
Dimitar Bojantchev
Robert Borger
Chuck Chamberlin
Douglas Cheney
Denis Cline
Tuck Colby
Rene de Wet
Todd Eichel
Jim Epting
Dwane Ferguson
Stephen G. Floyd
Steve Freedman
Hans Friedebach
John Michael Guidry
Stephen Hackney
Nick Hatzis
David B. Hill
Martin Hinshaw
Douglas Hornal
William Hounshell
Michael Hoyt
Scott Hunziker
John Hurst
Danford Jay
Ted Kapka
John Martin Keagy
Anthony I. Krause
Ronald Krohn
Raymond Landes
Kent Larson
Alan Levinson
Dean M. Johnson
Don Mack
Mark Marsiglio
Rick Matus
Randolph McClure
Walter Mercer
Daniel Mersich
Doug Morley
James Moss
J. Navarro
Dmitry Opolinsky
David Owen
Michael Parks
Kenneth Piller
Wayne Pratt
Jasean Rasnake
Bruce Riter
Frank Robinson
Darren Schehl
Chris schuldt
Dixon Smith
John L. Smith
Sidney R. Smith
Barrie Strachan
Reed Usrey
Jim Veazey
Noel Wade
Keith Walker
Charles G. Walker III
Matt Ward
Jeremy Whelchel
Simon Wippich
Jim Yares
Jeremy Zawodny

Associate Sponsors
James Adams
Moreno Aguiari
John Arnold
Allan Beiderman
Ian Brown
tim cantrell
Stephen Couzelis
Michael D'Angelo
Chris Dickson
Roger French
Frank Guarascio
Michael Haas
Alexander Sardo Infirri
Douglas J. Isern
Gregg Jackson
Steve Jones
Fazal Khan
Ryan R. Klems
Andrew Kullick
Rick Lafford
Jeff Lawson
Christian Meier
Ron Metcalf
Don Muncy
James Origliosso
John Pischl
Masoud Safa
Kenneth Schlehr
Mark Scott
Mark Sellers
Marcelo Silva
Dwight Small
Bernard Spaulding
Brian Stamper
John W. Stryker
Craig Wheeler
John Winter

VI ENGINES

Table of Contents

Prologue IX

Part I: Powerplant 101

Chapter 1   150-Year-Old Technology 3

Chapter 2   Energy and Efficiency 9

Chapter 3   Understanding EGTs 17

Chapter 4   Detonation and Pre-ignition 25

Part II: Top End

Chapter 5   Anatomy of a Cylinder 37

Chapter 6   Taking Cylinders to TBO 43

Chapter 7   Jug Economics 53

Chapter 8   Exhaust Valve Failures 59

Chapter 9   Preventing Exhaust Valve Failures 65

Chapter 10   Separation Anxiety 73

Part III: Bottom End

Chapter 11	Inside the Case	89
Chapter 12	Not-So-Plain Bearings	99
Chapter 13	Corrosion—Powerplant Enemy #1	109
Chapter 14	Cam Distress	117
Chapter 15	Making Metal	123
Chapter 16	Teardown Dilemma	129

Part IV: Key Systems

Chapter 17	Lubrication	137
Chapter 18	Ignition TLC	151
Chapter 19	Where Fuel and Air Meet	161
Chapter 20	Continental Fuel Injection	169
Chapter 21	Fuel Gone Wrong	185
Chapter 22	Turbosystems	197

Part V: Powerplant Management

Chapter 23	Leaning Basics	209
Chapter 24	Advanced Leaning	219
Chapter 25	Making Engines Last	227
Chapter 26	Temperature, Temperature, Temperature	239
Chapter 27	The Whys and Hows of Preheating	249
Chapter 28	The CamGuard Chronicles	257

Part VI: Condition Monitoring

Chapter 29	How Healthy Is Your Engine?	265
Chapter 30	Assessing Cylinder Condition	271

Chapter 31	Oil Analysis	
Chapter 32	Big Data	

Part VII: Troubleshooting

Chapter 33	The Art of Troubleshooting	313
Chapter 34	Flight Test Profiles	327
Chapter 35	Rough Engine	335
Chapter 36	Diagnosing a Temperamental Ignition	341
Chapter 37	High Oil Consumption	349
Chapter 38	Troubleshooting Turbosystems	357

Part VIII: Maintenance

Chapter 39	Oil Changes	373
Chapter 40	Spark Plugs	381
Chapter 41	The Dark Side of Maintenance	393
Chapter 42	Watch Your Language!	399
Chapter 43	A Little Dab'll Do Ya… In	405
Chapter 44	Fear and Balderdash	413
Chapter 45	The Perils of Cylinder Work	419

Part IX: Overhaul

Chapter 46	Maintenance Intervals	431
Chapter 47	Engine TBOs	439
Chapter 48	Deciding When to Overhaul	449
Chapter 49	Choosing an Overhaul Shop	457
Epilogue	Some Things Have Changed, Some Haven't	463

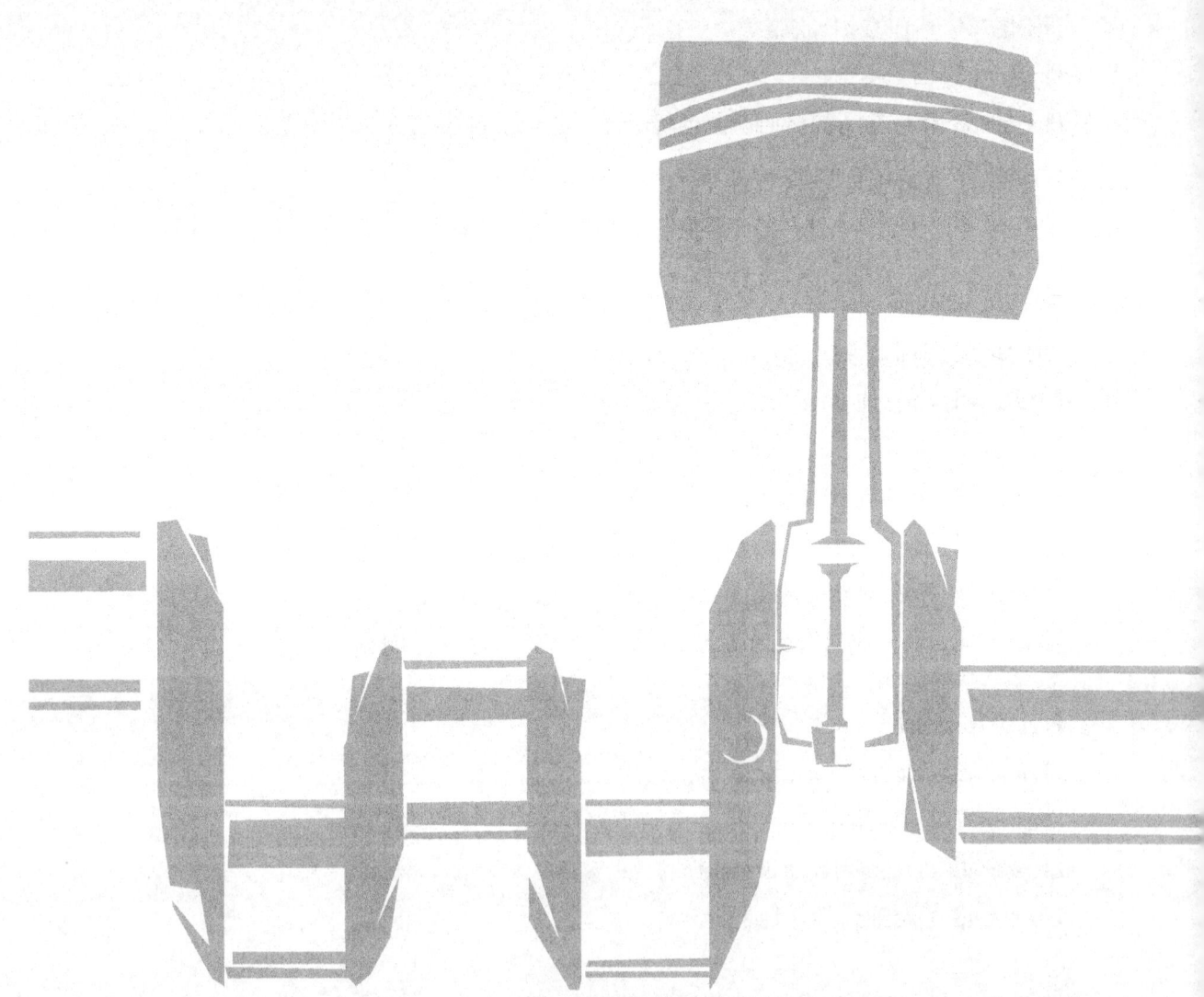

Prologue

This book is the second of a planned three-volume set.

The first volume, titled *Manifesto: A Revolutionary Approach to General Aviation Maintenance*, was published in July 2014. It's a pithy 112-page book articulating my philosophy of reliability-centered aircraft maintenance and proactive aircraft ownership, the credo that permeates all my writing, speaking and teaching activities, as well as the services offered by my company Savvy Aviation, Inc. The book has received rave reviews on Amazon.com (4.7 out of 5 stars), and remains its top-selling non-textbook title in the aircraft maintenance category.

This book is heftier, nearly 500 pages long, and contains far more grease-under-the-fingernails detail and real-life war stories. My purpose in creating this book (and its predecessor *Manifesto*) is to help piston GA owners become more knowledgeable about their aircraft, empowering them to operate in a safe yet cost-efficient fashion.

I've been a published aviation author since 1970, but almost everything I've written has been in the form of magazine articles, webinars and blog posts—all media that tend to be transitory and ephemeral. Hopefully these books will be more enduring. I plan to complete this series with a third volume compiling my articles about airframe maintenance, aircraft ownership, and GA flying. Stay tuned.

What's in this book

This book covers four-stroke spark-ignition piston aircraft engines inside and out. It starts with the origins of this 150-year-old engine technology, then moves on to the nuts-and-bolts detail in parts devoted to the top end (cylinder assemblies), bottom end (inside the crankcase), and major subsystems (lubrication, ignition, fuel and turbocharging).

The spotlight then shifts to powerplant management, including a discussion of combustion theory and what it can teach us about how to operate these powerplants to achieve optimum efficiency and maximum longevity. The controversial subject of leaning is covered in considerable detail, as is temperature control, preheating, and corrosion prevention.

Next is a series of chapters about how to use modern noninvasive condition-monitoring tools to assess engine health. These tools include borescope inspections, spectrographic oil analysis, scanning electron microscopy of oil filter contents, and digital engine monitor data analysis.

As loyal readers already know (and newcomers will discover), I'm something of a fanatic about troubleshooting. The next part of this book focuses on the art of diagnosing various common engine maladies—rough running, high oil consumption, ignition issues, etc. I talk about what mechanics could learn about this subject from their doctors—a methodology called "differential diagnosis"—and illustrate how it can be used to pinpoint the cause of problems without resorting to guesswork.

Next up is a series of chapters about engine maintenance. They discuss preventive maintenance that aircraft owners are permitted to do on their own reconnaissance without A&P supervision, as well as critical maintenance tasks (such as replacing cylinders) that even experienced professional aircraft mechanics often screw up. I'm a big fan of owners who take an active role in the maintenance of their aircraft, whether hands-on wrench swinging or deciding what work their A&Ps should or should not perform. Indeed, I was such a maintenance-involved owner long before I became an A&P/IA.

The final chapters address the costly event that every aircraft owner dreads most—engine major overhaul—and encourages readers not to think of it as an inevitability at a manufacturer-specified TBO, but rather as a condition-based decision to be made based on hard data and reasoned judgment. You'll learn how I took the engines on my turbocharged piston twin to more than twice the recommended TBO, and why you should aspire to do the same.

Acknowlegements

This book was originally intended to be an anthology of the engine-related magazine articles, webinars and blog posts that I've written over the past 25 years or so. I thought it would be simple enough to collect all these pieces and organize them into a coherent sequence. I somehow managed to sweet-talk my colleague Colleen Keller—A&P mechanic, aircraft owner, and Reno racer—into serving as my editor, a most fortunate choice indeed. Lynn Stuart (Lynn Stuart Graphic Design)—whose book design for *Manifesto* was nothing short of magnificent—agreed to do the cover design and typography for this book as well.

It soon became apparent to the three of us that this was going to be a big book—nearly twice our initial estimate of 250 pages—and a much larger effort than we originally thought. A lot of changes have occurred since the articles were published and the webinars conducted, so bringing everything up to date was a big job. There was also a lot of overlap between articles that needed to be identified and eliminated. Finally, many of the original photos, charts and other graphics didn't survive the conversion from color to grayscale and needed to be recreated from scratch.

All in all, this book turned out to be a year-long project, very much a cooperative effort between Colleen, Lynn and me. Colleen kept me honest and accurate. Lynn made me look good in print. But I wrote and final-proofread this book myself, so responsibility for any errors, omissions, miscalculations, blunders and glitches herein (and I'm sure you eagle-eyed readers will find quite a few) is mine and mine alone.

—Michael D. Busch, May 10, 2018

PART I
Powerplant 101

1
150-Year-Old Technology

Most of us are still flying (and driving) behind powerplant technology that dates from the 19th century.

The original four-stroke Otto-cycle internal-combustion engine was patented in 1862 by a Frenchman named Alphonse Beau de Rochas. More scientist than engineer, de Rochas never actually built an operational engine. The first working prototype was built by a German engineer named Nikolaus A. Otto, who was ultimately rewarded for his efforts by winning a gold medal at the Paris Exposition in 1867 and having the four-stroke cycle named after him. The first practical Otto-cycle engines were built by another, better known German engineer named Gottlieb Daimler, who together with his lifelong business partner Wilhelm Maybach built a one-cylinder automobile engine in 1885 and a two-cylinder engine in the now-classic "V" configuration in 1889. Daimler died in 1900, and in 1926 his company Daimler Motors Corporation merged with Benz & Co.—founded by two-stroke engine pioneer Karl Benz—to create Daimler-Benz AG.

The basic power-generating component of an internal-combustion engine is the cylinder assembly, whose major components are a cylinder, a piston, and a pair of valves

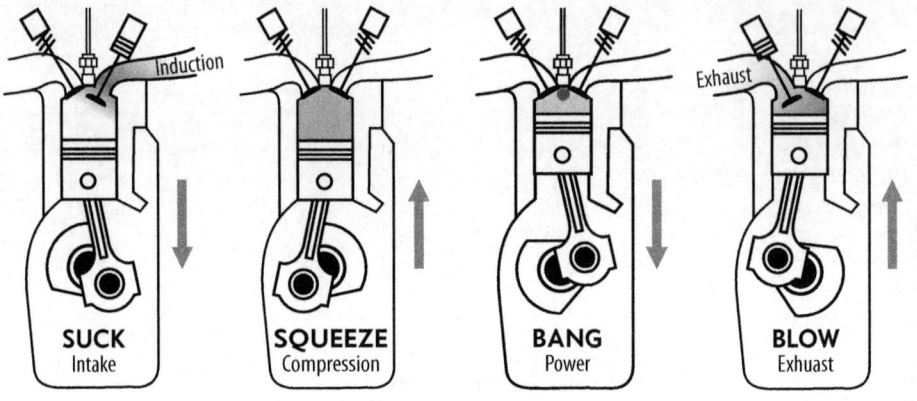

The four-stroke Otto cycle.

or ports (intake and exhaust). Each up or down movement of the piston within the cylinder is termed a "stroke."

Suck, squeeze, bang, and blow

An Otto-cycle engine employs an operating cycle composed of four strokes, with each successive stroke associated with a different phase of the cycle. The four phases are usually referred to as intake, compression, power, and exhaust—or colloquially, suck, squeeze, bang, and blow:

Suck: During the intake stroke, the piston moves away from the cylinder head with the intake valve open, creating a partial vacuum that sucks a combustible mixture (in our case, air containing atomized gasoline droplets) into the cylinder.

Squeeze: During the compression stroke, the piston moves toward the cylinder head with both valves closed, compressing the air-fuel charge into a much smaller volume, increasing its pressure and temperature, and making it more capable of combustion. The difference in volume of air-fuel charge between the start of the compression stroke (piston all the way down) and the end of the compression stroke (piston all the way up) is termed the "compression ratio." Most aircraft engines have very conservative compression ratios (between 7:1 and 8.6:1); automotive engines usually have compression ratios between 8:1 and 10:1, racing engines up to 12:1, and diesel engines

14:1 or more. The greater the compression ratio, the more efficient the engine at converting chemical energy into mechanical energy. (Piston aircraft engines aren't particularly efficient.)

Bang: During the power stroke, the air-fuel charge is ignited by an electrical spark (or by the heat of compression in diesel engines). Both valves remain closed, so the rapidly increasing pressure of the burning air-fuel charge drives the piston forcefully away from the cylinder head, converting chemical energy to mechanical energy. As the piston moves down in the cylinder and the volume of the air-fuel charge increases, its pressure and temperature decrease.

Blow: During the exhaust stroke, the piston moves toward the cylinder head with the exhaust valve open, allowing what remains of the spent air-fuel charge to exit the cylinder and be expelled through the exhaust system. Because piston aircraft engines are not very efficient, substantial energy remains in the exhaust gas as it exits the cylinder. In a normally-aspirated engine, this energy is simply wasted; in a turbocharged engine, some of the energy is used to spin a compressor and raise the pressure of the engine's induction air, allowing the engine to produce more power (especially at altitude).

The more, the ~~merrier~~ smoother

While the Otto cycle defines what's going on within a single cylinder assembly, most piston engines have more than one cylinder. That's because a fundamental limitation of the Otto cycle is that it only produces power 25% of the time. Consequently, the one-cylinder Otto-cycle engines commonly used on lawnmowers and small motorcycles tend to leave a lot to be desired in the smoothness and vibration departments.

The obvious solution is to have four cylinders arranged so that one is always in its power stroke at any given time; this approach results in a much smoother-running engine with far less vibration. Even greater smoothness is possible by adding additional cylinders and sequencing them so that one power stroke begins before the previous one finishes.

Numerous cylinder arrangements have been tried. Most automotive engines use either in-line (straight) or V-type layouts (for compactness), while most aircraft engines use either horizontally opposed or radial layouts (for improved air cooling).

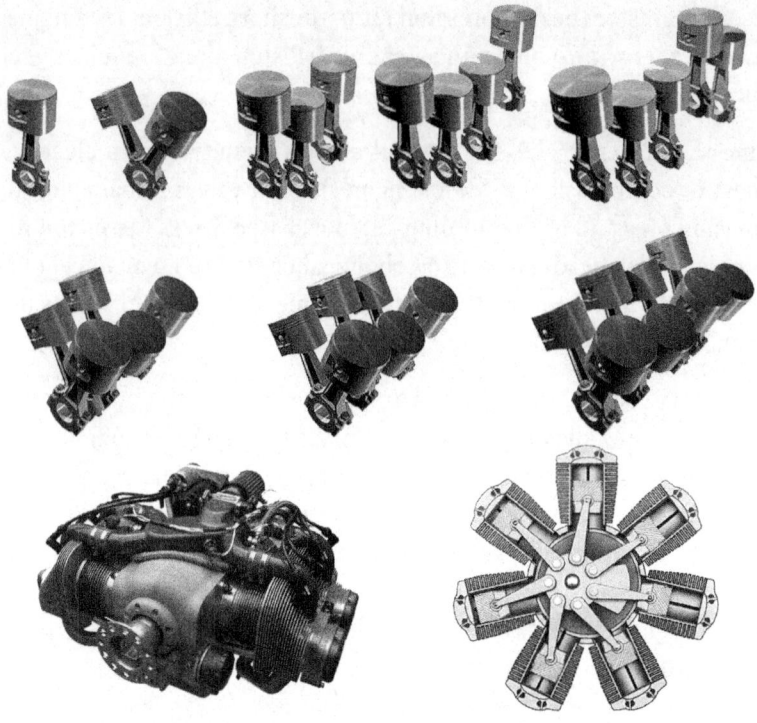

Otto-cycle four-stroke engines commonly have four or more cylinders arranged so that at least one is in its power stroke at any given time.

The most common configurations in piston-powered GA engines are 4 or 6 cylinders horizontally opposed.

Pressure and volume

Although the four-stroke Otto cycle is conceptually simple, what actually takes place inside the cylinder during each cycle is remarkably complex, as are the critical timing relationships of piston position, pressure, temperature, valve opening and closing, and ignition. The more you understand about the combustion event and timing relationships, the better job you will be able to do of managing your powerplant, optimizing your power and mixture settings, and troubleshooting any engine problems that may arise. With that in mind, let's explore the Otto cycle a bit more deeply.

An excellent tool for visualizing what goes on during the Otto cycle is a "P-V diagram" that plots combustion chamber pressure and volume. Look at the figure below and let's work through the four strokes of the Otto cycle:

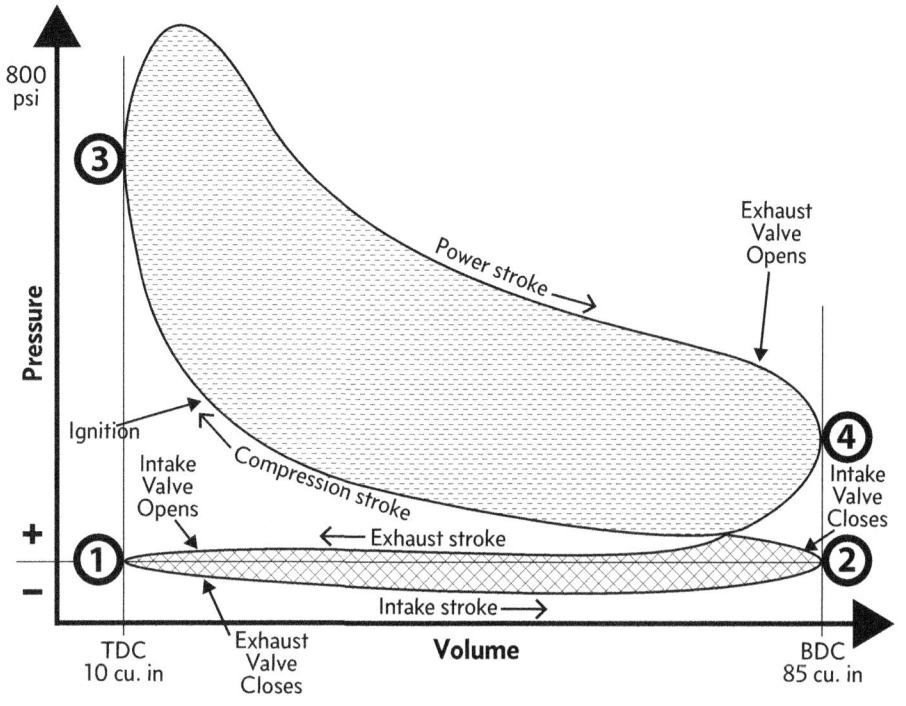

This P-V diagram plots pressure and volume of the Otto cycle.

Suck: Beginning at point (1) on the diagram, the piston starts at the top of its travel ("top dead center" or TDC) and moves to the bottom of its travel ("bottom dead center" or BDC). The intake valve is fully open, the exhaust valve closes, and the descending piston creates suction that draws the air-fuel charge into the cylinder.

Squeeze: At point (2), the piston reverses direction and moves from BDC to TDC. The intake valve closes, and the air-fuel charge is compressed—for example, from a volume of 85 cu.in. to 10 cu.in. (a compression ratio of 8.5:1)—causing the pressure and temperature in the combustion chamber to rise accordingly. As the piston approaches TDC—typically 20° to 25° of crankshaft rotation before it gets there—

the ignition system fires the spark plugs, and the air-fuel charge starts to burn, causing the pressure and temperature to increase even faster.

Bang: At point (3), the piston reaches TDC and reverses direction again, moving toward BDC. Meantime, the combustion of the air-fuel charge accelerates, reaching a maximum pressure and temperature at about 15° to 20° of crankshaft rotation after TDC. This is the point of peak internal combustion pressure (ICP), which is typically 800 psi in a normally aspirated engine and as much as 1,000 psi in a turbocharged engine. This high pressure pushes the piston down towards BDC rather forcefully: 800 psi pressing on a 5¼-inch piston produces more than 17,000 pounds of force. As the piston descends and the air-fuel charge expands, its pressure and temperature drop considerably as chemical energy is converted to mechanical energy. Shortly before the piston reaches BDC, the exhaust valve starts to open. Since the pressure in the cylinder is still considerably greater than outside ambient, exhaust gas starts flowing out the exhaust valve into the exhaust system in a process termed "blowdown."

Blow: At point (4), the piston reaches BDC and reverses direction once more, moving toward TDC. As the piston rises, it compresses the remaining fuel-air charge and forces it out the exhaust valve. Shortly before the piston reaches TDC, the intake valve starts to open, so that it can be fully open by the time the piston reaches point (1) and reverses direction to start the intake stroke. The brief period during which both intake and exhaust valves are open here is known as the "valve overlap interval."

Because of their low compression ratios, spark-ignition piston aircraft engines are unusually inefficient as Otto-cycle engines go. They typically convert only about one-third of the fuel's chemical energy to mechanical energy, and waste about one-half of it out the exhaust and the remaining one-sixth in radiated energy from cylinder fins and oil cooler. The EPA-mandated move to unleaded avgas won't help this one bit, since the octane-enhancing properties of tetraethyl lead (TEL) are what made it possible for spark ignition Otto-cycle gasoline engines to have higher compression ratios and greater efficiency and power output. Diesel engines with their much higher compression ratios represent our best hope for more efficient piston aircraft engines in the future.

2
Energy and Efficiency

Why are our piston aircraft engines so @#$%*! inefficient?

Our piston aircraft engines convert chemical energy into mechanical work, but they don't do it very efficiently. It turns out that only about one-third of the energy contained in the 100LL we burn winds up getting to the propeller and doing useful work to propel us through the air. The remaining two-thirds winds up getting lost between the fuel truck and the prop hub. At today's stratospheric avgas prices, that's pretty depressing.

Let's do the math

Consider a Continental IO-550 engine rated at 300 hp. If the fuel system is set up properly, fuel flow at maximum takeoff power is about 26.6 gallons/hour or 156 pounds/hour. How much chemical energy does that fuel provide?

We can calculate that. 100LL is rated at a "minimum lower heat value" of 18,700 BTUs per pound. Let's convert that figure into something more meaningful to pilots like you and me:

1. divide 156 pounds/hour by 3,600 seconds/hour to get .0433 pounds/second.

2. multiply by 18,700 (the thermal content of 100LL in BTUs per pound) to get 810 BTUs/second.

3. multiply by 1.414 (the horsepower equivalent to one BTU/second) to get 1,145 hp.

Does this mean that your IO-550 consumes 100LL with thermal energy equivalent to 1,145 hp, and yet produces only 300 hp of output power? Unfortunately, that's EXACTLY what it means—and that works out to a miserable thermal efficiency of 26%. Good grief!

Should an IO-550 really be drinking this much fuel? Well, we can calculate that, too:

1. At takeoff power, the engine is turning at 2,700 RPM. Since it's a four-stroke engine, each power cycle requires two crankshaft revolutions. Therefore, the engine is operating at 1,350 power cycles/minute.

2. The displacement of the engine is 550 cubic inches, or 0.32 cubic feet. Due to induction system losses, however, the engine's "volumetric efficiency" is only about 85%, so it "inhales" only about 0.27 cubic feet of air per power cycle.

3. Multiplying 1,350 power cycles/second times 0.27 cubic feet of air/cycle, we calculate that the engine should inhale 365 cubic feet of air/minute.

4. Sea level air under standard atmospheric conditions weighs 0.0765 pounds/cubic foot. Therefore, the engine breathes 27.9 pounds of air/minute.

5. Best power mixture requires an air-fuel ratio of about 12.5 to 1 by weight. Dividing 27.9 by 12.5, we get a fuel burn of 2.23 pounds of fuel/minute—or multiplying by 60, we get 134 pounds/hour or 22.3 gallons/hour calculated fuel flow at best power mixture.

The actual book fuel flow figure of 26.6 gallons/hour or 156 pounds/hour is higher than this calculated value because of the unusually rich mixture required to provide adequate detonation margins at full takeoff power.

What about LOP?

Surely engine efficiency is much better at cruise power settings with aggressively lean mixtures, right? Let's take a look. An IO-550 engine running at 65% power and operating "lean of peak" (LOP) uses approximately 13 gallons/hour or 78 pounds/hour. What kind of thermal efficiency does that represent? Repeating the calculations:

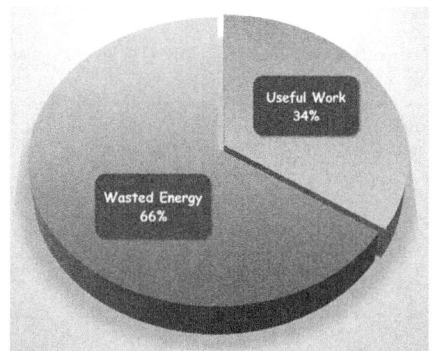

Our piston aircraft engines are woefully inefficient.

1. divide 78 pounds/hour by 3,600 seconds/hour to get .0217 pounds/second.

2. multiply by 18,700 (the thermal content of avgas in BTUs/pound) to get 405 BTUs/ second.

3. multiply by 1.414 (the horsepower equivalent to one BTU/second) to get 573 hp.

So even at LOP cruise, the IO-550 consumes 573 hp worth of go-juice in order to produce 195 hp (65% of 300), for an efficiency of about 34%. Definitely better, but certainly nothing to write home about.

Why so wasteful?

Here's one breakdown of efficiency losses (from *Performance of Light Aircraft* by John T. Lowry, Ph.D.):

Otto cycle efficiency: The thermodynamic efficiency of a four-stroke internal combustion engine is limited by the compression ratio (i.e., the ratio of cylinder volumes as the piston moves from bottom-dead-center to top-dead-center). The higher the compression ratio, the greater the efficiency. For an IO-550 with a compression ratio of 8.5:1, the Otto-cycle efficiency works out to about 57.5%.

Volumetric efficiency: As mentioned earlier, the ability of the engine to breathe in its full theoretical displacement of air during each power cycle is restricted by a variety of pressure losses at various points in the induction system: air filter, throttle body,

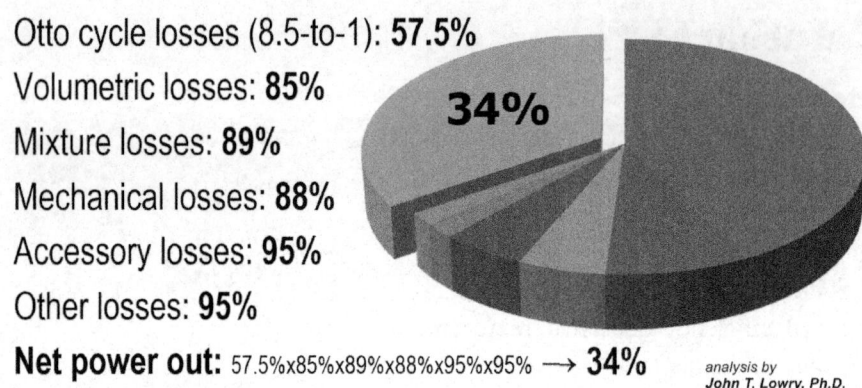

Functional breakdown of efficiency losses by John T. Lowry, Ph.D..

intake manifold, and intake valves. For most of our engines, volumetric efficiency is around 85%, bringing total efficiency down to 57.5% x 85% or 49%.

Mixture losses: Optimum fuel efficiency occurs at very lean mixture settings (so-called "best economy mixture") with an air-fuel ratio in the vicinity of 18-to-1 by weight. Best economy mixture occurs very LOP, however, and most pilots don't operate that lean. (Not to mention that many engines won't run smoothly that lean.) Many pilots operate rich of peak EGT in the vicinity of best-power mixture, at an air-fuel ratio around 12.5-to-1, which provides a fuel efficiency that's only 70% of optimum. Even if you operate slightly LOP (let's say at an air-fuel ratio of 16-to-1), your efficiency is just 89% of optimum, and that brings total efficiency down to 49% x 89% or 44%.

Mechanical losses: Friction losses involving the reciprocating and rotational parts inside the engine consume a significant amount of power that could otherwise be delivered to the propeller. Mechanical efficiency varies with engine speed (lower losses at lower RPM), but is typically around 88%, bringing total efficiency down to 44% x 88% or 38%.

Accessory losses: A certain amount of engine power is consumed driving accessories such as magnetos, fuel pumps, alternators, vacuum pumps, hydraulic pumps, air conditioning compressors, etc. Figure this robs 5% of the remaining power, bringing total efficiency down to 36%.

Other losses: This includes a grab bag of other inefficiencies including blow-by past the piston rings, unburned hydrocarbons in the fuel, humidity in the air, backpressure in the exhaust system, and so forth. Figure another 5% loss, bringing total efficiency down to 34% (which agrees with our earlier figure for an IO-550-B at 65% LOP).

Thermal and chemical losses

A quite different analysis (from *Fundamentals of Power Plants for Aircraft* by Joseph Liston) analyzes the various thermal and chemical losses suffered by a piston aircraft engine.

We've already seen that an internal-combustion engine is incapable of converting all the heat of combustion into mechanical energy, limited primarily by its finite compression ratio. The rest of the heat of combustion, as well as a small amount of additional heat generated by friction, is lost through the engine's exhaust and cooling systems.

There are also some chemical losses. In theory, the combustion of pure hydrocarbon fuel at stoichiometric mixture should produce nothing but carbon dioxide (CO_2) and water (H_2O). In reality, however, there's always some sulfur in the fuel, which is transformed by combustion to sulfur dioxide (SO_2) and sulfuric acid (H_2SO_4). If the mixture is a bit on the rich side, the exhaust also contains carbon monoxide (CO)

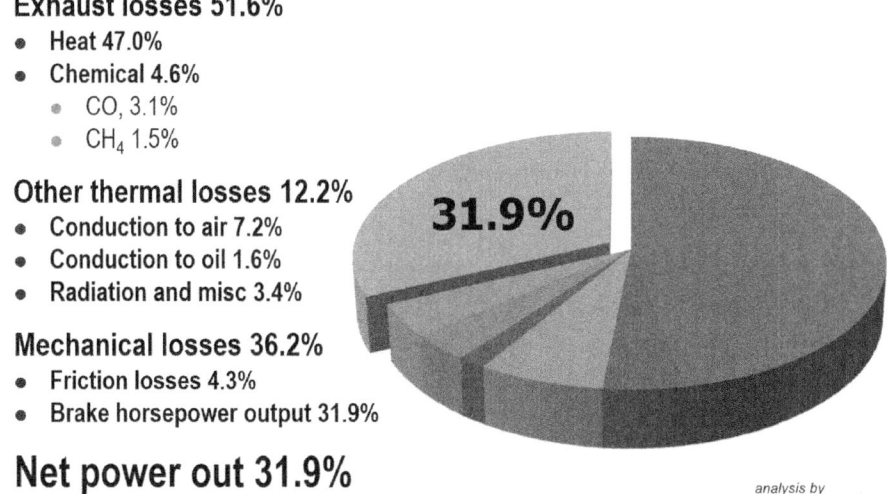

Thermal and chemical breakdown of efficiency losses by Joseph Liston.

which results from incomplete combustion, as well as some unburned carbon particles and some methane gas (CH_4).

Here's how Liston breaks this all down:

```
Fuel energy ....................... 100.0%
Exhaust .............................. 51.6%
    Heat..................... 47.0%
    Chemical................. 4.6%
    CO....................... 3.1%
    CH4...................... 1.5%
Other thermal ........................ 12.2%
    Conduction to air .......... 7.2%
    Conduction to oil .......... 1.6%
    Radiation and misc. ........ 3.4%
Mechanical ........................... 36.2%
    Friction losses............. 4.3%
    Brake horsepower output ... 31.9%
```

Again, this figure agrees pretty well with our earlier 34% figure for the IO-550-B at 65% LOP cruise.

Can we do better?

What, if anything, can we do to improve this dismal efficiency? Well, don't expect any miraculous improvements of large magnitude. But every little bit helps, and there are certainly a few areas where the potential exists for improvement:

Otto cycle efficiency: As we've seen, the basic thermodynamic efficiency of an internal combustion engine is a function of compression ratio. Unfortunately, high-compression engines have traditionally required high-octane gasoline in order to avoid detonation, and high-octane gasoline is fast becoming unobtainable because of the campaign to eliminate tetraethyl lead (TEL) from avgas. Consequently, the trend in recent years has been toward lower compression ratios that are compatible with low-lead or unleaded fuel. While this may be wonderful for the environment, it sure doesn't help the thermodynamic efficiency of our engines.

One bright light on the horizon is the prospect of moving from fixed-timed magnetos to sophisticated, computerized electronic ignition systems capable of protecting engines against detonation by varying ignition timing. The incorporation of variable ignition timing and detonation sensors should permit the use of higher compression ratios even with unleaded fuel. It may take a few more years before any such systems make it through FAA certification, but the prospects for improved efficiency are significant.

Even more exciting is the recent advent of certified Diesel engines for piston aircraft, which run on Jet A and have 18:1 compression ratios that offer much greater thermal efficiency than any spark ignition gasoline engine.

Volumetric efficiency: Small improvements in this area are possible through the use of tuned induction systems, large intake valves, venturi-style valve seats, ram recovery airscoops, and turbocharging. Auto engines have even gone to multiple intake valves per cylinder, but the weight and complexity might make this impractical for aircraft engines.

Mixture losses: Major strides have already been made in this area, partially through pilot education to encourage the use of lean mixture settings, and partially through improvements to engine instrumentation and mixture distribution to facilitate operation at or near best economy mixture (i.e., considerably LOP).

Mechanical losses: The biggest thing that can be done to reduce mechanical losses is for pilots to cruise at low RPM and high manifold pressure, rather than vice-versa. Small additional gains are possible through the use of high-lubricity synthetic oil to reduce friction losses, but the leading all-synthetic oil (Mobil AV 1) was pulled off the market in the 1990s due to its inability to control lead deposits, and even semi-synthetics like Aeroshell 15W-50 have lead-deposit problems, particularly in small-sump engines like the ones used in the Cessna TTx and Cirrus SR22. When the lead is ultimately removed from avgas, all-synthetic oils may come back in favor for piston aircraft engines.

Accessory losses: The conversion to electronic ignition systems may also provide some small benefits by eliminating the mechanical losses involved in driving dual magnetos, although this may be partially offset by the requirement for electronic-ignition engines to have dual alternators. The trend toward all-electric airplanes with no pneumatics or hydraulics may also help slightly.

For now, the best thing you can do to improve efficiency is to lean aggressively (considerably LOP if feasible), and to cruise at low RPM and high MAP rather than vice-versa. In the foreseeable future, further improvements may be possible through the use of variable-timing electronic ignition systems and installation of higher-compression pistons. Efficiencies in the area of 40% are possible, but don't expect much more than that from spark-ignition engines, at least any time soon.

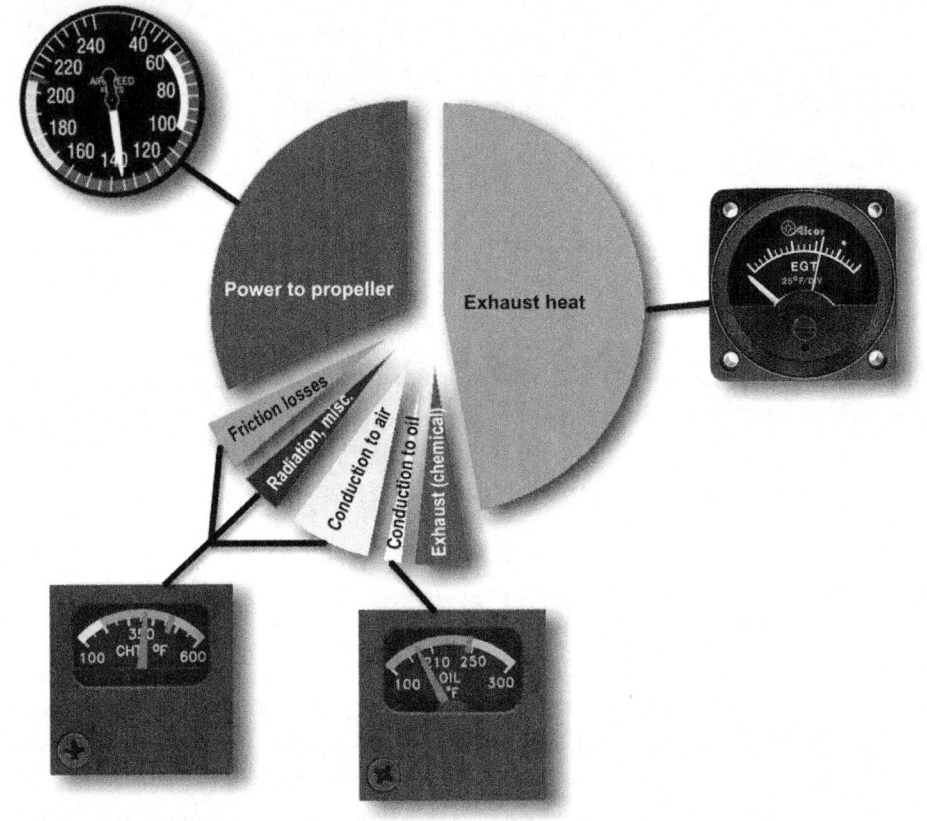

In the cockpit, we can see most of these losses on the EGT, CHT, and oil temperature gauges. Net power output affects what we see on the airspeed indicator.

3
Understanding EGTs

There are a lot of misconceptions about EGT. Let's see if we can clear some of them up.

These days, with the proliferation of modern digital engine monitors, pilots of piston-powered aircraft seem to be fixated upon exhaust gas temperature (EGT). Scarcely a day goes by that I don't receive a phone call or email asking some EGT-related question.

Pilots will send me a list of EGT readings for each of their cylinders and ask me if I think they look okay, whether I think their EGTs are too high, what maximum EGT limit I recommend, why their EGTs seem to be higher in the winter than in the summer, or why the EGTs on their 1972 Cessna 182 are so much higher than the ones on their friend's 1977 model. They'll voice concern that the individual cylinders on their engine have such diverse EGT readings, worry that the spread between the highest and lowest EGT is excessive, and ask for advice on how to bring them closer together. They'll complain that they are unable to transition from rich-of-peak (ROP) to lean-of-peak (LOP) operation without producing EGTs that are unacceptably high.

Each of these questions reveals a fundamental misunderstanding of what EGT measures, what it means, and how it is interpreted. Let me attempt to clear up some of this confusion by asking you, dear reader, to forget everything you thought you knew about EGT and start at the beginning.

A little history

When petroleum engineer Al Hundere introduced the very first EGT instrumentation for piston aircraft engines in the 1960s, he did something quite clever. The analog EGT gauges manufactured by his company Alcor (and widely installed by Beech, Cessna, Mooney, Piper, and other aircraft manufacturers of the day) had no absolute temperature markings, just a series of unlabeled tick-marks spaced 25°F apart, with an asterisk at 80% of full-scale. As the pilot leaned the engine, the EGT needle rose to a peak value on the meter, then started to fall off. The pilot would note where the needle peaked, then would richen until the needle dropped by the desired number of tick marks (e.g., three ticks for 75°F). The gauge provided the pilot no way to determine the *absolute* value of EGT (e.g., 1,475°F), but only its *relative* value (e.g., 75°F rich-of-peak).

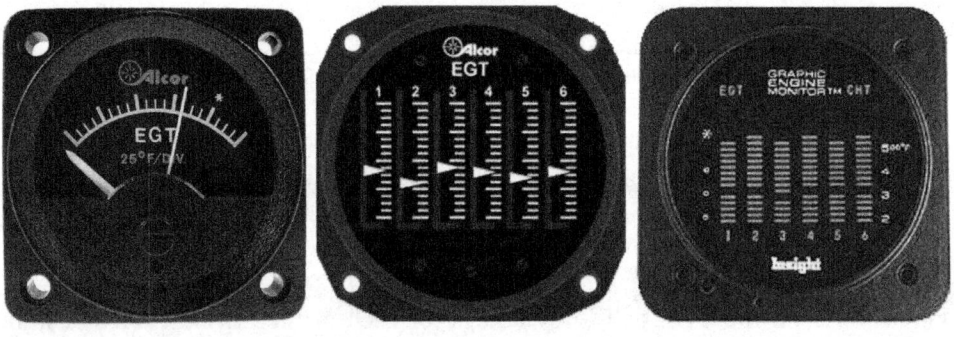

The original Alcor EGT gauge provided no absolute temperature information; the same was true of early engine analyzers and even the original Insight GEM 601/602.

Hundere understood that the absolute value of EGT is not particularly meaningful, and that presenting this information to the pilot would simply be a distraction. Since the Alcor EGT gauges provided no absolute temperature information, pilots never worried about whether their EGTs were too high or what the maximum EGT limit should be—and that was a good thing.

The first probe-per-cylinder engine analyzers introduced by Alcor and Bill Simkinson's KS Avionics were arrays of vertical analog meter movements, and also provided only relative EGT information to the pilot. When John Youngquist introduced the original Insight graphic engine monitor (GEM), its novel LED bar-graph display also provided only relative EGT.

Things started getting confusing when Electronics International introduced its "Ultimate Scanner" that provided digital (rather than graphical) readouts of EGT, and touted its 1°F accuracy as being far superior to the 25°F granularity of the GEM's bar-graph. (Never mind that the absolute EGT values it was reporting

E.I.'s US-8 Ultimate Scanner displays absolute digital EGT information, as does JPI's EDM-700, and subsequent generations of digital engine monitors have all done the same.

so accurately were essentially meaningless.) Not to be outdone, J.P. Instruments introduced its EDM-700 that featured both a GEM-like bar graph and an Ultimate Scanner-like digital readout on the same instrument. The EDM-700 was a smashing market success, and forced both E.I. and Insight to respond with similar products (the UBG-16 and GEM 610, respectively).

Now pilots were being presented with precise digital values of absolute EGT, scary temperatures in the 1300s, 1400s, 1500s and even 1600s Fahrenheit. Few understood what these temperatures meant, but most assumed that—as with most other temperatures in aviation (CHT, TIT, OAT, oil temp, etc.)—cooler was better and hotter was worse.

What EGT is not

In fact, however, absolute values of EGT are not particularly interesting for several reasons. The most important is that *indicated EGT is not a "real" temperature!*

To understand what I mean by this, I'd like you to conduct a thought experiment: Imagine that you're an EGT probe, located in an exhaust riser between two and four inches from the exhaust port of a cylinder, and think about what you would see.

As the figure to the right illustrates, you'd see nothing much two-thirds of the time—during most of the intake, compression, and power strokes—because the exhaust valve is closed and so no exhaust gas is flowing out of the exhaust port and past you (the EGT probe). During the one-third of the time that the exhaust valve is open, you'd see a constantly changing gas temperature that starts out very hot when the valve first opens but cools very rapidly (within a few milliseconds) as the hot compressed gas escapes and expands, and then ultimately is scavenged by cold induction air during the valve overlap period (at the end of the exhaust stroke and the beginning of the intake stroke) when both intake and exhaust valves are open simultaneously.

Now, all these gyrations are happening about 20 times a second, and you (the EGT probe) cannot possibly keep up them. So you wind up stabilizing at some temperature between the hottest and coolest gas temperature you see, and you dutifully report this rather arbitrary equilibrium temperature to the panel-mounted instrument, where it is displayed to the pilot as a digital value accurate to one degree. The temperature you report to the pilot is not exhaust *gas* temperature (which is gyrating crazily 20 times a second), but rather EGT *probe* temperature (which is stable but related to actual exhaust gas temperature in roughly the same fashion as mean sea level is to high tide).

To make matters worse, there are numerous factors that affect indicated EGT besides actual exhaust gas temperature. These include probe mass and construction (grounded or ungrounded), cam lobe profile, lifter leak-down rate, valve spring condition, probe installation location, and exhaust manifold topology, among others.

For example, the two front cylinders (#5 and #6) on the left engine of my Cessna T310R always indicate lower EGT than the other four cylinders. The exact same phenomenon also occurs on the right engine. This is not because those front cylinders produce cooler exhaust gas than their neighbors (they don't), but because the exhaust risers for those cylinders curve aft while the other four risers go straight down. Thus, the gas flow past the EGT probe is different for the front cylinders than for the others, and their indicated EGT is lower. This temperature anomaly is quite obvious on my digital engine monitor—and also quite meaningless.

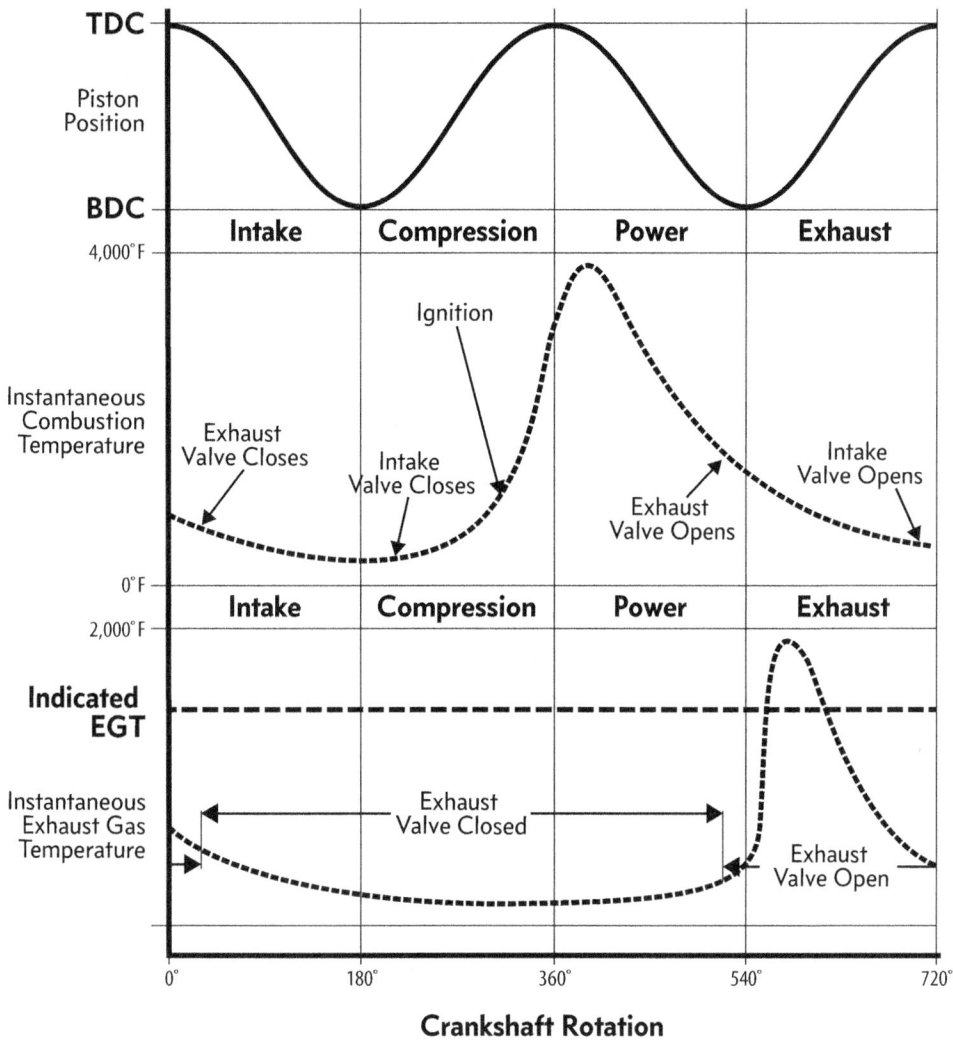

An EGT probe "sees" a very rapidly changing gas temperature during the one-third of the time that the exhaust valve is open, and nothing much during the two-thirds of the time that the exhaust valve is closed. Thus, EGT indicated by a digital cockpit gauge is not a "real" gas temperature.

What EGT means

Even if indicated EGT accurately reported actual exhaust gas temperature (which it doesn't), it's important to understand that exhaust gas temperature does not correlate with stress on the engine the way cylinder head temperature does. In fact, many things that increase engine stress (such as advanced ignition timing and high compression ratio) cause EGT to go down, while things that reduce engine stress (like retarded ignition timing and low compression ratio) cause EGT to go up.

Remember that CHT mainly reflects what's going on in the cylinder during the power stroke when the cylinder is under maximum stress from high internal temperatures and pressures, while EGT mainly reflects what's going on during the exhaust stroke after the exhaust valve opens and the cylinder is under relatively low stress.

High CHTs often indicate that the engine is under excessive stress, which is why it's so important to limit CHTs to a tolerable value (no more than 400°F for Continentals and 420°F for Lycomings). By contrast, high EGTs *do not* indicate that the engine is under excessive stress, but simply that a lot of energy from the fuel is being wasted out the exhaust pipe rather than being extracted in the form of mechanical energy.

For instance, a 1972 Cessna 182 with an O-470-R engine will typically have indicated EGTs that are 100°F hotter than those seen in a 1977 Cessna 182 with an O-470-U engine. The -R has a relatively low 7.0-to-1 compression ratio because it was certificated for 80-octane avgas, while the -U engine has a much higher 8.6-to-1 compression ratio because it was certificated for 100-octane. Because the high-compression -U engine is significantly more efficient at extracting heat energy from the fuel, it wastes less energy out the exhaust and thus its EGTs are cooler (despite the fact that the -U engine is more highly stressed than the –R).

High EGTs do not represent a threat to cylinder longevity the way high CHTs do. Therefore, limiting EGTs in an attempt to be "kind to the engine" is simply misguided.

DIFF vs. GAMI spread

Right behind the "high EGTs are bad" myth is the "identical EGTs are good" myth. Many pilots believe incorrectly that a flat-topped graphic engine monitor display (with all EGTs equal) is the mark of a well-balanced engine, and that unequal EGTs are a sign that something is wrong. This common misconception tends to be reinforced by digital engine monitors that display a digital "DIFF" showing the difference between the highest and lowest EGT indication.

As illustrated by the earlier anecdote about the front cylinders on my Cessna T310R, differences between absolute EGT values are both normal and benign. It is not uncommon for well-balanced fuel injected engines to exhibit EGT spreads of 100°F, and carbureted engines often have spreads of 150°F or more. In fact, EGT spreads are usually smallest near or just rich of peak EGT (the worst place to operate the engine), and often significantly greater at leaner or richer mixtures (that are much kinder to the engine).

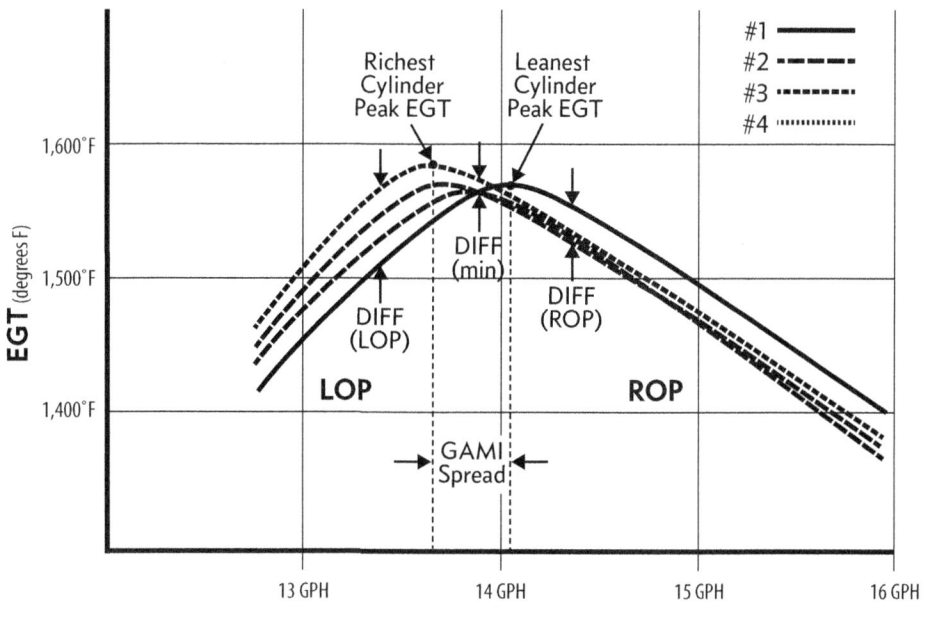

The spread between absolute EGT indications (depicted as "DIFF" on some digital engine monitors) is not important. What is important is the spread between fuel flows where the various cylinders achieve peak EGT (known as the "GAMI spread").

The mark of a well-balanced engine is *not* a small EGT spread ("DIFF"), but rather a small "GAMI spread"—defined as the difference in fuel flows at which the various cylinders reach peak EGT. Ideally, we would like to see this spread be no more than about 0.5 GPH (or 3 PPH). Experience shows that if the GAMI spread is much more than that, the engine is unlikely to run smoothly with LOP mixtures.

So what good is EGT?

The absolute value of EGT is not relevant – it's the relative values of EGT between cylinders, and how these values change as engine operating parameters are changed that is important. So what good is an EGT display in the cockpit? It's fantastic for troubleshooting engine issues. A wide variety of issues can be detected using EGTs, including malfunctioning magnetos, defective or fouled spark plugs, incorrect magneto timing, induction system leaks, and partially or completely clogged fuel nozzles. The art of troubleshooting with an engine monitor is a subject that's rich and fascinating, and we will explore it more in Part 7. CHT and EGT may also be used to detect abnormal combustion events, including detonation and pre-ignition that we'll discuss in the next chapter.

It's all relative

Al Hundere had it right after all: The only important thing about EGT is its *relative* value: how far below peak EGT and in which direction (e.g., 100°F ROP or 50°F LOP). ***Absolute*** values of EGT (e.g., 1475°F) are simply not meaningful and are best ignored. There is no such thing as a maximum EGT limit or red-line, and trying to keep absolute EGTs below some particular value—or even worse, leaning to a particular absolute EGT value—is simply wrongheaded. Don't do it. If you must fixate on those digital engine monitor readouts, fixate on something important like CHT.

Turbine Inlet Temperature (TIT)

Turbocharged engines are often equipped with a turbine inlet temperature (TIT) gauge or a TIT probe connected to the aircraft's digital engine monitor. Unlike EGT, absolute values of TIT are meaningful, because the TIT probe is mounted far downstream in the exhaust system where the gas flow past the probe is steady (not pulsed) and has relatively steady temperature (not constantly fluctuating). Observance of an absolute TIT red-line (typically 1650°F or 1750°F) is appropriate and prudent to obtain maximum useful life from the turbocharger. Brief excursions above TIT red-line for seconds or even minutes (e.g., when transitioning from ROP to LOP) are not harmful, but don't spend hours above TIT red-line if you want your turbocharger to live long and prosper.

4
Detonation and Pre-Ignition

Destructive detonation and pre-ignition events can ruin your engine in two minutes flat. Here's what's going on, and what you should do about it.

Although we often hear people describe what goes on inside the cylinders of an Otto-cycle engine as being an explosion—i.e., a violent, nearly-instantaneous event—it's not. The air-fuel charge does not explode when ignited by the spark plugs, but rather burns in an orderly fashion, starting at the spark plugs and progressing across the combustion chamber until it is quenched upon reaching the cylinder walls and piston crown when the air-fuel charge has been completely consumed and there's nothing more to burn. The combustion event takes a significant period of time—roughly 6 milliseconds or 90° of crank-shaft rotation, give or take.

The graphic on the next page illustrates this. Note that ignition occurs 20° to 25° of crankshaft rotation before the piston reaches top-dead-center (TDC), but that the peak combustion pressure doesn't develop until 15° to 20° after TDC. In an engine turning 2400 RPM, each crankshaft rotation takes 25 milliseconds, so the elapsed time from ignition to peak pressure (which takes about 1/8 of a rotation) is around 3 milliseconds.

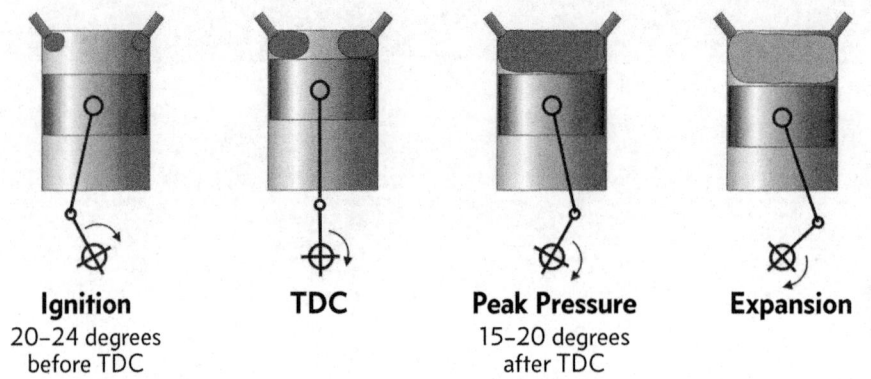

| Ignition | TDC | Peak Pressure | Expansion |
| 20–24 degrees before TDC | | 15–20 degrees after TDC | |

Normal combustion cycle.

It's crucial that peak pressure occurs well past TDC, because the geometry of the crank-shaft and connecting rod near TDC does not permit combustion pressure to be converted into useful work (i.e., crankshaft rotation), but rather simply generates excessive stress on the cylinder, piston, connecting rod and crankshaft. The graphic below attempts to dramatize this point.

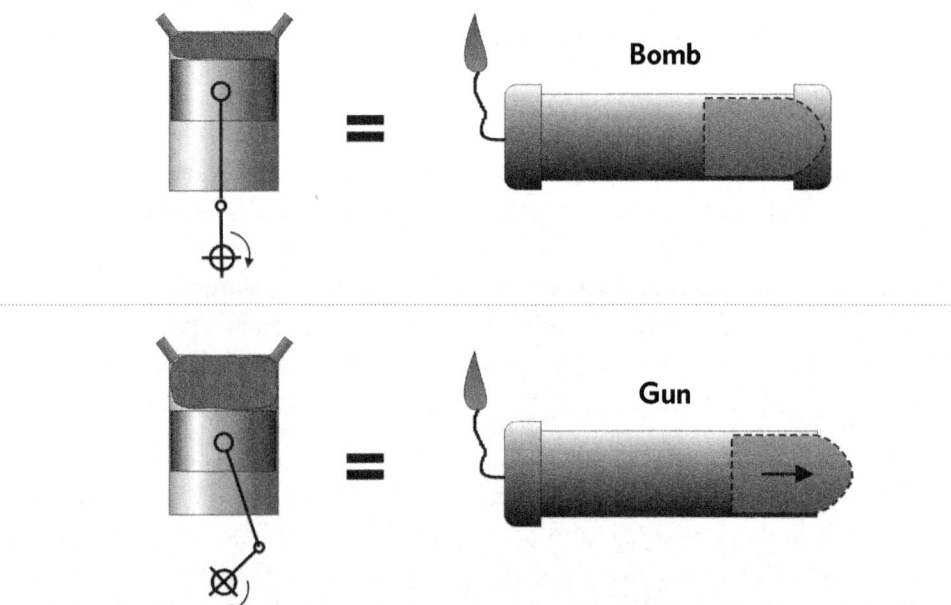

If peak pressure occurs at TDC, excessive stress on the drive train results.

Look at the plot of the pressure inside the cylinder as the combustion event progresses (below). If we crank the engine without supplying fuel or ignition (so there's no combustion), the compression stroke of the piston produces a modest pressure peak (perhaps 120 psi) right at TDC. On the other hand, if we provide a combustible fuel/air mixture and fire the spark plug at 22° before TDC, combustion causes the pressure to build to a much higher value—typically 800 psi in a normally-aspirated

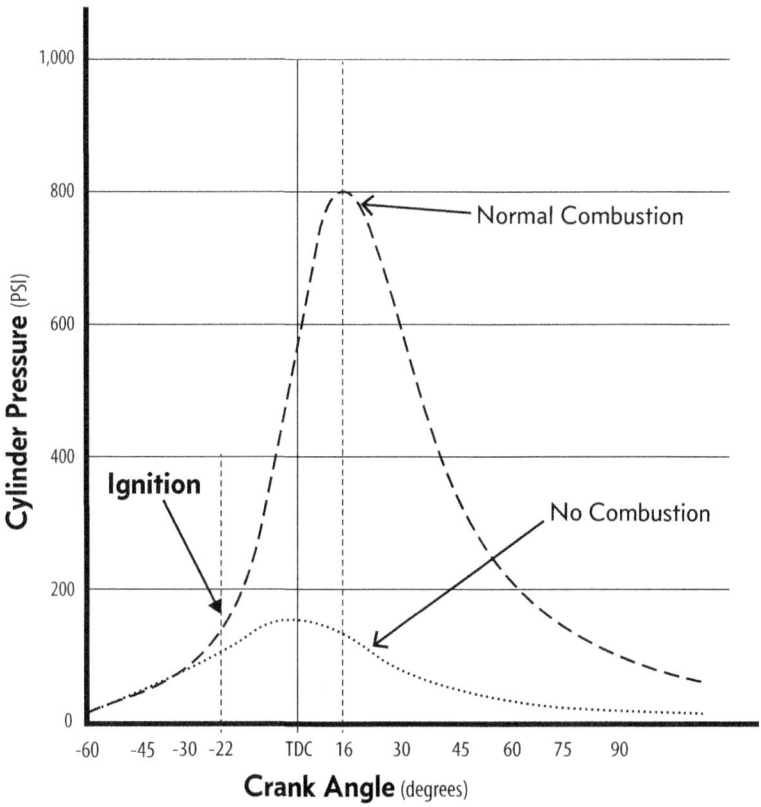

Cylinder pressure during a normal combustion event.

engine, and sometimes 1,000 psi or more in a turbocharged engine—that peaks 15° to 20° after TDC, just as the piston, connecting rod, and crank have reached a position where that pressure can be converted into crankshaft rotation with reasonable mechanical efficiency.

But if the combustion process moves too fast and the pressure peak occurs too early, the result can be excessive pressure, excessive temperatures, and unstable pressure pulses known as "detonation."

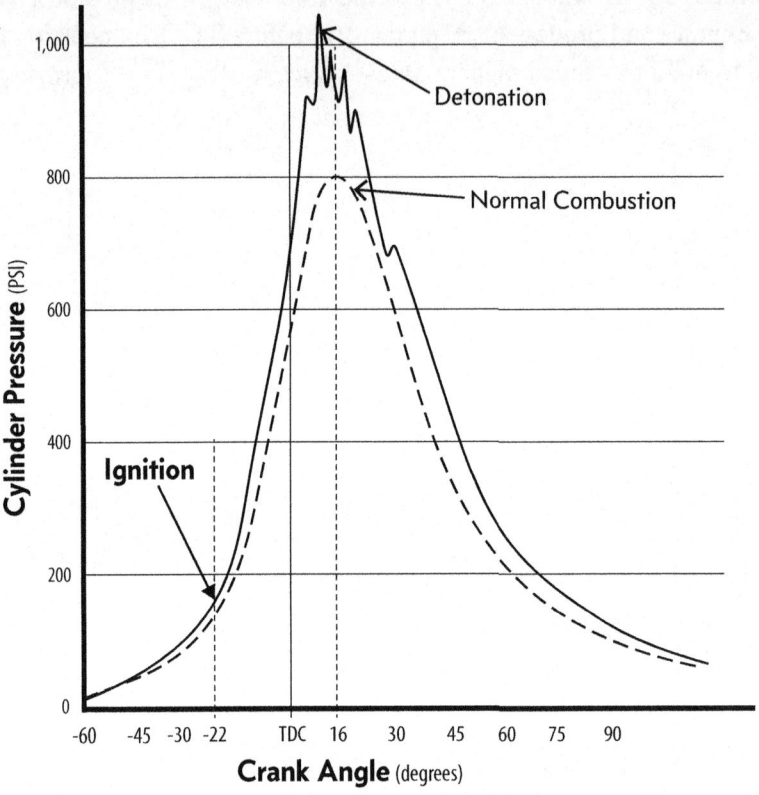

The ragged, spiked appearance of the upper trace is the characteristic pressure signature of detonation.

Detonation

Detonation is so damaging because when the piston is in the immediate vicinity of TDC, it can't move downward in the cylinder to relieve the pressure (and accomplish some useful work in the process).

In an automobile, we can usually hear detonation in the form of audible "knock." In an airplane, we can't—there's too much noise—but we can observe it on an engine monitor in the form of excessive CHT and depressed EGT.

Detonation is something that occurs near the peak pressure point in the combustion event, after the air-fuel charge has been ignited normally by the spark plugs. It is characterized by abnormal pressure spikes near the peak pressure point, caused by spontaneous combustion of end gas due to excessive temperature and pressure.

Contrary to what your CFI or A&P may have told you, detonation is not necessarily harmful. Many engines operate in light detonation quite regularly, and some can withstand moderate detonation for extended periods of time without damage. Detonation is not an optimum situation, but it's not necessarily destructive. The higher the specific output of the engine, the more likely it is to sustain detonation damage. An engine that produces 0.5 hp/in^3 (horsepower per cubic inch of displacement)—as is typical of most carbureted aircraft engines—can usually sustain moderate levels of detonation without damage, but highly-boosted turbocharged engines rated at 0.625 hp/in^3 or more can be damaged fairly quickly by detonation.

When detonation damage does occur, it typically manifests itself in the form of fractures (of spark plug electrodes and insulators, and sometimes piston rings and lands), pitting (typically of the piston crown), and/or heat distress (often piston skirt scuffing and piston corner melting).

Here is an example of a detonation event involving a Cirrus SR20 powered by a 200 horsepower Continental IO-360-ES engine. The plane was equipped with a snazzy Avidyne Entegra Multi Function Display (MFD) with an integrated engine monitoring system called "EMAX."

Everything looked fine until about two minutes after the pilot applied takeoff power, at which point the #1 cylinder's CHT began to climb rapidly compared to the other five cylinders. At the three-minute mark after brake release—with the aircraft at roughly 2,000 feet AGL—CHT #1 rose above 400°F and set off a high-CHT alarm on the MFD.

CHT #1's rapid rise—nearly 1°F per second—continued unabated until the piston and cylinder head were destroyed approximately five minutes after takeoff power was applied and two minutes after the CHT alarm was displayed. At that point, since the cylinder was no longer capable of combustion, CHT #1 started plummeting.

We can't be sure just how hot CHT #1 got because the Avidyne EMAX system "pegged" at 500°F. A reasonable guess is that the CHT peaked somewhere between 550°F and 600°F. No cylinder or piston can tolerate such conditions for very long, and this one obviously didn't.

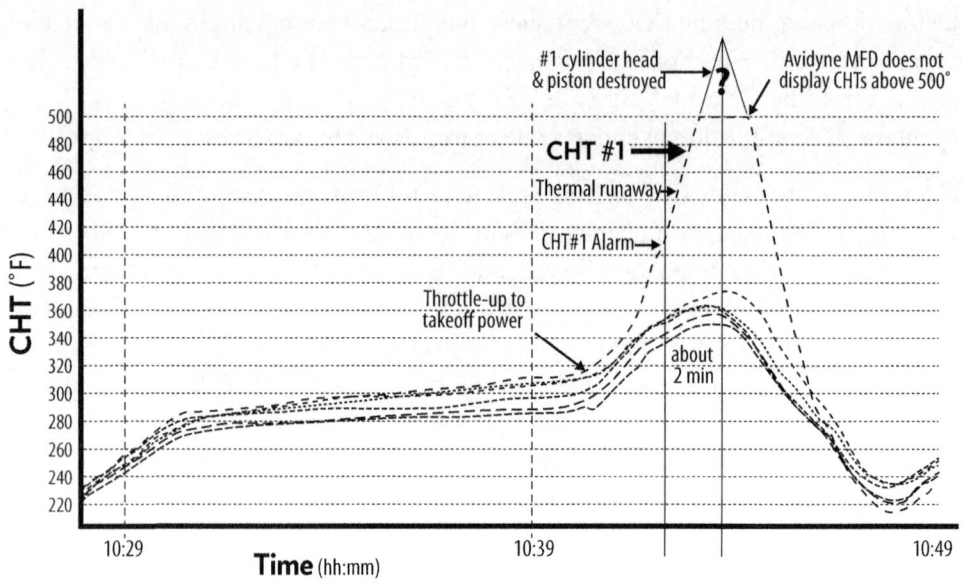

The CHT data downloaded from the EMAX system tells the short tale of this engine's demise.

Not long after CHT #1 went off-scale on the MFD, the pilot realized something was very wrong, and pulled the power way back. But he was a couple of minutes late, and the engine was already toast.

Here's what the #1 piston looked like after the event. Note the melted corners of the piston crown, the destruction of the top compression ring lands, and the severe metal erosion above the piston pin.

Much of the molten metal wound up inside the crankcase, contaminating the bearings and oil passages. Also note the severely hammered and gouged appearance of the piston crown, undoubtedly caused by loose chunks of metal flying around inside the combustion chamber. The cylinder head was found to have a big chunk of metal missing from it. Both spark plugs were destroyed by the event as well.

The engine was a low-time Continental factory engine, so the owner figured that the severe engine damage would be covered under Continental's warranty. I advised him not to bother filing a warranty claim, because I've never known Continental to give warranty consideration for a destructive detonation event. Continental considers this to be operational abuse, not a defect in materials or workmanship, and therefore not covered by warranty. (For what it's worth, I agree with their position on this.)

The owner didn't believe me and filed a warranty claim anyway. Continental promptly and unequivocally denied the claim, just as I told him they would.

What causes detonation?

There are a number of things that can cause or contribute to destructive detonation. One of the top offenders is advanced ignition timing. I'm astonished at how often we see engines with the magneto timing advanced several degrees from the engine specifications (E.g., 25° before TDC when the engine data plate calls for 22° before TDC.) Even a couple of degrees is enough to significantly reduce the detonation margin of the engine. Add a hot day and perhaps a cooling baffle that isn't quite up to snuff, and BOOM!

Owners should be particularly alert for mis-timed magnetos whenever maintenance is done that involves magneto removal or adjusting magneto timing. (More often than not, these occur during the annual inspection.) If mag timing is advanced, you'll notice that your EGTs are lower and your CHTs are higher than what you were seeing prior to maintenance. (Retarded timing results in the opposite: higher EGTs and lower CHTs.) If you notice this after the airplane comes out of maintenance, take it back to the shop and have the mag timing re-checked. It's a quick check and could save your engine (not to mention your gluteus maximus). Magnetos are required to be timed within one degree of the timing specified on the engine's data plate, and any error should be in the retarded direction.

Another common culprit is inadequate fuel flow on takeoff. When taking off from a near-sea-level airport—or from any elevation if you're flying a turbocharged airplane—you need to see fuel flow that's right up against the red-line on the gauge (or the maximum fuel flow shown in the Pilot's Operating Handbook (POH)). Unlike most other gauges on your panel, hitting red-line on the fuel-flow gauge (or even going a smidgen over) is a good thing. Takeoff fuel flow is a lot like tire pressure—a bit too much is a whole lot better than a bit too little. Anything less that red-line fuel flow on takeoff reduces the engine's detonation margin, and significantly less can reduce it enough to cause a catastrophic event.

Yet another cause is a partially clogged fuel injection nozzle. This can occur anytime, but most frequently occurs shortly after the aircraft comes out of maintenance because that's the most likely time for foreign material to get into the fuel system. (I've had two serious clogged-nozzle episodes in my airplane over the past 22 years, and both occurred shortly after an annual inspection.)

Pre-ignition

"Pre-ignition" is another abnormal combustion event that is often confused with detonation, but in fact is completely different. Pre-ignition is the ignition of the air-fuel charge prior to the spark plug firing. Anytime something causes the mixture in the chamber to ignite before the spark plugs fire, it is classified as pre-ignition. The ignition source can be an overheated spark plug tip, carbon or lead deposits in the combustion chamber, or (rarely) a burned exhaust valve—any of these things can act as a glow plug to ignite the charge prematurely.

Such a hot spot in the chamber can ignite the charge while the piston is very early in the compression stroke. The result: for a significant portion of the entire compression stroke, the engine is trying to compress a hot mass of expanding gas. This obviously puts tremendous mechanical stress on the engine and transfers a great deal of heat into the aluminum piston crown and cylinder head. Substantial damage is almost inevitable.

Holed piston resulting from pre-ignition.

Detonation causes a very rapid pressure spike near the peak pressure point for a very brief period of time. Pre-ignition causes tremendous pressure that is present for a very long time—possibly the entire compression stroke. Not only is pre-ignition far more damaging, but it's also much harder to detect. In fact, typically you find out about it only after the engine has been damaged catastrophically. The figure to the left shows an example of extreme damage caused by pre-ignition.

Engines can tolerate detonation for substantial periods of time, but there is no engine that can survive for very long when pre-ignition occurs. The engine will not run for more than a few seconds with pre-ignition. If you see a piston crown that looks sandblasted or a ring land that has fractured, it was probably caused by heavy detonation. If you see a hole melted in the middle of the piston crown, it was probably caused by pre-ignition. Other signs of pre-ignition are spark plugs with melted electrodes or insulators spattered with molten metal.

Detonation-induced pre-ignition

Although detonation and pre-ignition are two completely different phenomena, it is possible for heavy detonation to induce pre-ignition. If the engine is operating in heavy detonation for a significant period of time, the excessive temperatures and pressure spikes (which disturb the usual protective boundary layer) can cause spark plug electrodes and other things in the combustion chamber to overheat to the point where they start to glow red hot. At that point, the glowing item can cause pre-ignition and rapid destruction of the cylinder. Upon teardown, forensic analysis would reveal the tell-tale signs of both detonation and pre-ignition damage, although it's the pre-ignition that ultimately did the engine in.

Save your engine!

Regardless of whether its detonation or preignition that's threatening your engine, the solution is not rocket science. There are two simple rules that will almost always prevent these sorts of destructive events from occurring:

First, check your fuel flow gauge early on every takeoff roll. If the fuel flow is not at red-line or very close to it, abort the takeoff and sort things out on the ground. (The exception is takeoffs at high density altitudes in normally-aspirated airplanes; detonation is quite unlikely under those conditions.)

Second, set your engine monitor CHT alarm to 400°F or less for Continentals (I have mine set to 390°F), or 420°F or less for Lycomings. When the alarm goes off, do whatever it takes RIGHT NOW to bring the CHT back down below the alarm level: (1) Verify that the mixture is full-rich. (2) Turn on the boost pump if it isn't already on. (3) Open the cowl flaps if you have them. (4) If CHT climbs above 420°F (Continental) or 440°F (Lycoming), throttle back aggressively. Don't be shy about doing these things immediately, because you may only have a minute or two to act before your engine craters.

(Oh, and if your airplane isn't equipped with a digital engine monitor with CHT alarm capability, do yourself a favor and install one. Trust me, it'll pay for itself quickly.)

If you had a CHT excursion, when you get on the ground, put the airplane in the shop and have the spark plugs removed and inspected for damage, the cylinders borescoped, and the magneto timing checked. If takeoff fuel flow was short of red-line, have it adjusted before further flight.

PART II
Top End

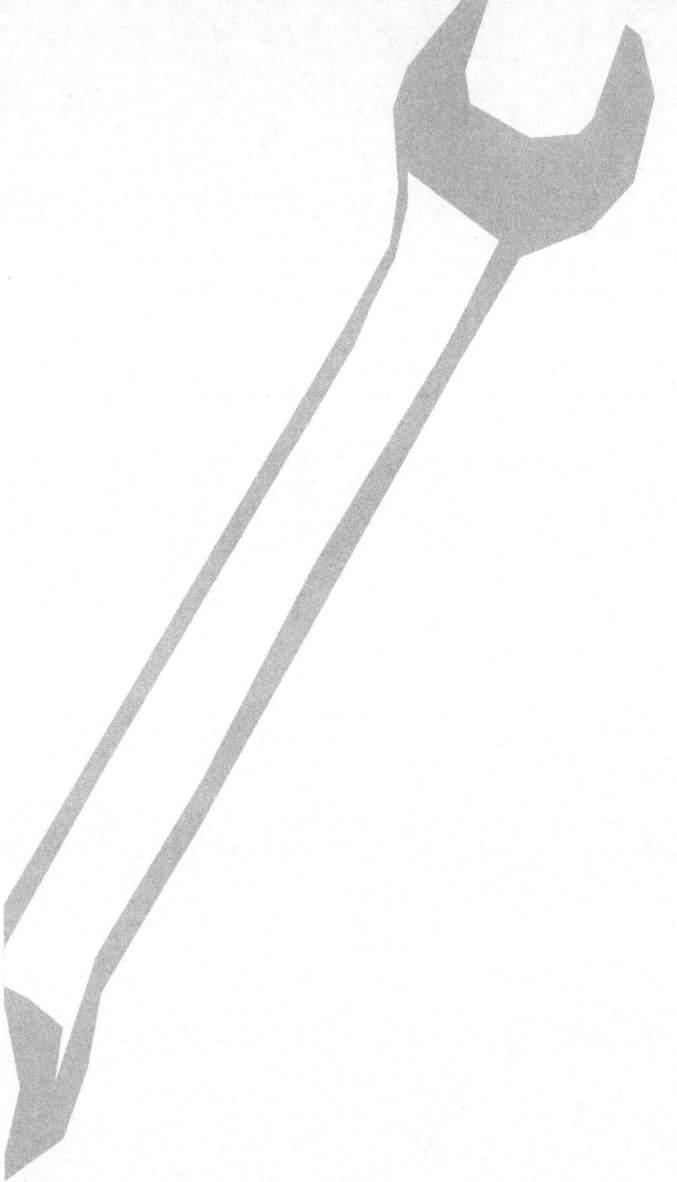

5
Anatomy of a Cylinder

The "jugs" of our engines are where most of the action is.

The "hot section" of our piston aircraft engines is comprised of bolt-on cylinder assemblies—four or six in nearly all horizontally-opposed GA engines. The cylinder assembly is where the action is. It contains the combustion event that converts chemical energy from fuel into mechanical energy that turns the propeller to produce thrust, and also turns various other accessories to produce electricity and fuel, oil, hydraulic and pneumatic pressure.

The cylinder assembly is also where that portion of the fuel's energy that cannot be converted to mechanical energy—and unfortunately, that's most of it—is dissipated in the form of exhaust gases and radiant heat. So while the rest of the engine typically never gets much hotter than your morning cup of coffee, cylinders get very hot indeed. They also cope with very high pressures—particularly when abnormal combustion events like detonation and pre-ignition occur.

Heat and pressure are the enemies of cylinder longevity. The hotter we allow them to run and the higher the peak pressures to which we expose them, the sooner they become troublesome and start letting us down.

In this chapter, we'll look at how cylinders are constructed and manufactured. Subsequent chapters will discuss how they wear out and fail and what aircraft owners and operators can do about it.

Gross anatomy

The cylinder assemblies used on horizontally-opposed GA engines are assembled from two major components: a forged steel barrel and a cast aluminum alloy head. Both are elaborately machined with cooling fins, and with mating threads where the head and barrel join.

During assembly, the head casting is heated in an oven to about 600°F and the barrel is chilled in a refrigerator to sub-freezing temperature. The valve seats and valve guides are also chilled. The heated head is removed from the oven, and the chilled valve seats and valve guides are quickly inserted into their respective bosses in the hot head casting. Then the still-hot head and the cold barrel are quickly screwed together.

When the temperatures equalize, the head-to-barrel junction winds up with an "interference fit" that will not come apart even at high temperatures and pressures. The valve seats and guides are also secured firmly to the head with an interference fit.

Compared to Continental cylinders, the head-to-barrel junction of Lycoming cylinders has more surface area and is more robust. That's one reason that Lycoming cylinders have a much higher CHT red-line temperature than Continentals do.

Cylinder barrel

The cylinder barrel in which the piston reciprocates must be strong, hard and light. It must withstand high temperatures and pressures, and must retain a film of oil that protects it against friction and wear. The barrel is made of chrome-nickel-molybdenum steel that is forged under high pressure for maximum tensile strength. The forging is then machined by a sophisticated CNC (computer numerical control) milling machine to create its exterior cooling fins and mounting flange, its smooth cylindrical interior working surface, and the male threads at its top end that mate with the female threads of the head casting.

Most cylinders have a working surface that is tapered slightly, with the end of the bore nearest the cylinder head a few thousands of an inch smaller in diameter than the skirt end. This "choke" is created to compensate for the higher operating temperature of the cylinder-head end of the barrel. When the cylinder is at full operating temperature, the bore becomes very nearly straight.

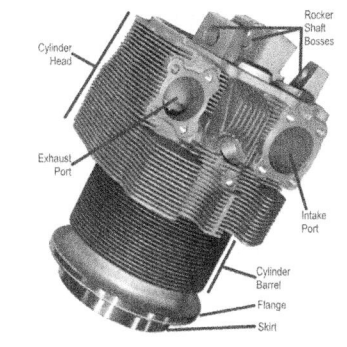

A cylinder is assembled from two major components—a forged steel barrel and a cast aluminum head—that are screwed together with an interference fit.

The cylinder barrel incorporates a machined mounting flange near its base. After the barrel and head are mated, the flange is drilled with eight close-tolerance mounting holes to accommodate the hold-down studs and through-bolts that secure the cylinder to the crankcase. The skirt portion of the cylinder barrel extends beyond the mounting flange part-way into the crankcase, making it possible to use a shorter connecting rod and reduce the external dimensions of the engine.

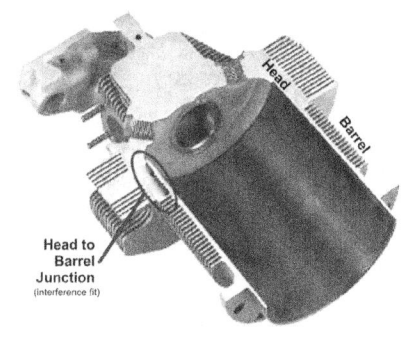

This cutaway of a cylinder shows the head-to-barrel junction.

After machining, the barrel is "through-hardened" by means of a heat-treating process. The interior working surface may be further "case hardened" for additional durability and/or corrosion resistance by means of one of three alternative processes:

- **Nitriding** involves masking off all but the interior working surface, then baking the cylinder barrel in an oven for about 40 hours at a temperature just under 1,000°F in an atmosphere of pure ammonia gas (NH3). During the baking process, nitrogen combines with the steel on the unmasked working surface to form an extremely hard layer about .015" thick.

- **Chrome plating** involves masking off all but the interior working surface, then immersing the cylinder barrel in a bath of chromic acid (CrO_3) and electroplating the working surface with a thick coating of metallic chromium. Once the necessary thickness of chromium has been applied, the polarity of the electric current is reversed to create channels or fissures in the surface. Such channels are necessary to make the chrome surface oil-wettable and allow it to retain a film of oil necessary for lubrication.

- **Nickel-carbide plating** is similar to chrome plating, except that the cylinder's working surface is electroplated with a layer of nickel that is permeated with tiny particles of silicon carbide (artificial diamonds). The silicon carbide particles give the nickel surface the necessary hardness and oil wetability.

Both Continental and Lycoming factory steel cylinders are manufactured with nitrided barrels. Continental NIC3 cylinders and ECi Titan cylinders (no longer in production) are nickel-carbide plated. Superior Millennium cylinders are through-hardened (heat treated) but not case-hardened. Chrome plating is often used as a repair technique to restore worn cylinder barrels to new dimensions. ECi repaired cylinders using nickel-carbide plating under the tradename "CermiNil." Harrison Engine Service in LaPorte, Indiana offers both traditional channel chrome plating and silicon-carbide-impregnated chrome plating trade named "Nu-Chrome."

Once the barrel's working surface has been nitrided or plated with chrome or nickel, it is honed to a precise micro-finish to provide an optimal bearing surface for the piston rings. The surface roughness must be precisely controlled at the micro-inch level. If it is too smooth, it will not hold an oil film sufficient to provide necessary lubrication (especially during initial break-in); if it is too rough, it will result in accelerated wear of the piston rings and cylinder wall.

The outer surface of the steel barrel is painted to protect it from corrosion. Continental cylinders are painted gold, while Lycoming cylinders are painted blue-gray. Both paints will change color if the barrel is subjected to overheating, something IA's are trained to look for during inspection.

Barrels that are plated with chrome or nickel-carbide are painted with a color code between the flange and lower cooling fin for easy identification. Chrome-plated cylinders have an orange band, while nickel-carbide cylinders have two silver stripes. The absence of such a color code denotes an unplated (through-hardened or nitrided) steel barrel.

Cylinder head and valves

The cylinder head is made of an aluminum alloy sand casting. The rough casting is then extensively machined using CNC equipment to create a hemispherical combustion chamber, cooling fins, intake and exhaust ports and flanges, and bosses (holes) for the valve guides and seats, rocker shafts, spark plugs, and fuel nozzles.

The machined head casting is treated with an Alodine chromate conversion coating to make it corrosion-resistant. The spark plug bosses are reinforced with threaded steel inserts called Heli-coils. Stainless steel studs are inserted at the exhaust port flanges to secure the exhaust risers.

Valve guides and seats are press fitted into bosses in the cylinder head casting while it is still hot, creating an interference fit. Valve guides support and guide the valves, while valve seats provide a durable sealing surface and protect the relatively soft aluminum head casting from being eroded by the repetitive hammering of valve closure.

In both Continental and Lycoming engines, intake valve guides are made of bronze (a copper-aluminum alloy), while exhaust valve guides are made of Ni-resist (a high-nickel cast-iron alloy). In engines with steel or chrome-plated cylinders, elevated nickel in oil analysis reports is an unambiguous marker for accelerated exhaust valve guide wear. (In engines with nickel-carbide-plated cylinders, elevated nickel is ambiguous—it can be caused either by cylinder barrel wear or by exhaust valve guide wear.) Intake valve seats are forged from aluminum bronze, while exhaust valve seats are forged from chrome-molybdenum steel.

The valves themselves are poppet valves consisting of a cylindrical stem with a face at one end and a tip and the other. Continental uses simple flat-headed valves for both intake and exhaust. Lycoming valves are more exotic (and expensive): intake valves are tulip-style, while exhaust valves are mushroom shaped with hollow stems filled with metallic sodium for improved heat transfer.

The valves are held closed by a pair of concentric steel helical coil springs. The two paired springs are made from different wire diameters and have different pitch in order to help reduce resonance effects. The valve springs are held in place by a pair of specially-shaped retaining washers. The lower retainer seats against the cylinder head casting, while the upper retainer is secured by split washers called "keepers" that fit into a groove machined into the valve stem near its tip. Exhaust valves are

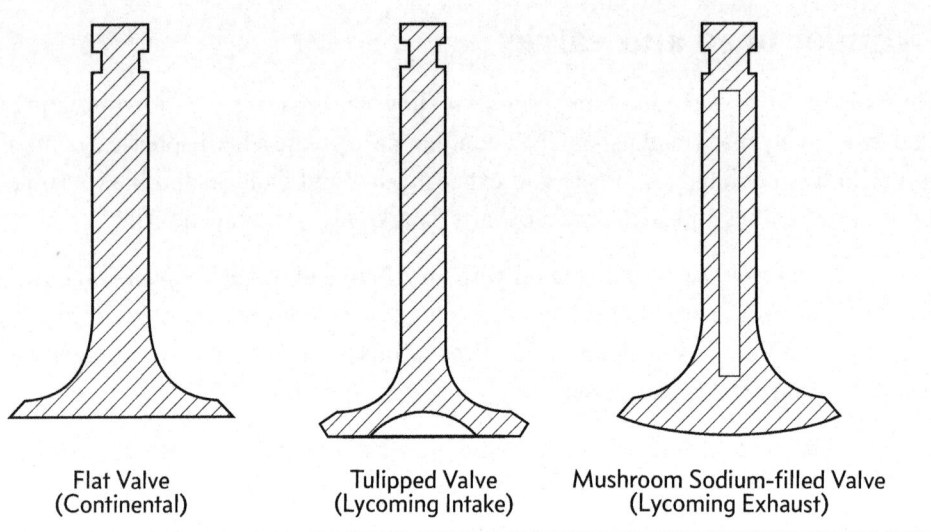

Valves come in different shapes and sizes.

equipped with "rotator caps" that impart a slight rotation to the valve every time it opens. (Continental and Lycoming exhaust valves rotate roughly one revolution per minute during cruise operation.)

The valves are opened by rocker arms that pivot on rocker shafts secured by mounting ears cast into the cylinder head. The rocker arms are operated by pushrods and hydraulic lifters (tappets) that ride on the lobes of the cam. Valves, springs, rocker arms and shafts are lubricated and cooled by engine oil and protected by a cast aluminum rocker box cover that bolts to the top of the cylinder head.

6
Taking Cylinders to TBO

Has the industry forgotten how to build a decent cylinder?

Lots of aircraft owners are frustrated by short cylinder longevity. Many say they had new or factory rebuilt engines whose cylinders started losing compression after 500 hours or so. Some old-timers gripe, "My cylinders always used to make it to TBO—why can't they make cylinders like they used to?" Some have replaced their Continental or Lycoming factory jugs with PMA replacements from Superior or ECi, swearing angrily that they'd never buy another factory cylinder.

The "jug du jour" syndrome

There always seems to be a new-and-improved cylinder technology that promises to solve the problem once and for all. But we've heard this song before.

In the early 1990s, carbide-impregnated chrome-plated barrels were all the rage. Sold under the trademarks Cermicrome and Nu-Chrome, they were heavily promoted and touted by leading overhaul shops … until owners started reporting precipitous drops in compression and sharply increased oil consumption after 500 hours or so. Today, Cermicrome is long gone and Nu-Chrome is rarely used.

In the late 1990s, the same leading overhaul shops that had previously pushed Cermicrome cylinders were now pushing Superior Millenniums with their lovely-to-look-at investment-cast heads and through-hardened steel barrels. Initial reports were extremely promising, but then we started hearing about Millenniums wearing out after 500 hours. To add insult to injury, Superior cut its cylinder warranty from two years to one, and later discontinued their investment-cast cylinders altogether, claiming the sandcast ones were just as good. Ultimately, the FAA issued a costly Airworthiness Directive (AD) requiring that all investment-cast Superior cylinders be relegated to the scrap heap.

Every cylinder manufacturer—Continental, Lycoming, ECi, Superior—has had issues at various times.

In the early 2000s, ECi's nickel-carbide-plated Titan jugs were touted as the final solution to cylinder problems. Not only is nickel-carbide more durable than through-hardened or nitrided steel, but it's also immune to rust. Certain high-profile overhaul shops (notably RAM Aircraft) started using ECi Titans exclusively. Then thousands of ECi Titans started developing cylinder head cracks at low hours and it turned out that the head castings were flawed. ECi fixed that problem in 2006, but the cylinders were slapped with an AD in 2016 that relegated most of them to the scrap heap alongside the investment-cast Millenniums.

Sandcast Superior Millenniums then became the cylinder of choice until they started developing head cracking problems, too, and Superior recalled a bunch. Superior tried to fix that problem with an improved cylinder head casting, but got tangled in another costly AD prompted by cylinder head-to-barrel separation on some sandcast cylinder assemblies.

So, believe it or not, right now as I'm writing this chapter in January 2018, the "jug du jour" seems to be factory cylinders from Continental and Lycoming…you know, the ones that thousands of angry owners said they'd never buy again! By most accounts, Continental seems to have squared away most of their cylinder problems, and is now delivering ones that are pretty darn good, including carbide-impregnated nickel-plated "NiC3" cylinders that are ideal for operators who face high corrosion risks.

Flawed anecdotal evidence

Every cylinder manufacturer—Continental, Lycoming, ECi, Superior—has had issues at various times. It can be a mistake to make decisions based upon stories you hear and things you read about what brand of cylinder seems to be winning the longevity race. That's because the cylinder manufacturers are constantly changing and improving their products (or at least trying to improve them), and most of the stuff you read and hear is basically ancient history.

Think about it. The average owner-flown piston airplane flies 100 hours or less a year. If you hear that such an aircraft needed a complete top overhaul at 500 hours SMOH, it's hard to know whether the problem was with the cylinders themselves or with how the airplane was flown. But even if you assume that the cause was a manufacturing defect rather than an operational problem, the presumed defect was present in cylinders that were probably manufactured no less than five years ago. That probably tells you almost nothing about the quality of cylinders that same manufacturer is building today.

Over the past two decades, Continental has had problems with cylinder honing microfinish, oil control ring tension, valve-to-guide concentricity, and excessively thick application of manganese phosphate coating. ECi had problems with plating adhesion, plasma-coated compression rings, and head casting geometry. Superior has had problems with heat treating of head castings and certification issues. Lycoming had an emergency AD in 2017 because of non-conforming piston pin bushings. Many aircraft owners were affected by these problems, and are hopping mad at one or more of these firms.

Of course, some airplanes fly a lot more than 100 hours a year. I can think of quite a few air tour operators and freight haulers that put 500 hours on their jugs in less than a year, and a small handful fly nearly 1,000 hours a year. But those guys almost always make it to published TBO (and often beyond) without having to replace a cylinder.

Premature cylinder wear seems to be a problem predominantly for owner-flown airplanes that take many years to amass 500 hours. Do you suppose there's a lesson there?

Barrel hardness—not the whole story

Each new cylinder technology—chrome plating, nitriding, through-hardening, carbide-impregnated chrome, carbide-impregnated nickel—has sought to provide a harder, more durable barrel surface that would be more resistant to wear and therefore last longer. It's an appealing idea, and one that's easy to sell.

However, this fixation with barrel hardness diverts attention from what I believe to be the biggest culprit in poor cylinder longevity: lubrication failure. After all, surface hardness doesn't really matter much if there's no metal-to-metal contact between the cylinder barrel and the piston rings.

That's precisely the way things are supposed to work. The cylinder wall is supposed to be well-coated with lubricating oil, constantly replenished and distributed by the oil control ring and held in place by the surface roughness of the cylinder's crosshatch hone pattern (or in plated cylinders, by porosity or impregnated carbide particles).

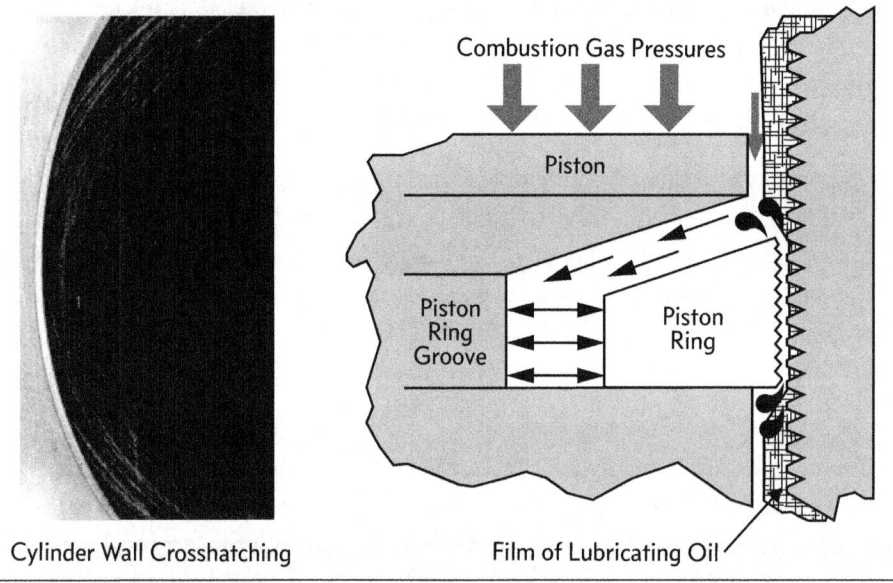

Cylinder walls are honed in a crosshatch pattern to maintain an oil barrier with the piston rings. Hydrodynamic lubrication prevents metal-to-metal contact.

As the piston travels up and down in the cylinder, the rings are supposed to "hydroplane" on the oil film, never touching the metal surface of the cylinder wall. This is known as "hydrodynamic lubrication" and is absolutely essential to cylinder longevity. So long as the oil film on the cylinder wall is not breached by the piston rings, there's no metal-to-metal contact, and therefore no appreciable wear. It's only when the film is breached that barrel wear becomes an issue.

So long as you have sufficient upper-cylinder lubrication to prevent metal-to-metal contact between the rings and barrel, any cylinder can make it to TBO and quite possibly well beyond. But if the oil film is breached on a regular basis, even the hardest barrel surface will wear out prematurely.

Why lubrication fails...

Unfortunately, a certain amount of metal-to-metal contact is inevitable. Otherwise, our cylinders would last practically forever. Let's examine some of the things that can cause the oil film to be breached:

Inadequate film thickness: To keep the rings separated from the cylinder walls, the barrel must be coated with an adequately thick oil film along the entire length of the piston's stroke. Unfortunately, it's not easy to get enough oil up to the top part of the barrel. The piston's oil control ring is responsible for replenishing the oil film and spreading it up and down the cylinder walls. But the compression rings constantly scrape the oil away, and the combustion process constantly consumes some of the oil by carbonizing it.

Proper upper-cylinder lubrication depends pivotally on subtle technical characteristics of the cylinders and rings, notably the tension applied to the oil control ring by its expander spring, and the depth and contour of the hone pattern on the cylinder wall. If the oil control ring tension is too great, the oil film won't be thick enough. If the hone pattern is not sufficiently rough (or worn or corroded away), the oil applied to the cylinder walls won't adhere.

The worst loss of oil film thickness occurs during engine start—particularly on the first start after days or weeks of disuse when the pre-existing oil film has had a chance to strip off. Since gravity causes oil to run off during periods of inactivity, the greatest loss occurs at the top of the barrel, resulting in asymmetrical wear that ultimately causes the cylinder bore to become out-of-round.

Low oil viscosity: Hydrodynamic lubrication depends on two factors: the viscosity of the lubricating fluid and the relative velocity of the moving parts. The slower the parts are moving with respect to one another, the more viscous the lubricant must be in order to keep them apart. For cylinder barrels and piston rings, the critical point occurs at top-dead-center where the rings slow to a stop and reverse direction. If metal-to-metal contact is going to occur between the rings and cylinder wall, that's where it'll happen.

Oil's viscosity varies with temperature—the hotter the oil gets, the less viscous it becomes. In this case, we're not so much interested in the oil temperature gauge as we are in the CHT gauge—that's the best proxy for the temperature of the oil adhering to the upper cylinder walls. Operate at excessive CHTs and your jugs will wear out more quickly.

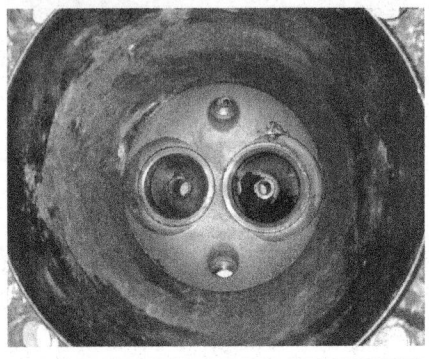

Aircraft that fly irregularly and live in a corrosive environment often lose cylinders prematurely to rust.

High peak combustion pressure: Compression rings are held against the cylinder walls by combustion chamber pressure. The rings have a slight trapezoidal taper that allows gas pressure to press the rings outward. If the peak pressure gets too high, the pressure on the rings can breach the oil film and cause metal-to-metal contact. This can be especially severe in highly boosted turbocharged engines; they are hard on cylinders.

Corrosion: Yet another culprit in lubrication failure is corrosion of the cylinder walls. This problem most affects aircraft that fly irregularly and are based in a corrosive environment near the coast or in a humid climate. Corrosion typically develops when the cylinder's protective oil film breaks down during extended periods of disuse. Once corrosion pits start to develop, they disrupt the critical hone pattern required to retain a consistent oil film on the cylinder walls. Disruption of the oil film in the pitted area causes metal-to-metal contact with the rings that causes the pits to enlarge. This process degenerates, causing progressively higher oil consumption and lower compression readings.

Making cylinders last

While there's probably not much that can be done about the microfinish roughness of your cylinder hone patterns or the expander spring tension of your oil control rings, most of the factors that contribute to lubrication failure are under your control. Give them the attention they deserve and your cylinders have a good chance of going the distance. Ignore them and you could be facing a top overhaul sooner than you'd like.

Fly regularly: The single most important thing you can do to maximize cylinder longevity (and cam/lifter longevity, for that matter) is to fly your aircraft regularly. It doesn't really matter whether you fly 100 or 500 hours a year. What matters is that the airplane is flown every week or two in order to keep the cylinders coated with oil and corrosion-free.

If you can't fly at least once a week, it really helps to hangar your aircraft, unless you're fortunate enough to be based in a dry mountain or desert climate where corrosion is not a big problem. I also recommend using thick single-weight oil (at least during the warm months of the year) to minimize oil stripping and maximize corrosion protection.

Watch your CHTs: The hotter your cylinders run, the less viscous the oil that adheres to them. When cylinder head temperatures exceed 400°F or greater, oil viscosity can decline to the point where hydrodynamic lubrication is marginal at best. Try to keep your CHTs in the 300s (°F) whenever possible. Using a higher airspeed during climbout, paying attention to cowl flap settings (if applicable), and using conservative power settings are all helpful in this regard (as is operating lean-of-peak if that's your cup of tea).

A digital engine monitor that measures CHT on all cylinders is essential to engine longevity.

Probe-per-cylinder CHT instrumentation lets you see if some cylinders are running hotter than they should. I fly behind two Continental engines, and I set my engine monitor CHT alarm to go off at 390°F. Lycomings tend to run about 20°F hotter, so an alarm threshold of 410°F would be appropriate. If you have trouble keeping your CHTs below that, careful attention to your baffles and baffle seals and fuel flow and ignition timing should correct the problem.

Watch your power settings: Equally important is avoiding excessive peak combustion pressures that can rupture your upper cylinder oil film and cause accelerated barrel wear. Reduced power settings will help, but what's even more important is to avoid the MAP/RPM/mixture combinations that produce the most damaging pressure peaks. To make a long story short, the worst pressure peaks occur at high MAP, low RPM and roughly 50°F ROP.

Don't be quick to yank that jug!

Another closely related problem is that I see lots of cylinders being replaced unnecessarily because of allegedly low compression readings. Most owners break into a cold sweat when a differential compression test produces any number below 70/80, and entirely too many mechanics have recommended pulling cylinders simply because they measured in the low 60s.

This makes no sense at all. Continental Manual M-0 (available on the Continental Motors website) describes the proper procedure for performing a cylinder leakage check. It requires the use of a master orifice tool to establish the go/no-go threshold for each particular differential compression test gauge each time it is used. For most gauges I've checked, this threshold is in the low 40s (typically 41/80 to 44/80). Thus, a cylinder that measures 50/80 is perfectly okay and should not be pulled on the basis of compression alone.

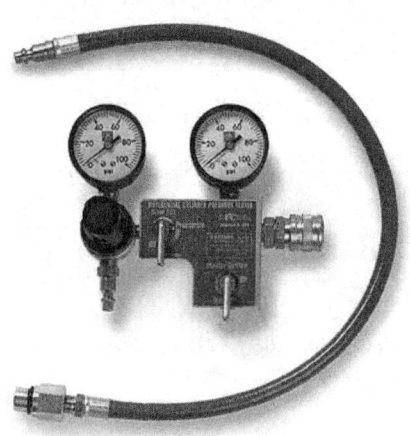

A low compression test reading on a cylinder does not necessarily mean the cylinder must come off.

Furthermore, even if a cylinder exhibits compression below the no-go threshold, Continental Manual M-0 still does not authorize yanking it off the engine. It directs the mechanic to perform a borescope inspection of the cylinder to determine the reason for the low compression reading. If the borescope inspection fails to identify any specific problem, then M-0 says that the cylinder should remain in service and the aircraft flown for a minimum of 45 minutes, after which the compression check should be repeated.

It is quite common that the re-test performed after flying the aircraft for a while produces considerably different results. For example, an owner based in Phoenix, Ariz., encountered a rough-running engine while flying in California. He put his aircraft in a shop there and asked them to try to determine the cause of the roughness. The shop ran numerous tests, including a compression test, and found that one cylinder measured 40/80. They told the owner that they would have to replace this cylinder "before they could let him fly the aircraft" and recommended that he do a complete top overhaul.

The owner consulted with me. I recommended that he NOT let the shop replace any cylinders (since low compression cannot cause the engine to run rough), and instead fly the airplane back to Phoenix so the owner's normal shop could work on it. The owner demanded that the California shop release his aircraft (over their loud protests), and flew it back to Phoenix, where his local mechanic repeated the compression test and found that the cylinder in question measured in the mid-50s.

In another case, a Cirrus SR22 owned by one of my company's clients had a cylinder measure 38/80 during an annual inspection in Florida. The shop wanted to pull the jug, but we insisted on a borescope inspection which revealed no anomalies. We arranged to have the owner fly his airplane for an hour after which the shop repeated the compression test of that cylinder while it was still hot from the flight. The compression reading on the re-test was 72/80. The jug was fine. It was the test that was flawed.

Compression tests are overrated

This underscores something that many owners and mechanics don't understand: In the absence of evidence of a burned valve, low compression does not represent a safety hazard. No airplane ever fell out of the sky simply because it had an engine with low compression cylinders.

Many years ago, I had dinner with Continental's longtime head of engineering, John Barton (now retired), who related a story to me that illustrates just how overrated the compression test is. It seems that Continental engineers were trying to determine just how much deterioration in compression it would take to cause an engine not to make full rated takeoff horsepower. To do this, they put an IO-550 engine on a dynamometer test stand and performed a series of full-power test runs. Between each run, they filed the compression ring gaps oversize to intentionally increase leakage and reduce compression readings.

At 60/80 in all cylinders, the engine put out full rated power on the dyno; there was no measurable horsepower decrease at all. They repeated the test with all compressions at 50/80, and the result was identical. They repeated it once more with all compressions at 40/80, and once again the engine made full rated power.

As the Continental engineers continued the experiment with ever-lower compressions, the crankcase pressure due to excessive blow-by became so high that it threatened to blow all the oil overboard through the breather. But they never did detect any measurable horsepower decrease.

The moral of the story is that in the absence of unacceptable oil consumption or obviously oil-fouled plugs, a low compression reading due to leakage past the rings almost never justifies yanking a cylinder.

Who's to blame?

When a cylinder wears out prematurely at 500 hours, it's easy to blame the manufacturer, and easy to get mad when the manufacturer declines to replace the cylinder under warranty because the cylinder took five years to reach 500 hours and the warranty is only good for a year or two.

The difficult truth is that it's often not the manufacturer's fault. The engine probably didn't fly often enough. It might have been run hotter or harder than it should. Maybe the cylinder was pulled when it didn't really need to be.

Hard as it may be to accept, we've met the enemy… and frequently it is us.

7
Jug Economics

Does it make economic sense to extend the life of a geriatric engine by doing a "top overhaul" or replacing a few soft cylinders? Let's do the math.

I had just wrapped up the 2002 annual on my Cessna T310R. The inspection turned up only one item of significant bad news, and that surfaced on the very first day of the inspection when I did a brief pre-maintenance flight to warm up the engines, pulled the airplane into the hangar, removed the top cowls and top spark plugs, and did a hot compression check.

Despite the fact that the TSIO-520-BB engines were at 1402 SMOH—two hours past Continental's published TBO of 1400 hours—almost all of the cylinders still measured in the mid-70s. Not too shabby.

But the #4 cylinder on the right engine was clearly in deep kimshee. The compression tester read 52/80, and from the roar I heard when I put my ear to the tailpipe, it didn't take rocket science (or a borescope) to figure out that I had a fried exhaust valve. No question that jug was coming off.

NOTE: In 2002, Continental's guidance was that no leakage past the exhaust valve was acceptable. That changed radically in early 2003 when Continental revised their guidance to say that a compression reading of 52/80 was quite acceptable.

So now what?

Some folks might say that with an engine at "manufacturer's recommended TBO" and an unairworthy cylinder, the time had obviously come to do a major overhaul or exchange it for a factory rebuilt. Sorry, I don't agree. The way I looked at it, the engines were running great, not making metal in the filters or oil analysis, not leaking, not burning oil, and 11 out of 12 jugs were just fine, thank you. With any luck at all, these engines had at least a few hundred more hours left in them—and at the 120 hours/year I averaged, that was at least several more years of flying.

No, my plan was to replace the soft jug and keep on trucking. But replace it with what?

Given the high time on the engines, I had no illusions that this cylinder would still measure up within service limits and be able to be re-valved and reinstalled. I had to assume that the geriatric jug was trash and couldn't reasonably be reconditioned.

If the "bottom end" of the engine is in good shape, there's really no reason to consider a major overhaul.

At the same time, it didn't seem to make sense to pony up big bucks for a new cylinder assembly from Continental or Superior. After all, it might not have been realistic to expect to get another 1,000 hours out of these engines and thus get my money out of a new cylinder. I figured that all I really needed was a jug that would take me for another 500 hours or so until major overhaul.

So I phoned up the owner of a top-notch engine shop in California and asked him if he could fix me up with a decent serviceable jug that would take me to overhaul time. He said he was sure he could. I shipped him my soft jug (which he confirmed was "beyond economic repair" just as I suspected), and he shipped me back a "continued-time" cylinder assembly with about 500 hours on it. The cost was just over half of what a new cylinder assembly would have cost. No guarantees, of course, but it was pretty good bet that this jug would last me until overhaul time.

Fast forward: That half-price continued-time jug served me well for another 1600 hours until I majored the engine.

Pushing TBO

If you read my first book "Manifesto," you know that I am a strong believer in overhauling "on condition" rather than at some arbitrary number of hours. The engine—not the tach or Hobbs—will tell you when the time has come to overhaul. So long as the engine isn't making metal, isn't leaking, isn't burning excessive oil, and isn't exhibiting any other disturbing symptoms or alarming trends, there's no reason not to keep on flying.

Published TBO should be thought of as an actuarial statistic, much like human life expectancy figures. Some engines won't make it to published TBO. Some will happily go for hundreds or even thousands of hours past it (unless arbitrarily euthanized). It's clearly a bad idea to push an engine that's obviously tired or sick, but it's a crime to retire one that's elderly but still spry.

Cylinders are expendable. They bolt on and they're relatively inexpensive and easy to change. (We'll quantify that shortly.) If the "bottom end" of the engine—case, crank, cam, main bearings, and gears—is in good shape, there's really no reason to consider a major overhaul.

Ah, but how do you know if the bottom end is in good shape? Most of the time, you have to rely on indirect indicators—oil filter inspection and spectrographic oil analysis being the most important.

But if you're unlucky enough to have to yank a jug (as I did), you wind up with a big hole in the side of your engine. That gives you a unique opportunity to peer inside and inspect what's inside the crankcase—at least the stuff visible through the hole—and get another data point on the condition of your bottom end.

When I yanked the #4 cylinder off the right engine at the 2002 annual, I conducted a careful inspection of the cam lobes, tappets, and everything else I could see and feel. There was no sign of cam or lifter spalling, no visible corrosion, and nothing else that would put up a red flag. Combined with the clean oil filters and good oil analysis reports, I came away reassured that my right engine was not likely to give up the ghost any time soon, despite the 1402 SMOH on the hourmeter.

Doing the math

Doing a major overhaul or exchanging an engine for a factory rebuilt is really expensive. In 2018 dollars, you can figure $30,000 to $50,000 depending on what kind of airplane you fly. Therefore, if you can put off the overhaul by changing a cylinder, it's almost always a good decision to do so. In round numbers, a new cylinder "pays for itself" if it can extend the life of your engine by 100 hours; anything more is gravy. An owner who flies 100 hours a year could change out a cylinder every year and still be money ahead, compared to doing a major overhaul sooner rather than later.

If you do what I did and install continued-time reconditioned cylinders instead of new ones, the break-even point becomes about 50 hours. At that price, it's a no-brainer.

My actual experience with the two engines on my Cessna T310 was that I wound up changing out just three cylinders (out of 12) over a 14-year period during which I flew about 1,600 hours beyond published TBO. As you might imagine, my decision to extend the life of my engines by installing continued-time cylinders about once every five years saved me a small fortune compared to overhauling or exchanging the engines at TBO.

What stuff costs

To do the math yourself for your own airplane, you need to know what stuff costs so you can do the calculations. Fortunately, the Internet puts all this stuff right at your fingertips.

In today's world, the cost of a factory rebuilt engine and the cost of a first-rate field overhaul aren't much different. There are pros and cons of rebuilts and field overhauls, but costwise it's about a push. You can look up the price of any Continental or Lycoming rebuilt engine on the Air Power Inc. site (www.factoryengines.com). For prices on new cylinders from Lycoming, Continental, or Superior, check Air Power's parts site (www.factorycylinders.com). The cylinder prices quoted are for complete cylinder assemblies that include valves, rings, piston, rocker shafts, and even gaskets. Reconditioned cylinders (which cost between half and two-thirds of new jugs) are worth considering if you have a high-time engine and are trying to buy a few hundred more hours as cheaply as possible (as I was).

Unless you do your own maintenance, labor can be a significant cost factor in cylinder replacement calculations. Changing a cylinder involves removing baffles and dropping the exhaust and induction plumbing on the affected side of the engine. As a general rule-of-thumb figure around four hours of labor to change one jug, and add two more hours for each additional jug changed on the same side of the engine at the same time. These figures can vary a fair amount depending on make and model—my Cessna Turbo 310 twin is exceptionally easy to work on, but some high-performance singles are considerably tougher due to poorer maintenance access.

What about top overhauls?

A "top overhaul" generally refers to replacing all the cylinders at once—usually around mid-TBO. Many an owner whose engine has one or two soft cylinders has been talked into replacing them all, on the theory that if one or two cylinders are bad, the others can't be far behind.

My experience suggests that this isn't often the case. The fact that the #4 cylinder on my right engine burned an exhaust valve doesn't suggest to me that more cylinders are likely to follow suit any time soon. The fact is that in the 3,300 hours that I'd been operating my airplane, I've only ever had one other burned exhaust valve—and believe it or not, that occurred on the #4 cylinder of the right engine (about five years and 600 hours earlier). If I'd replaced the other five (or eleven) cylinders then on the theory that they would soon suffer the same fate, I would have wasted tens of thousands of dollars needlessly. In most cases, there's no reason to replace a cylinder unless there's actually something wrong with it.

> My decision to extend the life of my engines by replacing a cylinder about every five years saved me a small fortune.

A top overhaul for most engines has a break-even point of roughly 500 hours, give or take. So if your cylinders are all in serious trouble (an unlikely scenario) but you're convinced that the bottom end is good for another 500 hours or more, a top overhaul can be cost-justified.

Every case is different, and without a crystal ball it's impossible to know with certainty what the best course of action will be. But in general, I'd advise an owner with one or two bad jugs to replace only those jugs. On the other hand, if three out of four (or four out of six) are bad, then the handwriting would seem to be on the wall and replacing all jugs might be a good decision.

With regular use, careful powerplant management techniques, and a little luck, any cylinders can make it to TBO. From time to time, some won't—and that's why they make the things bolt-on units that are relatively easy to change. But before you change 'em, do the math.

8
Exhaust Valve Failures

Exhaust valves are the most heat-stressed components in your piston aircraft engine, and the most likely to fail prior to TBO.

I suffered my first (and only) in-flight exhaust valve failure in the mid-1990s. The engine started running very rough (as you might expect of a six-cylinder engine that was only running on five cylinders). After I landed, I noticed that the manifold pressure at idle was several inches higher than normal, confirming that something was definitely wrong with the engine.

I put the airplane in the hangar, removed the top cowling and the top spark plugs, and performed a differential compression test. Five of the cylinders measured just fine, but one measured 0/80 with a hurricane of air blowing out the exhaust pipe. It was clear that this jug was going to have to come off. Once I wrestled the cylinder off the engine and looked at the exhaust valve, it was pretty obvious that something was missing. A fragment of the exhaust valve face had broken off and departed the premises for parts unknown. Luckily, it opted to depart through the wastegate and to spare the turbocharger turbine wheel from destruction.

This exhaust valve failed in-flight, shutting down the cylinder.

I sent the jug out for repair. It came back with a new exhaust valve and guide, and with some dressing to the valve seat. I installed the cylinder back on the engine, where it happily operated for another 25 years and 3,200 hours until the engine was overhauled at 220% of TBO.

Hot, hot, hot!

Exhaust valves are the most heat-stressed components in your engine. They live their lives exposed to hideously high temperatures, while oscillating back and forth through a valve guide largely without benefit of lubrication (since they're too hot for engine oil to tolerate without coking). Frankly, it's astonishing that they last as long as they do.

During the peak pressure and temperature portion of each combustion event, gas temperatures in the combustion chamber approach 4,000°F, far hotter than the exhaust valve could withstand. Fortunately, the valve is closed during this time, so the heat energy absorbed by the valve face is quickly transferred through the valve seat to the cylinder head, where it is absorbed by the head's large thermal mass and dissipated by its cooling fins. This "heat sink" arrangement is essential to the survival of the valve. Without it, the valve face would overheat and self-destruct quite rapidly.

As the combustion event subsides, the exhaust valve opens. By this time, the gas temperature in the combustion chamber has transferred much of its heat energy to the piston (converting it to mechanical energy), so the exhaust gas that flows past the valve and out the exhaust port starts out at less than 2,000°F and cools very rapidly as the combustion chamber pressure drops. This is a good thing, because when the exhaust valve is open it loses its primary heat sink (the valve seat), and the only way the valve can dissipate heat is through the valve stem to the valve guide. This secondary heat path is a bit more effective in Lycoming engines (with their sodium-filled valve stems) than it is in Continental engines (which use solid-stem valves).

Cutaway of a cylinder's exhaust port, showing the exhaust valve, seat, guide, and cylinder head.

At the end of the exhaust stroke, the exhaust valve closes, once more making firm contact with the valve seat and establishing the primary heat sink arrangement in preparation for the thermal assault of the next combustion event.

How exhaust valves fail

Exhaust valve problems often cause aircraft owners to suffer from pangs of guilt. "Why did the valve burn? What did I do wrong?" Mechanics often contribute to such guilty feelings by telling owners that their exhaust valve burned because the engine was leaned too aggressively. This is almost always wrong.

Most exhaust valve problems are caused by excessive valve guide wear. Some guide wear is normal and inevitable, given that the guide is softer than the chrome-plated valve stem that passes through it, and that the two are in constant relative motion without benefit of lubrication. But if the guide wears excessively, it cannot hold the valve face perfectly centered in the valve seat. That's when problems begin. If the valve face and seat are not perfectly concentric, then one spot on the valve face will not seal properly against the seat when the valve is closed during the combustion event. This causes two bad things to happen. First, the heat path from the valve face through the seat and head is disrupted, interfering with the ability of that spot on the valve face to shed heat. Second, tiny amounts of extremely hot combustion gas leak past the spot that isn't sealing properly. The result is a "hot spot" on the valve face.

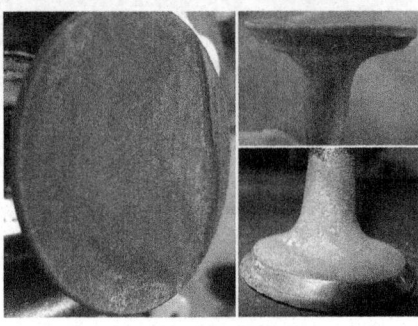

A badly burned exhaust valve. Note the hot spot (left panel, 2-4 o'clock), the warping (top-right panel), and metal erosion (bottom-right panel). This valve was only hours from complete failure.

Once the exhaust valve develops a hot spot, things can deteriorate rather quickly. Metal starts eroding from the hot spot, causing its seal against the valve seat to get worse, interfering with the heat path even more, and allowing increasing amounts of leakage during the hottest part of the combustion event. When the hot spot gets hot enough, the valve face will start to warp, further degrading the seal and increasing the leakage. Deterioration progresses at an ever-increasing pace until the hot spot gets so hot that the valve ultimately sheds a chunk of metal, at which point compression goes to zero and the cylinder shuts down. (Colloquially, we say the engine "swallowed a valve.")

Bottom line: Once the hot spot develops, the valve is doomed—it's not a question of whether it will fail, only when.

Why they fail prematurely

Any exhaust valve will fail if it remains in service long enough. In a perfect world, the valve, guide and seat will survive to TBO or beyond. In the real world, that isn't always the case.

Several factors can contribute to premature exhaust valve failure. If the guide is not properly machined (reamed) during cylinder manufacture, overhaul or repair to hold the valve perfectly concentric with the seat, then a hot spot can develop relatively quickly. For example, there is considerable evidence that Continental had some valve concentricity issues on cylinders they manufactured during the late 1990s and early 2000s, resulting in an epidemic of burned exhaust valves at 500 to 700 hours. Continental changed its manufacturing procedures and these problems seem to have gone away.

Another factor involves how the valve seat is ground, and how wide the contact area is between the valve and the seat. If the contact area is too wide, there may not be enough pressure between the valve and seat to cut through carbon deposits that form on the valve seat (particularly when the engine is operated at low power and/or rich mixture). If the contact area is too narrow, then the heat transfer path from the valve to the seat is compromised and the valve runs too hot (particularly at high power settings and lean mixtures). Grinding the seat to obtain the optimal contact area can be more of an art than a science.

If the engine is operated with a rich mixture (particularly during taxi and other ground operations), lead, carbon, and other unburned combustion byproducts can build up on the portion of the exhaust valve stem that projects into the exhaust port when the exhaust valve is open. When the valve closes, this deposit build-up is dragged into the lower portion of the valve guide, and often causes accelerated guide wear ("bellmouthing"), particularly in Continental engines that use relatively soft valve guides. As we've seen, accelerated valve guide wear generally leads to valve hot spots ("burned valves") and ultimately to valve failure ("swallowed valves").

In most Lycoming and some Continental engines that use relatively hard valve guides, the deposit build-up on the valve stem makes it difficult for the valve to close fully. This can also cause leakage past the valve, resulting in hot-spotting and ultimately valve failure. If the situation gets bad enough, the result in a stuck valve that won't close. (The same problem can be caused by valve guide corrosion in engines that sit unflown for long stretches of time.) The first symptom of this condition is usually "morning sickness" where the engine runs very rough when first started, but smooths out as the engine comes up to operating temperature. If the problem isn't addressed promptly, it can lead to an in-flight stuck valve that can have quite serious consequences: bent pushrod, damaged cam, even snapping the head right off the valve if the piston strikes the head of the stuck-open valve. Stuck valves are quite common in Lycomings and Continental O-200/O-300 engines; they are quite rare in big-bore Continental engines.

So, contrary to popular belief, to the limited extent that pilot leaning procedure contributes to burned, stuck, and swallowed exhaust valves, such issues are far more likely to be caused by excessively rich mixtures (particularly during ground operations) than by lean mixtures. I operate my engines brutally lean during ground ops, and lean-of-peak EGT during all phases of flight other than takeoff and initial climb. This assures the cleanest and coolest operation, which is the optimum prescription for long valve life.

During the late 1980s and early 1990s, Continental switched to a new, ultra-hard "nitralloy" exhaust valve guide in an attempt to reduce guide wear. Unfortunately, some of these guides weren't properly chamfered and developed a sharp edge that chiseled the chrome plating from the valve stems and allowed the valves to wobble, burn, and ultimately fail. That was the reason for my exhaust valve failure in the mid-1990s. As is true more often than not, my valve failure was not caused by pilot error but by manufacturing error. Interestingly, the other 11 nitralloy valve guides in my engines didn't have this problem and kept my valves healthy for more than twice TBO.

9
Preventing Exhaust Valve Failures

Early detection can save you from a costly exhaust valve failure.

In the last chapter, we discussed how exhaust valves fail. Now let's look at how we can prevent them from failing by monitoring exhaust valve condition, detecting incipient valve problems, and dealing with them before in-flight failure occurs.

My in-flight exhaust valve failure described in the previous chapter occurred "back in the bad old days" before we had the sophisticated engine monitoring tools that we have today—specifically spectrographic oil analysis, borescope inspections, and digital engine monitors. Nowadays, there's really no excuse for such an in-flight failure because we have the technology to detect these problems early. Anyone who experiences an in-flight exhaust valve failure today just isn't paying attention.

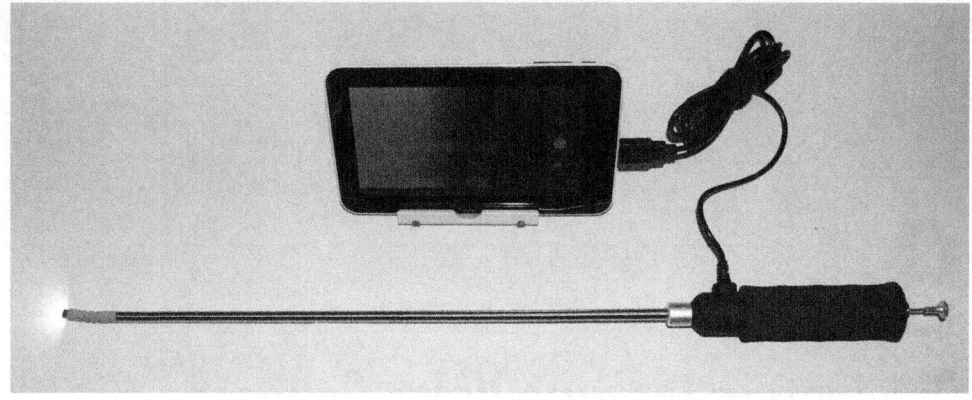

This inexpensive ViVidia VA-400 borescope is excellent for inspecting exhaust valves.

Borescope inspections

In my opinion, regular borescope inspections should be our first line of defense against exhaust valve failure. The borescope is an optical probe or a subminiature digital camera that can be inserted through a spark plug boss. It is used to perform a direct visual inspection of the combustion chamber, including the valves, cylinder head, cylinder barrel and piston crown. The borescope permits a quick, inexpensive, unambiguous determination of whether the exhaust valve is operating normally, or whether it is starting to burn or stick.

Note the symmetrical appearance of the exhaust valve (left), showing the valve is operating normally.

Here's an image of normal valves in a Continental cylinder. The smaller valve on the left is the exhaust valve. Note the pattern of exhaust deposits on the face of the valve. The deposits are quite minimal, indicating that this cylinder has been running a nice, lean, clean-burning mixture. More importantly, the deposit pattern is almost perfectly symmetrical—similar to a bullseye—showing that this valve is operating at the same temperature all the way around the circumference of its face, with no hot spots. That's exactly the way a healthy exhaust valve should look.

Compare that with this exhaust valve. Look at the highly asymmetrical pattern of exhaust deposits on the face of this valve. The cylinder has been running rather rich, causing thick deposits to form around most of the circumference of the valve face. But the valve has an extreme hot spot in the 8 to 10 o'clock position, so hot that is has burned off almost all the exhaust deposits in this area. This valve is in very serious shape, and wouldn't have survived very many more hours without failing.

This exhaust valve is in serious trouble, and doesn't have much longer to live. There's an obvious hot spot in the 8 to 10 o'clock position.

The borescope inspection is the gold standard for evaluating exhaust valve condition. Unlike the differential compression test (which has proven to be inconsistent and unreliable), the borescope provides a clear, unambiguous indication of whether or not the exhaust valve is healthy. If the valve has a symmetrical appearance under the borescope, it's fine. If the appearance is asymmetrical (lopsided), the valve is in distress and needs to be replaced. Simple as that.

The only problem with borescope inspections is that they need to be done regularly, and often enough to ensure that a distressed valve is detected before it fails in flight. How often is that? My research indicates that a well-trained inspector can generally detect a hot spot on an exhaust valve 100 to 200 hours before the valve fails in flight. If you have your cylinders borescoped at intervals of 100 hours or less, you can be fairly confident that a burned exhaust valve will be detected before it fails in flight.

By happy coincidence, 100 hours is just about the right interval for cleaning, gapping and rotating spark plugs. If the top cowling and the top spark plugs are removed, then doing a borescope inspection is a no-brainer, and shouldn't take more than an extra 30 minutes. In fact, any time a spark plug is removed from a cylinder for any reason, it would be maintenance malpractice not to stick a borescope in the hole and look around.

Engine monitor analysis

These days, an increasing proportion of the piston aircraft fleet—including most high-performance aircraft—are equipped with digital engine monitors that display and record per-cylinder EGT and CHT data and often numerous other parameters as well. The digital engine monitor should be our second line of defense against exhaust valve failures. Although it can't give nearly as much advanced warning of valve distress as the borescope, the engine monitor's compelling advantage is that it monitors the engine continuously, and doesn't need to be scheduled.

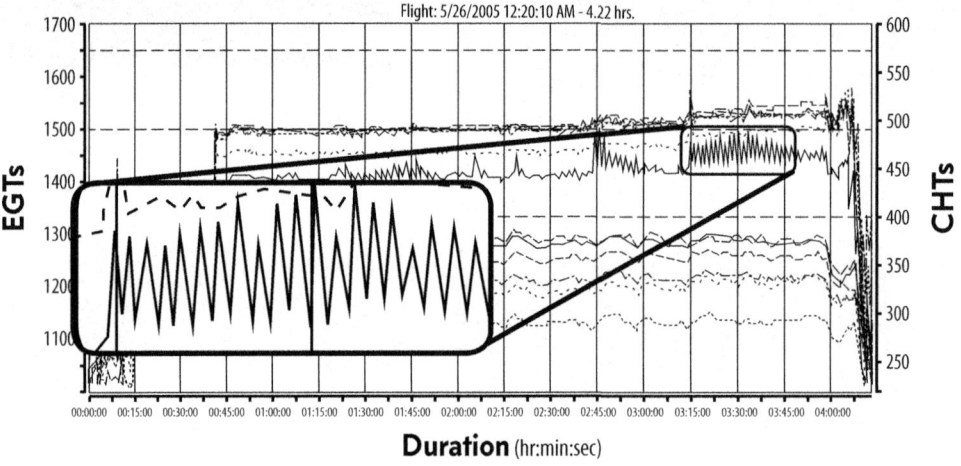

A very slow, rhythmic EGT oscillation (often on the order of one cycle per minute) is the unique signature of a failing exhaust valve. Inset: Cylinder #5 EGT.

Look at this graph of engine monitor data, and focus on the anomalous appearance of the EGT trace for cylinder #5 (highlighted). At first glance, this might seem like it's coming from a failing probe or a loose connection. But closer examination reveals that the EGT oscillations are not rapid and random (as one would expect from a bad probe or connection), but rather very slow and almost perfectly rhythmic. As the inset shows, the EGT oscillated exactly 10 times in each 15-minute period, and for a total of exactly 20 cycles in 30 minutes—a bit less than one cycle per minute. A failing probe or connection can't do that. In fact, there's only one thing that can do that: a failing exhaust valve.

Most engines employ a "rotator cap" mounted at the tip of the exhaust valve stem that causes the valve to rotate a fraction of a degree every time the valve cycles open and closed. The purpose of this valve rotation is to prolong valve life by equalizing the heat load around the circumference of the valve face, and helping to keep the valve and seat clean and free of deposits. The rate of valve rotation varies with engine RPM and rotator cap construction. For most Continental engines its about one revolution per minute at typical cruise RPMs, a bit faster for most Lycoming engines.

Consequently, if you notice a slow, rhythmic variation of EGT with a frequency on the general order of one cycle per minute, there's only one phenomenon that can possibly account for that EGT variation: exhaust valve leakage. Your response to such an observation should be to schedule a borescope inspection of the offending cylinder as soon as possible. In all likelihood, the borescope will reveal that the exhaust valve has an obvious hot spot, and the cylinder will need to come off for replacement of the exhaust valve and guide and dressing of the valve seat.

On the following page, look at the progressive deterioration of the #2 exhaust valve in a Bonanza's Continental IO-520 engine over a period of five months. Note how the EGT variation becomes increasingly obvious, regular and rhythmic as the exhaust valve deteriorates. Also note the frequency: almost precisely one cycle per minute. The valve was literally crying out for attention. Ultimately the owner noticed the problem and pulled the cylinder before the valve failed in-flight.

The engine monitor will not give nearly as much advance warning of a failing exhaust valve as the borescope, but my research suggests that it will give something on the order of 25 hours lead time before failure—provided the pilot is paying attention and knows what to look for.

Note that this EGT signature depends on the fact that the exhaust valve is rotating during engine operation. A few engines don't use rotator caps (notably Lycomings with solid-stem exhaust valves); also, certain failure modes (e.g., stuck valves) may prevent the valve from rotating. Therefore, the slow rhythmic EGT variation may not be present in every failing exhaust valve scenario, but it will be present in most of them.

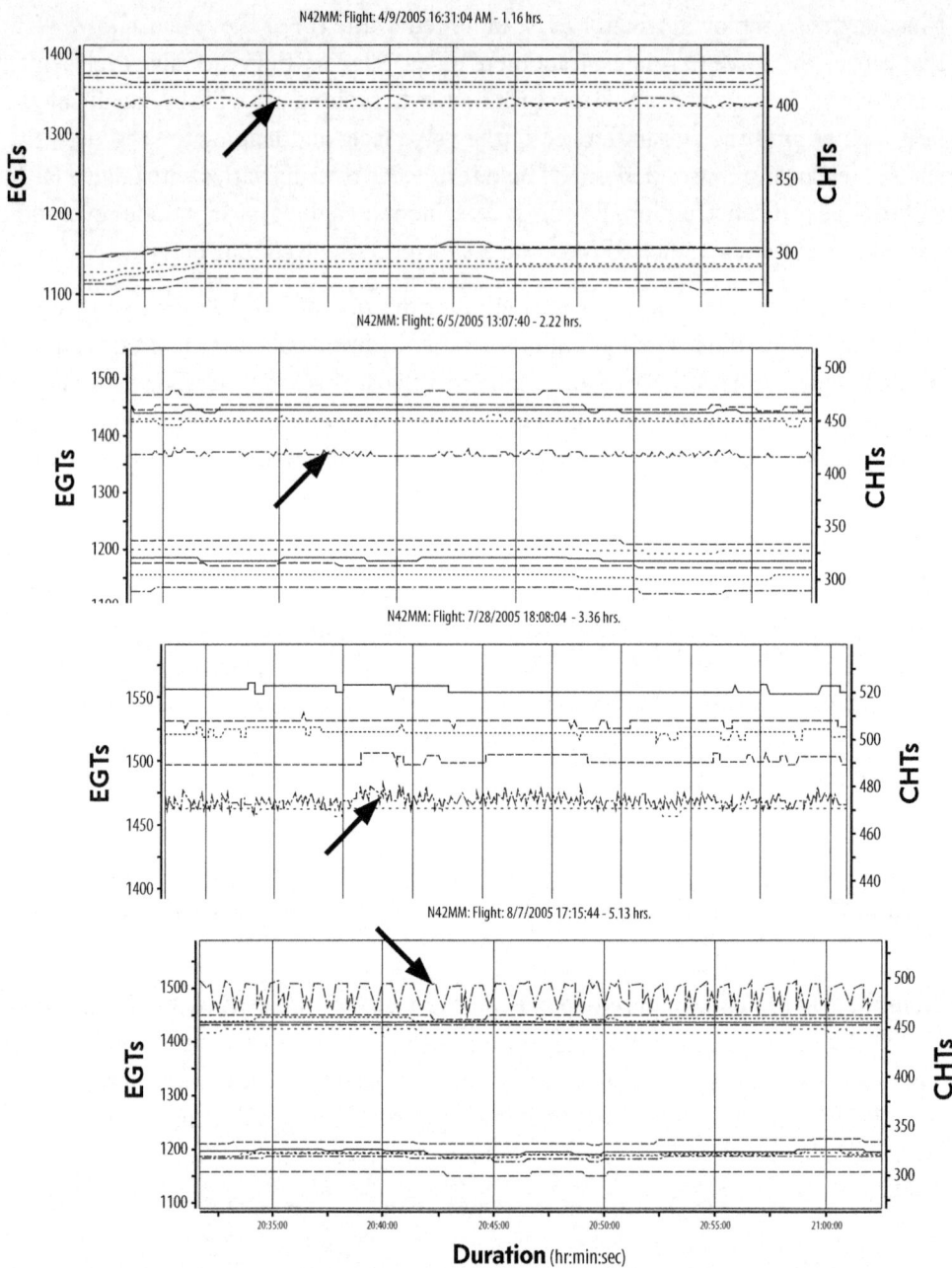

Successive data dumps show progressive deterioration of #2 exhaust valve over a five-month period. Note the precisely rhythmic EGT variation at almost exactly one cycle per minute.

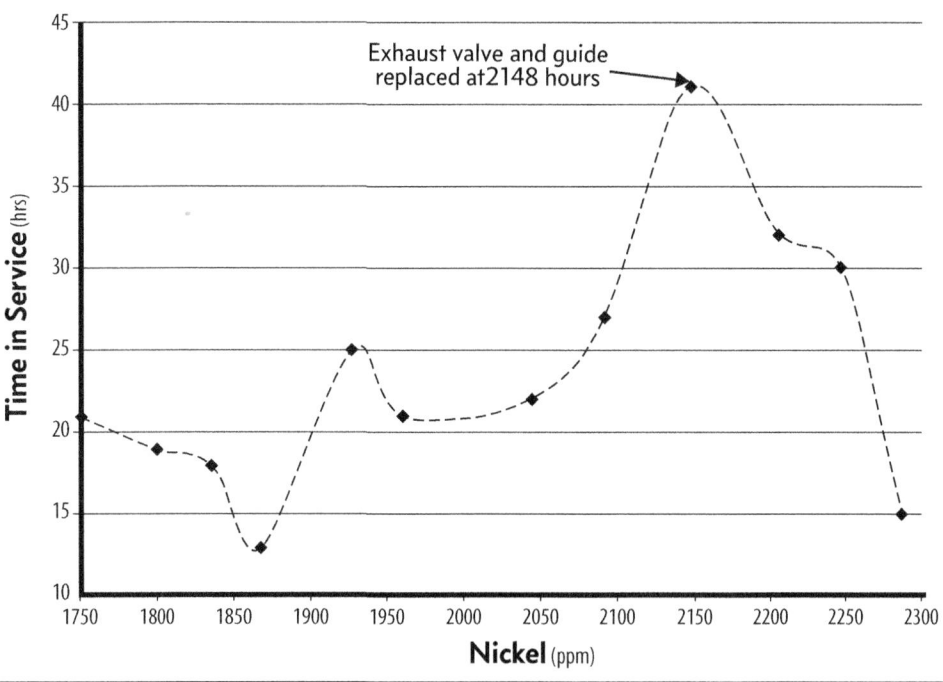

Oil analysis started showing elevated nickel (exhaust valve guide material) long before the failing exhaust valve was detectable by other means.

Oil analysis

A third line of defense against exhaust valve failure is spectrographic oil analysis. We've seen that exhaust valve failure is usually caused by accelerated valve guide wear. Exhaust valve guides are made of a high-nickel alloy, so accelerated guide wear usually shows up in oil analysis as increased nickel.

The figure above shows 550 hours' worth of nickel readings in the left engine of my Cessna 310. This engine had traditionally "made" about 14 parts per million (ppm) of nickel during the first 1500 hours since major overhaul, which is about normal for a six-cylinder Continental engine. Then the nickel readings started acting strange, rising into the low 20s, then the high 20s, giving warning of accelerated exhaust valve guide wear.

Unfortunately, oil analysis cannot identify which cylinder is the culprit. But the increasing nickel readings put me on notice that there was likely to be a failing exhaust valve in my future. Consequently, I started watching the engine monitor carefully, and started doing more frequent borescope inspections. Eventually, the #3 exhaust valve was found to be developing a hot spot. At 2,148 hours SMOH, the #3 cylinder came off and the exhaust valve and guide were replaced. Nickel then started to decline as the new valve and guide broke in, and after a few oil changes it dropped back to a healthy 15 ppm.

By using oil analysis, engine monitor data, and regular borescope inspections, exhaust valve distress can be detected reliably long before the valve reaches the point where it will fail. Nowadays there's really no excuse for suffering the kind of swallowed valve episode that I experienced two decades ago. We have the technology to eliminate these failures.

10
Separation Anxiety

Fatigue failure of a cylinder can ruin your day.

In past chapters, I've talked about the remarkable longevity I've achieved with the cylinders on my Continental engines. But not every aircraft owner is quite so lucky, and a few seem to be downright unlucky. Check out this email I once received from a Cessna 182 owner:

> I was flying westbound at 6000 ft MSL (3300 AGL) from Bismarck, N.D. to Lewistown, Mont. in good VFR weather. After about one-hour of flight, while flying over the Dakota Badlands, about 20 miles from the Montana border and about 400 miles from the real mountains, I heard a loud "pop" and developed a slight engine vibration. Fortunately, the engine continued to run at reduced power. The manifold pressure, oil pressure and temperature remained within limits. Outside of the initial "pop" there was not excessive noise or vibration, although the engine was rough. In spite of the normal manifold pressure I lost about 15 knots of airspeed, but was able to maintain altitude.
>
> Fortunately, I had chosen to follow I-94 instead of flying direct. There really is not much civilization out here once you get away from the Interstate. But

I-94 provides a 500-mile-long emergency runway. I routinely keep track of emergency landing spots. It was only about 20 miles back to Dickinson, Mont. (KDIK) so I immediately turned back. The engine ran okay and I made a precautionary simulated engine failure approach and landing at Dickenson and taxied in.

The #1 cylinder head on this Continental O-470 engine suffered a catastrophic fatigue failure in flight. The #3 cylinder suffered a similar failure about two hours later!

After landing I noticed a lot of oil on the ramp and all over the aircraft. I had lost about 3 of the 9 quarts of oil. It made a real mess at the ramp. When we removed the top cowling we noted that the #1 cylinder head had separated, as you can see in the accompanying photograph.

Fortunately, Milt Purvis at Dickinson Air Services jumped right on the problem.

With the help of FedEx, we had a replacement cylinder on the aircraft the next day. I test-flew the aircraft for about 0.3 hours and it was inspected after landing and found okay. We loaded up and continued on our way to Montana. After takeoff I climbed to altitude in the traffic pattern, made sure everything looked good and turned on course toward Montana, again following I-94.

About 40 miles from Dickinson, it happened again! Same symptoms! This time I was only about 6 miles from the Beach, N.D. (population 900) airport (an unattended airport along I-94 about 3 miles from the Montana border). Upon removing the cowling we found a similar cylinder head separation on the number 3 cylinder.

The Continental O-470 engine is about 400 hours from TBO. It had been running exceptionally well during the 30 months and 300 hours that I have owned the airplane, although oil consumption was relatively high—a quart every 3-4 hours. The aircraft had spent much of its prior life in Florida and showed corrosion on the cylinder fins. I do not know if the cylinders were new or overhauled on the last engine major.

The engine is currently being replaced in Beach by the folks from Dickinson (60 miles away) with a gold seal engine from Western Skyways.

I have not seen a failure similar to mine on a Continental O-470 engine, much less two failures within two hours! I do not know if the failure was from fatigue or corrosion. In hindsight, since all cylinders had the same history, I probably should have bit the bullet and replaced the engine after the first failure.

Actually, cylinder head separations are not all that uncommon—although *two separations in two hours on the same engine* certainly is extraordinarily rare. This owner/pilot clearly did an outstanding job of keeping his cool and handling the in-flight emergencies well. He also did a great job of flight planning to make sure he always "had an out" nearby in case of trouble.

Fatigue facts

There are two major causes of catastrophic engine failure: wear and fatigue. Failures caused by wear are almost always preventable, because wear events are usually detectable well in advance of failure using standard condition-monitoring techniques like oil filter inspections, oil analysis, compression tests and borescope inspections.

In contrast, fatigue failures usually happen suddenly, with little or no warning. They almost never "make metal" beforehand that could be detected through oil filter inspections or oil analysis. Sometimes a sharp-eyed IA will catch an incipient fatigue crack in a crankcase or cylinder head during an annual inspection. That may be adequate for crankcases, but once-a-year inspections are simply not adequate to detect head cracks reliably before failure occurs. Sometimes we get lucky and catch head cracks before failure, but sometimes we aren't so lucky—as our intrepid Skylane owner learned the hard way while flying over the Dakota Badlands.

Fatigue is the progressive structural damage that occurs when metal is subject to cyclic stress. The process starts with a microscopic flaw (called the *initiation site*) that widens slightly with each stress cycle. As the part continues to undergo repetitive stress, the tiny crack begins to grow more rapidly. After many stress cycles, the crack grows to critical length at which point crack growth accelerates dramatically and becomes unstable, and complete failure of the part is inevitable.

Engineers use a graph called an "S-N curve" to characterize the fatigue properties of a particular material. The S-N curve plots the magnitude of repetitive stress S (in pounds per square inch) against the average number of repetitive stress cycles N the material can endure before it fails.

The S-N curve reveals a substantial difference in fatigue characteristics between ferrous metals (like iron, steel and titanium) and non-ferrous metals (like aluminum, magnesium and copper). Ferrous metals exhibit a "fatigue limit" stress below which they can endure an infinite number of repetitive stress cycles without failing. Non-ferrous metals have no fatigue limit, and will always fail eventually if subjected to enough stress cycles.

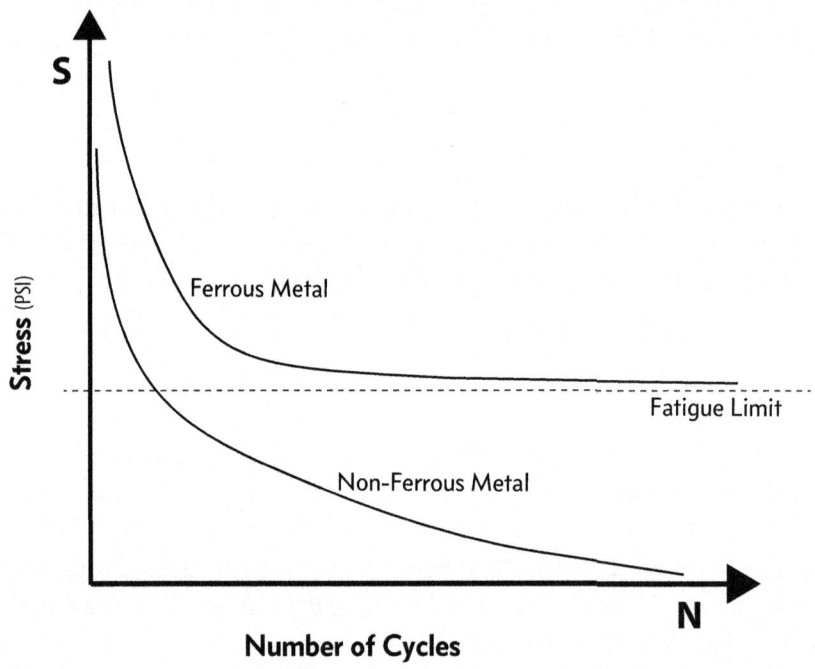

The S-N curve shows how fatigue life of a metal part varies with the magnitude of the repetitive stress applied. Ferrous metals have a "fatigue limit" stress below which their fatigue life becomes infinite; non-ferrous metals have no fatigue limit and will always fail eventually.

Piston engines and fatigue

Fatigue failures can occur to various critical components of a piston aircraft engine, including the crankcase, crankshaft, connecting rods, pistons, and cylinders. The crankshaft, connecting rods and cylinder barrels are made of steel, and are engineered to operate well below their fatigue limit, so that they have an infinite fatigue life—at least in theory. We do see a small number of fatigue failures of these steel parts, but only when the parts are improperly manufactured (e.g., bad steel), improperly assembled (e.g., incorrect torque), or stressed beyond their design limits (e.g., prop strike). History shows that fatigue failures of steel parts due to improper manufacture or assembly usually happen rather quickly after the engine enters service—typically within the first 200 hours, and often quite a bit less. They are known as *infant-mortality* failures.

On the other hand, the crankcase, pistons and cylinder heads are made of aluminum alloy, so they have no fatigue limit below which they're immune from fatigue failure. If a crankcase, piston, or cylinder head remains in service long enough and endures enough repetitive stress cycles, it WILL fail due to metal fatigue.

If the engine is operated within its design limits, the useful fatigue life of a crankcase, piston or cylinder head is something far in excess of recommended TBO—generally at least several times TBO. Pistons are always replaced at major overhaul, which is why we don't often hear of the fatigue failure of a piston (although I did have one piston fail from fatigue at about 225% of TBO due to a hidden manufacturing flaw). Crankcases are normally reused at major overhaul, so crankcase cracks are not uncommon. Fortunately, the crack growth rate in crankcases tends to be quite slow, so a careful inspection at each annual inspection is usually sufficient to detect crankcase cracks long before they reach critical length, and catastrophic fatigue failure of crankcases is almost unheard of.

Historically, cylinders have often been overhauled and reused at engine major overhaul, so it's not uncommon to see cylinders with heads that have been in service for many TBOs. Such high-time cylinder heads are thought to be at far greater risk of fatigue failure than are first-run heads that have been in service for one TBO or less. Furthermore, crack growth in a cylinder head can progress quite rapidly, so a once-a-year inspection is not sufficient to assure that fatigue cracks will be caught before the head fails catastrophically.

Head-to-barrel separations are not limited to Continental engines. Here's a Lycoming cylinder that lost its head. (NTSB photo)

Nowadays, most top-notch engine shops insist on installing new cylinders at major overhaul, which greatly reduces the likelihood of cylinder head fatigue failure. But it's not a guarantee. There have been bouts of infant mortality fatigue failures of cylinder heads due to improper manufacturing. At various times, cylinders manufactured by Continental, ECi and Superior have been subject to massive and costly ADs trigged by head-to-barrel separations.

The Superior Millennium AD

In April 2008, the FAA issued a Notice of Proposed Rulemaking (NPRM) announcing their intention to issue an AD against all investment-cast Millennium-brand cylinders manufactured by Superior Air Parts and installed on Continental IO-520, TSIO-520 and IO-550 engines. The concern was cracks in the area of the exhaust valve and separation of cylinder heads from barrels. The proposed AD would require repetitive 50-hour compression and soap-bubble testing of all investment-cast Millennium cylinders with more than 750 hours in service, and mandatory retirement of such cylinders at Continental's published engine TBO. The cost of compliance to aircraft owners was estimated by the FAA at $12,400,000.

Many of us in the industry (including me) were outraged by this NPRM. First, there had only been 24 reported failures of these Superior Millennium cylinders out of a population of some 8,000 cylinders, giving a failure rate of 0.3%—lower than the historical failure rate of Continental factory cylinders. Second, most of the reported failures occurred in Alaska-based airplanes operated by a single commercial operator in operations involving extraordinarily high numbers of maximum-performance takeoffs and landings per hour. Finally, these investment-cast Superior Millennium cylinders were—in the judgment of many piston aircraft engine experts—the best built, most durable, most efficient cylinders that had ever been offered for big-bore Continental engines. The FAA was proposing to legislate all of them out of existence because of a relatively trivial number of failures experienced almost exclusively by one operator in Alaska. Good grief!

Head-to-barrel separation of #5 cylinder on a Continental TSIO-520 engine.

Predictably, a hue and cry arose from the industry. Hundreds of comments were submitted to the FAA's rulemaking docket, almost all of them opposing adoption of the draconian proposed AD. Nevertheless, the FAA would not be deterred. On August 5, 2009, it issued the Final Rule on AD 2009-16-03 tolling the death knell for Superior investment-cast cylinders, and penalizing thousands of aircraft owners whose only crime was deciding to spend a bit more money to install top-of-the-line cylinders on their engines. If you ask me, this AD was a real travesty.

Next in the boresight: ECi Titans

On February 24, 2012, the National Transportation Safety Board issued Safety Recommendation A-12-7 to FAA acting administrator Michael Huerta, expressing great concern over 29 cylinder head fatigue failures of ECi Titan cylinders manufactured between 2003 and 2009 and installed on Continental IO-520, TSIO-520 and IO-550 engines. The NTSB recommended that the FAA issue an AD requiring repetitive inspections of these cylinders every 50 or 100 hours, and mandatory retirement of the cylinders at Continental's published TBO, but the FAA rejected this recommendation and—after a bitter four-year fight—mandated a far more expensive remedy. AD 2016-16-12 became effective in September 2016, and required removal from service of up to 30,000 ECi cylinders based on time-in-service, at an estimated cost in excess of $30 million. And to what end?

According to figures furnished to the FAA by ECi, the lion's share of the affected cylinders fell into three groups. One group of 7,797 cylinders manufactured in 2003 and 2004 installed on Continental IO-520 and TSIO-520 engines had 15 reported failures for a failure rate of 0.19%. Another group of 12,339 cylinders manufactured from 2005 to 2008 installed on TSIO-520 engines has had 9 reported failures for a failure rate of 0.07%. A third group of 5,232 cylinders manufactured from 2005 to 2008 installed on IO-520 engines had no reported failures and a perfect record to date. These failure rates are extraordinarily low—lower than those seen in investment-cast Superior Millenniums, and lower than the historical norms for Continental OEM cylinders. There was never a documented case of the failure of an ECi Titan cylinder causing an accident or injury.

By any historical measure, these ECi Titans had proven themselves to be extremely reliable—but apparently not reliable enough for the FAA. I can't help but wonder just how low a failure rate it would take to satisfy the Feds. Is it technically possible to build a totally failure-proof cylinder? Even if it is, could we afford to buy them? Do you suppose that the $30 million this AD against ECi Titans cost aircraft owners might have been better spent on something else…perhaps recurrent training to make those owners safer, more proficient pilots?

Why do they fail?

Rare as these head-to-barrel separation failures are, it's important to understand what makes them happen. I had the opportunity to discuss the matter at considerable length with ECi's legendary chief engineer Jimmy Tubbs—who probably knows more about head-to-barrel junctions than anyone else on the planet—and to study the results of research studies performed by his engineering group in San Antonio. As a result, I've become convinced that the key to preventing these failures doesn't lie in more FAA rulemaking or in building more robust cylinders, but rather in educating the pilots and aircraft owners who fly behind these cylinders.

Any aircraft cylinder—whether made by Continental, Lycoming, Superior or ECi—will fail eventually if allowed to remain in service long enough. Cylinder heads are made of aluminum alloy, and like all non-ferrous metals they have a finite fatigue life. A fundamental principle of Metallurgy 101 is that any non-ferrous metal will ultimately fail from fatigue if subjected to enough repetitive stress cycles. Fatigue failure of an aluminum alloy cylinder head is not a matter of "if," only a matter of "when."

If these cylinders are treated right, they can last for a long, long time. Case in point: 9 of the 12 cylinders on my 1979 Cessna Turbo 310 saw continuous service for 33 years and made it to 3.4 times Continental's published TBO without cracking. Why did my cylinders last so long while others failed catastrophically before attaining even one TBO?

I've long been convinced that the answer to cylinder head longevity lies in keeping the heads cool. The strength of aluminum alloys drops sharply as temperature increases. At 400°F, the head casting loses half of its strength; by 500°F, it loses three-quarters

At 400°F, the head casting loses half of its strength; by 500°F, it loses three-quarters of its strength.

of its strength. That's why I'm obsessive about keeping my CHTs well-controlled. I always try to keep all CHTs at 380°F or less. The CHT alarm on my engine monitor is set to go off at 390°F, and when it does, I take immediate action to bring the CHT down: I lower the nose if I'm climbing, open the cowl flaps, richen the mixture if I'm ROP, lean more if I'm LOP...whatever it takes.

But ECi's research suggested that this loss of tensile strength at elevated CHT was only part of the story. To understand the rest of the story, we need to know something about the anatomy of the junction that attaches the cylinder's aluminum head casting to its steel barrel.

Anatomy of the junction

The head-to-barrel junction is a threaded interface, with internal threads machined into the head mated to external threads machined onto the barrel. During manufacture, the head casting is heated in an oven to about 650°F, while the barrel is chilled in a refrigerator. The hot head and the cool barrel are then quickly screwed together until the top of the barrel bottoms out against the abutment surface of the head casting. As the temperatures of the head and barrel equalize, the head shrinks and the barrel expands to create an "interference fit" that locks the two together permanently.

Before talking to Jimmy, I'd been under the mistaken impression that it was the threads that held the head and the barrel together. It turns out that's wrong. The real strength of the head-to-barrel junction lies in the non-threaded area above the threads, variously known as the "seal band" or "friction band" or "shrink band." It is primarily the friction of the interference fit in the seal band area that gives the junction its strength. In fact, ECi's specs provide for a slightly tighter interference fit at the seal band than at the threads (by a few thousandths of an inch), because their testing revealed that this creates a more robust junction.

The "tightness" of the interference fit is a function not only of manufacturing tolerances, but also of operating temperature. As the cylinder temperature rises, both the head and barrel expand. But because the coefficient of expansion of aluminum is about twice that of steel, the head expands twice as much as the barrel. Thus, as CHT increases, the interference fit of the head-to-barrel junction loosens.

If the CHT becomes high enough, the seal band can start to lose its grip, transferring stress to the threaded portion of the junction. When that happens, the top thread

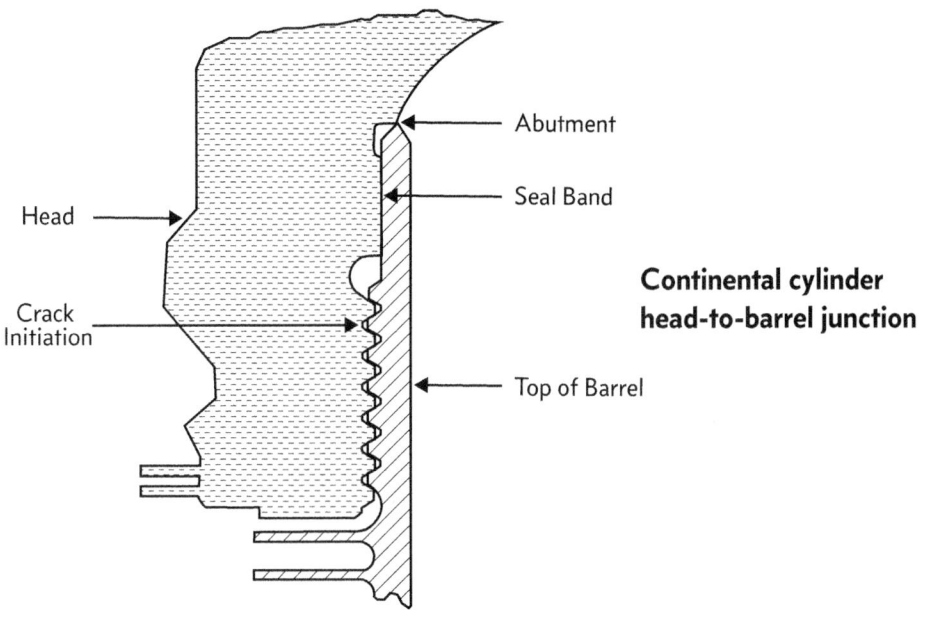

The strength of the Continental cylinder head-to-barrel junction lies in the interference fit of the non-threaded "seal band." If the seal band loses its grip, crack initiation occurs at the topmost thread.

takes the lion's share of the stress, and ultimately a crack will initiate at that location that can grow over time to the point that the junction fails catastrophically.

How hot does the CHT have to be before this happens? There are two major factors that affect the junction's tolerance for high CHT. We've already mentioned one of them: The amount of interference fit machined into the parts during manufacture. A second factor identified by ECi's studies—and a surprising one—is that junction strength is adversely affected by high blow-by past the compression rings, particularly if the new cylinder is not properly broken in and the barrel becomes glazed before break-in is complete. High blow-by appears to weaken the head-to-barrel junction because the associated hot gases heat up the junction and impair the interference fit without being reflected on the CHT gauge (which measures cylinder head temperature at a location quite distant from the junction). Thus, a high-blow-by cylinder will have a hotter junction temperature (and thus a weaker junction) than a low-blow-by cylinder with the same indicated CHT.

Operational issues

In addition to time-in-service and manufacturing errors, there are a number of operational issues that can affect the useful fatigue life of engine parts. One of the most important is corrosion. Corrosion creates surface pits that can serve as initiation sites for fatigue cracks and greatly foreshorten useful fatigue life. In fact, serious corrosion can result in fatigue failure of ferrous metal parts even though they are operated within their design fatigue limit and should theoretically be immune from fatigue failure.

> Head-to-barrel separations failures could be drastically reduced--perhaps even eliminated altogether--if we could persuade pilots to keep their CHTs under control.

Fatigue life is also profoundly affected by the magnitude of the repetitive stress cycles. Because aluminum engine parts are normally operated in a very flat portion of the S-N curve—low S (stress), high N (cycles)—a small increase in stress can result in a large decrease in fatigue life. If we're talking about cylinder heads, stress S is a function of peak internal cylinder pressure, which is affected both by power settings and mixture management. We know, for example, that peak internal cylinder pressure is maximum at a mixture setting of roughly 50°F rich of peak EGT (ROP), and is considerably lower at richer mixtures (e.g., 125°F ROP) or leaner mixtures (e.g., 25°F LOP). Thus, cylinders operated at 50°F ROP (as recommended in many POHs) are more likely to suffer fatigue failure than cylinders operated substantially richer or leaner.

Detonation, pre-ignition, and advanced ignition timing can all result in abnormally high peak internal cylinder pressures, and can drastically reduce the fatigue life of cylinder heads and pistons. Any time spark plug or borescope inspection reveals the tell-tale signs of detonation or pre-ignition, replacement of the affected cylinder and piston is prudent.

Preventing these failures

After reviewing all the data, I can't help but conclude that many head-to-barrel separation failures would be drastically reduced—perhaps even eliminated altogether—if we could persuade pilots to keep their CHTs under control, and teach them how to break in new cylinders properly. A big part of this is pilot education: so many pilots see that the green arc on their CHT gauge extends all the way up to the red-line at 460°F for Continental engines and 500°F for Lycomings, and think that there's nothing wrong with running their CHTs well up into the 400s. Wrong! We need to teach the critical importance of proper temperature control.

Another huge part of the problem is lack of proper CHT instrumentation. There are still way too many airplanes flying with nothing but the original factory CHT gauge that instruments only one cylinder and leaves the remaining three or five cylinders unmonitored. To make things worse, this legacy CHT instrumentation is seldom calibrated and often wildly inaccurate, sometimes displaying CHT values as much as 50°F less than what it actually is.

I'm convinced that if every piston-powered aircraft was equipped with a modern probe-per-cylinder digital engine monitor that set off an alarm anytime any CHT rose above 390°F for Continentals or 410°F for Lycomings, cylinder head-to-barrel separation failures would become largely a thing of the past.

Was it just bad luck?

Why did the Skylane owner's O-470 engine suffer two head separations in the space of two hours? It's impossible to say for sure, but we can certainly make some educated guesses.

It is clear from the Skylane owner's photos that the engine compartment suffered from substantial corrosion while the aircraft was based in Florida. It also sounds more likely than not that the cylinders were reconditioned and reused when the engine was last major overhauled. The reconditioned cylinders may have been dimensionally restored to new fits and limits, but there's no way to restore the fatigue life of a cylinder head. As the saying goes, "metal never forgets."

The particular O-470 engine involved turned out to be a high-compression version of the engine (O-470-U) with an 8.6-to-1 compression ratio (similar to IO-470, IO-520 and IO-550s). It therefore had substantially higher peak internal combustion pressures than the older O-470-R and -S variants that were designed with a 7.0-to-1 compression ratio to run on 80/87-octane avgas. High compression ratio engines are more efficient, but place higher stresses on cylinder heads and other engine parts.

We have no way of knowing how the engine was operated during the life of those cylinder heads, but since the aircraft didn't have an engine monitor and the owner didn't mention his leaning procedure, it seems likely that the engine was operated "by the book" at around 50°F ROP, which we know is the highest-stress condition. If this aircraft had been equipped with an engine monitor and the pilot had been trained to use it properly, there's a good chance that the cracked head might have been detected prior to the point where complete head separation occurred.

The fact that the #3 cylinder suffered a head separation just two hours after the #1 cylinder failed is certainly extraordinary. It makes me wonder whether perhaps the engine suffered some sort of detonation or pre-ignition event that caused the cylinders to be stressed beyond their design limits. With no engine monitor data available, it's impossible to know for sure.

PART III
Bottom End

11
Inside the Case

A peek at the vital organs hidden away inside your engine's ribcage.

Reciprocating aircraft engines come in a variety of different cylinder arrangements—radial, inline, V, and opposed—but most engines used in piston general aviation are horizontally opposed four- and six-cylinder engines, and those are the ones we'll focus upon here. The horizontally opposed design has a higher horsepower-to-weight ratio than other cylinder arrangements, better vibration characteristics, and a flat profile well suited to streamlined cowlings.

The engine has two banks of cylinders—left and right—directly opposite each other, with a single crankshaft between them. The crankshaft is enclosed in a crankcase, which also contains the camshaft, connecting rods, bearings, gears, and various other components that are collectively referred to as "the bottom end" of the engine. If something goes wrong with any of these bottom-end components, it becomes necessary to remove the engine from the aircraft and "split the case" to gain access to them.

Most piston general aviation engines are horizontally opposed.

Crankcase

Think of the crankcase as the engine's ribcage. Besides supporting itself, the crankcase must support all the other components of the engine—internal ones like the crankshaft and camshaft, and external ones like the cylinders and engine-driven accessories. It must provide a liquid-tight enclosure for the lubricating oil that bathes all of the engine's components when the engine is running. It must also incorporate provisions for attaching the powerplant to the airframe.

Crankcases of piston GA engines are made of cast aluminum alloy that is both lightweight and strong. Strength is important because the crankcase is subjected to tremendous reciprocating loads from the cylinders, as well as centrifugal and internal forces of the crankshaft's bending moments. In addition, the propeller places thrust forces on the crankcase, and potentially severe centrifugal and gyroscopic twisting loads during abrupt maneuvering. The crankcase must be stiff enough to withstand all these loads without major deflection.

The case consists of two halves—left and right—held together by a series of long through-bolts that pass through both case halves just above and below the crankshaft, plus a series of short flange bolts that connect the top and bottom flanges. Each half of the crankcase contains transverse bearing supports, one for each main bearing that supports the crankshaft. These bearing supports are an integral part of the crankcase structure, and add considerable strength and rigidity to the crankcase in addition to supporting the crankshaft main bearings and the camshaft. The surfaces of the crankcase halves—both the bearing supports and the top and bottom flanges—are known as the parting surfaces, and must be machined perfectly flat to ensure that the case halves can be assembled in a structurally sound and liquid- and gas-tight fashion. That's because, for structural reasons, the parting surfaces must be in direct metal-to-metal contact with no gaskets to seal the parting seam.

The two halves of the crankcase are perfectly machined to provide a gasketless seal.

The machined crankcase surfaces on which the cylinders are mounted are called pads or decks. They must include a suitable means of fastening the cylinders to the crankcase. Most opposed engines use a combination of hold-down studs fitted into threaded holes in the crankcase, and long through-bolts that run horizontally all the way though the crankcase and serve both to clamp the left and right crankcase halves together at the bearing supports, as well as to help attach a pair of opposing cylinders to the crankcase. Each cylinder has a mounting flange with holes to accommodate the hold-down studs and through-bolts. Each cylinder also has a skirt that extends a considerable distance inside the crankcase. Both Continental and

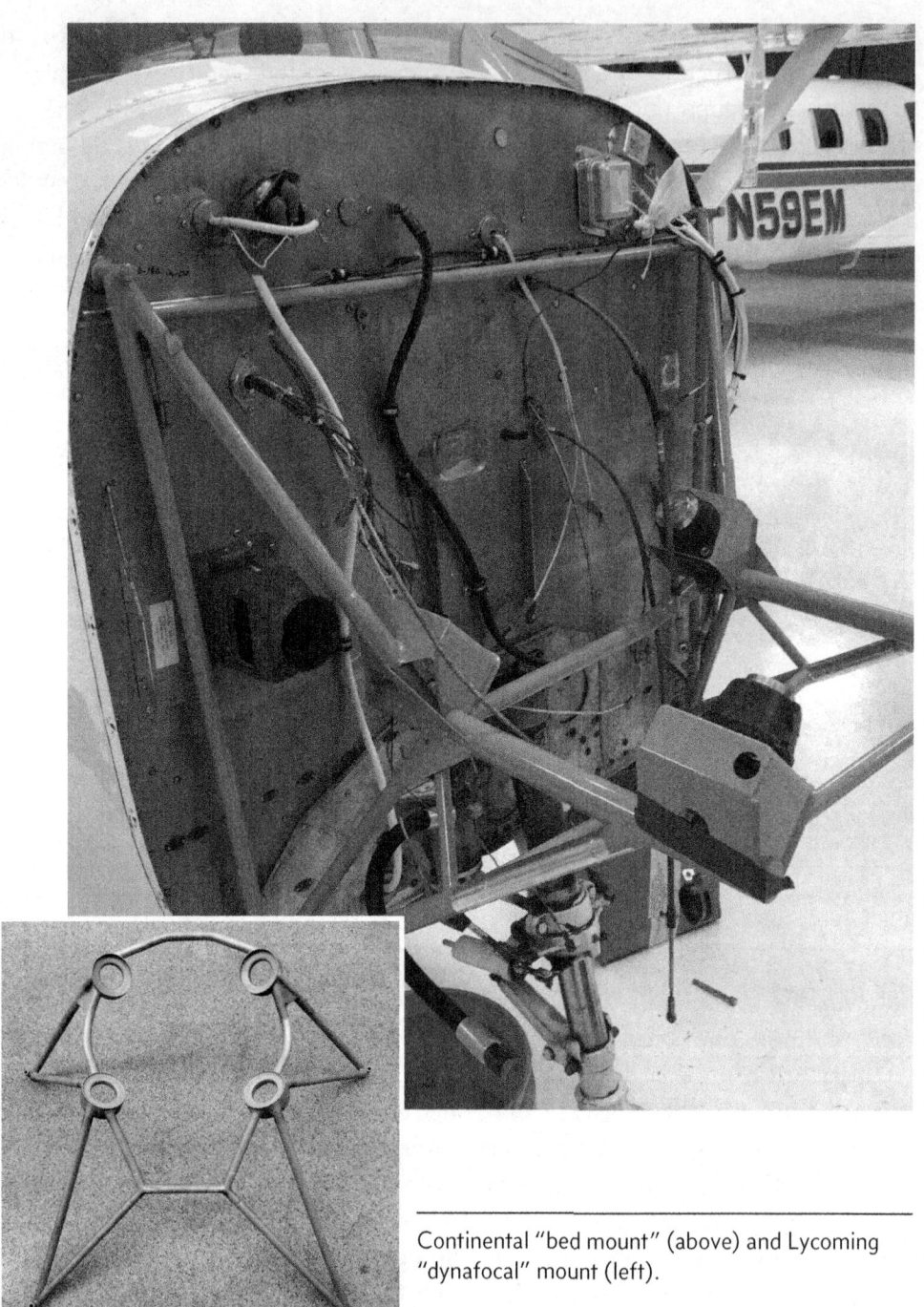

Continental "bed mount" (above) and Lycoming "dynafocal" mount (left).

Lycoming engines use a total of eight threaded fasteners—six hold-down studs and two through-bolts—to attach each cylinder base flange to the crankcase deck. Both also use large cylinder-base O-rings to provide a liquid-and gas-tight seal between the cylinders and the crankcase.

The crankcase incorporates mounting points for attaching the powerplant to the airframe. Most Continentals use a "bed mount" with four or six attachment points on the bottom of the case, while most Lycomings use a "dynafocal mount" with four attachment points on the rear of the case. These mounting points are designed to accommodate elastomeric shock mounts that absorb engine vibrations and prevent them from being transmitted to the airframe.

The crankshaft extends out of the crankcase both front and rear. The exposed portion of the crankshaft at the front of the crankcase incorporates the propeller mounting flange or splines. The nose portion of the crankcase incorporates a large bearing designed to withstand the centrifugal and gyroscopic propeller loads, and a front crankshaft seal to provide a liquid- and gas-tight seal. The exposed portion of the crankshaft at the rear of the crankcase accommodates a large crankshaft gear that drives the camshaft and various engine accessories.

Accessory case and gear trains

A cast aluminum accessory case is bolted to the rear of the crankcase to enclose the engine's gear trains and to provide mounting pads for accessories such as magnetos, pumps (fuel, oil, hydraulic, pneumatic), starter, alternator, tachometer drive or generator, propeller governor, etc.

Gear trains inside the accessory case drive the camshaft, fuel and oil pumps, and other accessories. The gear ratios are designed to drive each component at the proper speed. For example, the camshaft turns at exactly one-half crankshaft speed, so the camshaft gear is twice the diameter—and has twice as many teeth—as the crankshaft gear that drives it. Magnetos in four-cylinder engines turn at crankshaft speed, but magnetos in six-cylinder engines turn at $1\frac{1}{2}$ times crankshaft speed.

Crankshaft

If the crankcase is the engine's ribcage, the crankshaft is its backbone. It's a massive piece of high-strength chromium-nickel-molybdenum steel, forged for strength, machined and polished for smoothness, and case-hardened for durability. It's by far the costliest component of the engine—a new one could set you back between $10,000 and $30,000.

The crankshaft's primary function is to transform the reciprocating motion of the pistons and connecting rods into rotary motion that can drive the propeller. It's called a "crankshaft" because it's a shaft composed of multiple "cranks" (also known as "throws") along its length—one per cylinder in horizontally opposed engines. These are formed by forging offsets into the shaft before it is machined. Each of these cranks

The crankshaft converts piston motion into prop rotation.

is composed of a polished "crankpin" offset from the main axis of the crankshaft by a pair of "cheeks." These cheeks typically extend to the opposite side of the main axis from the throw to provide counterweights that keep the crankshaft in balance.

The crankshaft is supported in the crankcase by a main bearing at each end and between each throw. It is machined with polished journals that mate with the main bearings that are mounted on the crankcase's bearing supports. The main bearings and journals define the rotational axis of the crankshaft. The crankpins are offset from this axis by a distance that determines the "stroke" of the pistons and connecting rods. The stroke of the piston is twice the offset of the crankpin.

The main journals—as well as the crankpins—are typically bored hollow to reduce crankshaft weight, and nitrided to provide a very hard and long-wearing bearing surface. (Nitriding is a case-hardening process whereby the crankshaft is baked in a high-temperature oven for many hours in an atmosphere of ammonia gas; the nitrogen from the ammonia combines with the polished surfaces of the crankshaft to create a hardened outer shell about .015" thick.) A diagonal oil passage running from each main journal to its adjacent crankpin provides lubrication to the crankpin from oil that is pumped into the main bearing by the engine's oil system.

The crankshaft is kept in balance by means of counterweights that are simply extensions of the cheeks on the opposite side of the crankshaft from the throw. During manufacture or overhaul, the crankshaft is placed on two knife edges, and small amounts of metal are removed from the cheeks until the shaft has no tendency to turn toward any one position during the test.

Some crankshafts—notably those on six-cylinder engines—also incorporate movable counterweights called harmonic dampeners that help relieve the crankshaft of torsional stresses caused by the power pulses applied to the crankshaft throws, particularly the aft most ones that are furthest way from the propeller. These harmonic dampeners are attached to blades forged into the cheeks of the aft crankshaft throws by means of steel pins and oversized hanger bushings in a fashion that allows them to move and act as a pendulum. The distance the pendulum can move determines its resonant frequency, and this is carefully tuned to cause the pendulous counterweights to oscillate out of phase with the power pulses, thus absorbing some of the pulsatile energy and stress-relieving the crankshaft.

Connecting rods

The connecting rod transmits force between the reciprocating piston and the rotating crankshaft. It's easily the most highly stressed component of the engine. It must be strong enough to remain rigid under load and yet be light enough to minimize reciprocating inertia—the directional reversal that occurs at the start and end of each stroke of the piston.

The connecting rod has a "big end" that attaches to the crankpin and is fitted with a two-piece bearing, and a "small end" that attaches to the piston via the piston pin that rides in a press-fit bushing. The big-end bearing boss is sliced transversely to create a removable cap that is attached to the rod with a pair of high-strength close-tolerance rod bolts and nuts. The assembly torque of these highly stressed rod bolts is so critical that most Lycoming engines have the bolts torqued to a specified amount of bolt stretch (measured with a special large micrometer) rather than relying on a conventional torque wrench. After torqueing, the rod nuts are usually safetied with a cotter pin, although many Continental engines now use self-locking Spiralock nuts instead.

Camshaft

For an Otto-cycle reciprocating engine to operate properly, each valve in each cylinder must open and close at the proper time. Intake valves start to open just before the piston's intake stroke begins, and start to close just after the intake stroke ends. Exhaust valves start to open just before the exhaust stroke begins, and start to close just after the exhaust stroke ends. There's a brief time when both valves are open simultaneously—known as the valve overlap period—and an extended time when both valves are closed simultaneously—during most of the compression and power strokes.

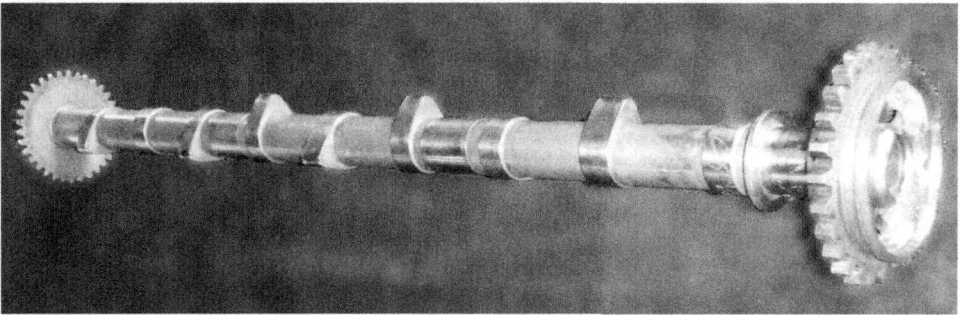

The lobes of the camshaft operate the cylinder valves

The camshaft is responsible for operating the valves and controlling the timing of their openings and closings. It is gear-driven from the crankshaft with a 2-to-1 reduction gear train that causes it to rotate at exactly one-half crankshaft speed. The camshaft typically has one eccentric cam lobe for each exhaust valve on each cylinder, and one double-wide cam lobe for each pair of intake valves on opposing cylinders. Thus, there are six cam lobes in a four-cylinder engine (four exhaust, two intake), and nine cam lobes in a six-cylinder engine (six exhaust, three intake). The cam lobes are highly polished and case hardened to provide a very hard and durable wear surface, using a process called "carburizing" (like nitriding except that the baking is done in an atmosphere of carbon monoxide instead of ammonia, creating a carbon-rich hardened wear surface about .015 inches thick).

Each cam lobe is carefully profiled to provide the necessary valve lift, duration and timing. A cam follower—also referred to as a tappet or lifter—rides on the cam lobe and converts its profile into reciprocating motion that controls the opening and closing of the valve. A pushrod transmits this motion to a rocker arm at the top of the cylinder assembly, and the rocker presses on the tip of the valve stem to cause the valve to open at the proper time. Valve springs surrounding each valve stem cause the valve to close when the pressure from the cam-lifter-pushrod-rocker linkage is relieved. Valve springs come in concentric pairs with different resonant frequencies in order to eliminate spring-surge vibrations that might otherwise interfere with their ability to hold the valve firmly closed against the valve seat.

Most horizontally opposed engines use hydraulic tappets that automatically compensate for any slop in the linkage, eliminating the need for valve lash adjustments. Such tappets contain a plunger, spring and ball that cause the tappet to pump up with engine oil to the necessary length to achieve zero lash. The hydraulic tappets also supply engine oil through the hollow pushrods to provide lubrication to the rockers and cooling to the valve stems and springs.

12
Not-So-Plain Bearings

There's a lot more to engine bearings than meets the eye.

According to Miriam-Webster, a bearing is "a machine part in which another part turns." Most aircraft have lots of them.

Wheels spin on their axles with the help of tapered roller bearings. Magnetos, alternators, generators and starter motors incorporate ball bearings to support their rotors. The landing gear trunions on my Cessna 310 pivot on needle bearings. Variable-pitch propeller blades are supported by large-diameter ball bearings. Turbine engine rotor shafts spin in ball and roller bearings. All these bearings consist of inner and outer "races" with spherical or cylindrical rolling elements between them. Such "rolling-element bearings" do a superb job of supporting a shaft in precise position while permitting it to rotate with very little friction.

But tear down a Continental or Lycoming engine and you won't find bearings like those. The bearings in which the crankshaft, crankpins, camshaft, rocker shafts and piston pins run have no races, balls, rollers, needles or other moving parts. They're just

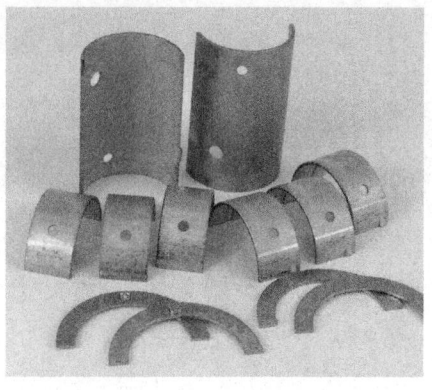

Main crankshaft bearings. The elongated ones at the top are front mains, and the ones at the bottom are thrust washers.

curved pieces of metal—known variously as "plain bearings" or "sleeve bearings" or "bushings"—that rely on sliding elements rather than rolling ones. Plain bearings are usually constructed of two semicircular halves called "shells"; one-piece plain bearings are usually called "bushings."

There's a good reason that reciprocating engines use plain bearings: They reciprocate! This means that the crankshaft, crankpin, piston pin, rocker shaft and camshaft bearings are subject to continuous sharp cyclic loads. Rolling-element bearings don't handle such loads very well, because they concentrate loads into very tiny contact regions between the rolling elements and the races, resulting in extremely high pressures. If ball bearings were used in a reciprocating engine, the result would be peened races, flat-spotted balls, and consequent short bearing life.

By contrast, plain bearings tolerate such cyclic loads better because they distribute the loads over a much larger area, so the pressure is greatly reduced. They also do a better job of accommodating shaft flexing, minor misalignments, and wide temperature swings. That's why nearly all reciprocating engines—from one-cylinder motorcycle engines to giant marine diesels—use plain bearings instead of ball or roller bearings.

These plain bearings and bushings look simple, but they aren't. There's a lot more to them than meets the eye.

Lubrication

When I had the engines in my Cessna 310 torn down for overhaul in 1990, I made a point of paying a visit to the engine shop to survey the damage before the engine was put back together. The engines had accumulated 1900 hours over 11 years, and I remember being rather astonished at the appearance of the main and rod bearings; they looked nearly brand new!

How can a plain bearing with no rolling elements withstand the torturous environment of a high-performance turbocharged reciprocating aircraft engine for 1900 hours and 11 years without showing any significant signs of wear? There's a two-word answer: hydrodynamic lubrication.

Plain bearings rely on hydrodynamic lubrication to prevent metal-to-metal contact between the rotating journal and the stationary bearing. Pressurized oil is pumped continuously into the gap between the journal and the bearing. This gap is only about .002" wide—about the thickness of a human hair. Rotation of the journal within the bearing, together with the viscosity of the oil, creates a dynamic wedge of high-pressure oil that keeps the parts separated. So long as the bearing gets adequate oil pressure and the journal rotates rapidly enough, there is no metal-to-metal contact and therefore no wear on either the bearing or the journal.

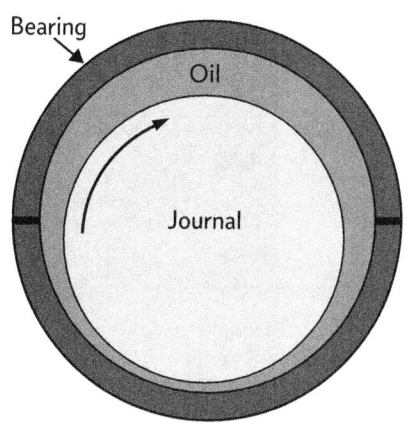

Plain bearings rely on hydrodynamic lubrication to prevent metal-to-metal contact between the rotating journal and the stationary bearing.

Does this mean that plain bearings can last forever? Actually yes, provided the engine is run continuously (as it might be in a test cell) with an uninterrupted oil supply and uninterrupted journal rotation. Unfortunately, that's not what happens in the real world. We start up our engines, run them for an hour or three while we're flying from point A to point B, and then shut them down. It's mainly those pesky startups and shutdowns that limit the useful life of plain bearings.

When we first crank the engine, there's no oil pressure and the crankshaft cranking speed is pathetically slow. The conditions for hydrodynamic lubrication simply do not exist. Consequently there is metal-to-metal contact between the journal and the bearing, and wear is inevitable.

This startup wear is mitigated in several ways: The mating journal and bearing surfaces are polished as smooth as possible. The bearing surface is made of a material that has low sliding friction against the steel journal. Anti-wear additives in the oil

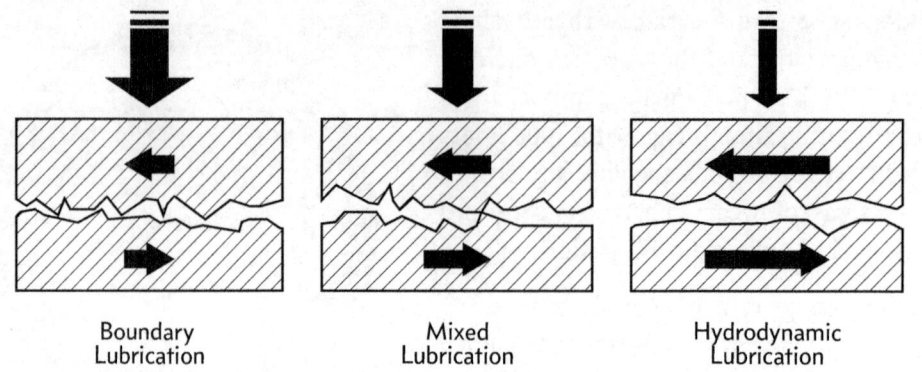

Hydrodynamic lubrication depends on adequate oil pressure and journal rotation speed. If those aren't present (as at engine startup and shutdown), the parts must depend on boundary lubrication.

Oil supply passages (black arrows) in each main bearing saddle in the crankcase provide a flow of pressurized oil to the main bearings.

react with the metal to form a thin protective film on the surfaces that further reduces friction through a phenomenon known as "boundary lubrication."

Getting oil to the bearings

Once the engine is running, pressurized oil passes through "galleries" (often misspelled "galleys"), which are internal passageways machined into the crankcase halves to conduct oil to key components such as main bearings and hydraulic lifters. Oil passages in each main bearing saddle connect to the galleries. Each main bearing shell has an oil supply hole that lines up with the oil passage in the saddle.

Getting oil to the crankpin bearings in the big ends of the connecting rods is a bit trickier. The crankshaft is machined with diagonal passages to conduct oil from the main bearings to the crankpin journals.

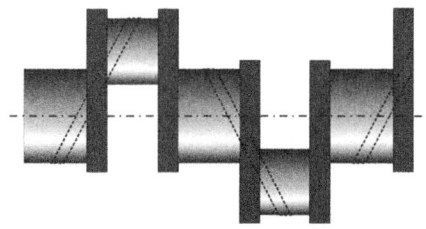

The crankshaft is machined with diagonal oil passages to conduct oil from the main bearings to the crankpin journals in order to lubricate the big-end rod bearings.

Oil pumped into the main and rod bearings is extruded from the edges of the bearings and returns to the engine's oil sump or tank. The rapidly rotating crankshaft flings this oil in all directions, filling the crankcase with a dense oil mist that provides "splash lubrication" to other engine components like the cam lobes, lifter faces, piston pin bushings and cylinder barrels.

Material properties

Plain bearings must have a running surface that will slide against steel journals with low friction and be highly resistant to galling (i.e., adhering). Because the bearing is exposed to high cyclic loads and high temperatures, its running surface must have good fatigue strength and retain that strength under high heat. It also needs to be resistant to corrosive attack by moisture and acids. Yet another key property is called "embeddability" which means that the bearing's running surface is soft enough to allow small particles of dirt, metal or other foreign material to become embedded in

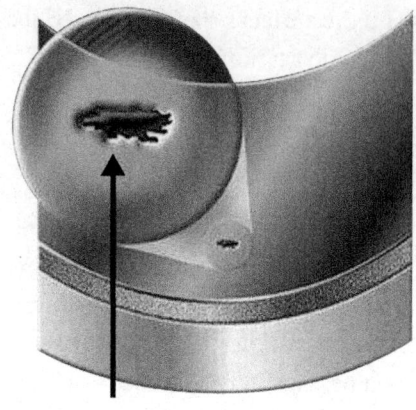

"Embeddability" means the ability of a bearing's running surface to capture particles of dirt, metal or other foreign materials.

the bearing so it doesn't scratch the crankshaft journal or jam the tiny oil clearance zone between the journal and the bearing.

Most bearings in piston aircraft engines have a running surface made of a family of alloys known as "babbitt" or "white metal" that are about 90% tin combined with small quantities of antimony and copper. (The term "babbitt" derives from Isaac Babbitt, who invented the original version of this bearing alloy in 1839 in Taunton, Massachusetts.) Babbitt offers exceptional slipperiness and embeddability, but its fatigue resistance and temperature strength deteriorate rapidly if more than a few thousandths of an inch thick. Consequently, most piston aircraft engine bearings use a layered construction referred to as "trimetal" although they actually have four or five layers.

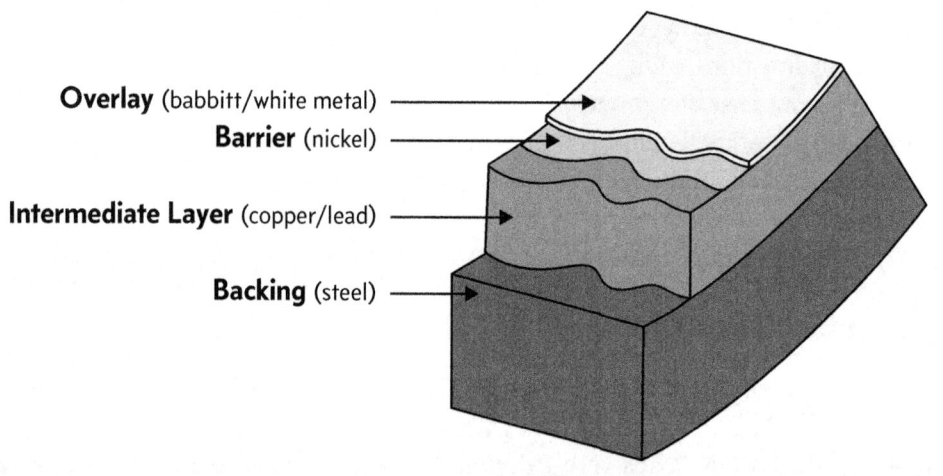

A trimetal bearing consists of a steel backing, an intermediate layer made of a copper/lead alloy, and a thin babbitt overlay.

Trimetal bearings

The bearing shell starts out with a semi-circular steel backing, to which is bonded an intermediate layer of copper/lead alloy designed to provide the necessary cushioning, fatigue strength and temperature conduction. A thin overlay of babbitt is then electroplated over the intermediate layer in order to provide the required surface properties (slipperiness, embeddability, corrosion-resistance). A micro-thin layer of nickel is deposited between the intermediate layer and overlay in order to prevent tin from migrating from the babbitt into the copper/lead alloy. Sometimes, another micro-thin layer of pure tin is deposited on top of the babbitt overlay to protect it from corrosion prior to installation in the engine.

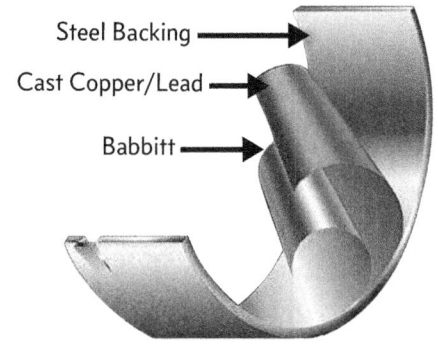

Trimetal bearing construction.

It can be useful to think of the construction of a trimetal bearing as being like a bed: The steel backing is like the box spring (providing support), the intermediate layer is like the mattress (providing cushioning), and the overlay is like a silk sheet (thin, smooth, slippery and comfortable).

The surface properties (slipperiness and embeddability) of the copper/lead intermediate layer are not nearly as good as those of the babbitt overlay, but they're generally adequate to prevent the bearing from self-destructing suddenly in the event the overlay is worn away. If that happens, the engine will start "making copper" that will be apparent in oil analysis and (if it gets bad enough) as visible nonferrous metal in the oil filter. That means it's teardown time.

NOTE: During the period 1995 through 2001, Lycoming used bearings with an intermediate layer composed of aluminum/tin alloy instead of copper/lead alloy. These aluminum bearings proved to have lower fatigue strength and a higher failure rate, so Lycoming stopped using them in February 2002.

Securing the bearings

The trimetal bearing shells are mounted in semicircular "saddles" in the crankcase, and secured firmly in place by means of a preload referred to as "bearing crush." The steel backing of the bearing shell is slightly taller (by a few thousandths of an inch) than the crankcase saddle in which the shell is installed.

When the engine is assembled and the through-bolts are torqued to spec, the bearing shell is firmly locked into the saddle with an interference fit.

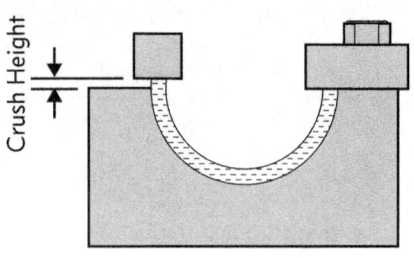

Bearing shells are slightly taller than their supporting saddles, creating a preload (called "bearing crush") when the through-bolts are torqued to spec, firmly locking the bearing shells in place.

IMPORTANT: If this bearing crush is lost—due to improper crankcase machining at overhaul or improper through-bolt torque during cylinder installation or replacement—the bearing can shift in its saddle, creating misalignment of the bearing's oil supply hole with the saddle's oil passage and causing impaired oil flow to the bearing. This condition—referred to as a "spun" or "displaced" bearing—can result in catastrophic engine failure.

Failure modes

Although trimetal bearings usually have excellent longevity when operated within design limits and provided adequate lubrication, premature bearing failure does occur. Here's a rod bearing whose babbitt overlay is almost completely worn away, leaving the copper/lead intermediate layer fully exposed.

The rod bearing shown at the top of the next page is badly contaminated with large amounts of foreign material (mostly metal particles) embedded in its babbitt overlay. Catastrophic failure (thrown rod) was probably not far away.

The rod bearing shown at the bottom of the next page exhibits severe damage due

A severely worn bearing with the babbitt overlay almost completely gone.

to cyclic overload, most likely destructive detonation or preignition in the associated cylinder. The extreme peak combustion pressures created loads on the connecting rod that exceeded the ability of hydrodynamic lubrication to prevent metal-to-metal contact between the rod bearing and crankpin. Undoubtedly the cylinder head and piston also sustained damage. You can tell all of these are rod bearings (rather than main bearings) because they don't have oil supply holes. As a general rule, rod bearings are more highly stressed and susceptible to damage and failure than main bearings.

The (not so) plain bearings in your piston aircraft engine will generally provide trouble-free service to TBO and often well beyond. (The ones in my Cessna 310's engines still looked great at 225% of TBO.) To accomplish that, all they really need is a steady supply of clean oil at a decent pressure (to provide good hydrodynamic lubrication), and some anti-wear additives to provide good boundary lubrication at engine start. Contamination is the bearing's main enemy, so make sure you have a good full-flow oil filter capable of trapping any particulate matter larger than about .001 inches (25 microns).

Inspect the filter regularly and watch your oil analysis reports for elevated copper or tin. If the filter is clean and the oil report looks normal, then you have nothing to worry about.

A badly contaminated bearing with large amounts of foreign material embedded in its babbitt overlay.

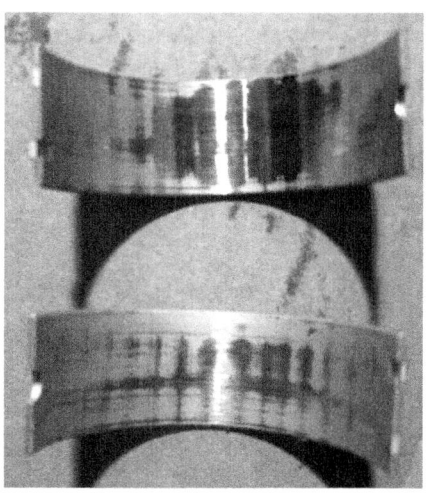

This bearing exhibits severe overload damage, probably due detonation or preignition.

13

Corrosion—Powerplant Enemy #1

If your engine fails to make TBO, it's probably not because you wore it out, but because you let it rust out.

An aircraft owner—let's call him Sam—once contacted me to ask if I could recommend any engine overhaul shops in the Pacific Northwest. I responded with several specific shop recommendations, and a dialogue ensued during which Sam told me the unfortunate circumstances that led to his query.

Sam had recently purchased his first twin. After a lengthy search, he located an aircraft that appeared to be in pristine condition in a hangar in northern New Jersey. The aircraft had excellent logs, mid-TBO engines, and had received its last half-dozen annual inspections at a well-known maintenance facility in the Midwest with a sterling reputation. Shortly after its annual inspection, the owner stopped flying (for whatever reason) and put the aircraft up for sale.

By the time Sam found the airplane, it had been sitting unflown in its northern New Jersey hangar for about eight months. Sam had the plane ferried back to the same well-known Midwest shop that had performed the annual, this time for a prebuy examination. The airplane had flown only seven hours since the annual eight months earlier, almost all of that ferry time from the Midwest shop to the New Jersey hangar and back.

During the prebuy, the shop cut open the oil filters and didn't find any metal. It also borescoped the cylinders and said they looked fine. The airframe was opened up and inspected for corrosion, but nothing significant was found. Sam was satisfied, purchased the airplane, and flew it home to Seattle.

Uh oh!

Everything seemed fine until just 30 hours later when Sam noticed that he had a problem keeping the props synchronized during a flight from Seattle to San Diego. Then, on the return flight to Seattle, he lost a vacuum pump. (Welcome to the joys of twin ownership!)

Back in Seattle, Sam asked his mechanic to replace the vacuum pump and check into the prop-sync problem. While inspecting the prop governor, the sharp-eyed A&P noticed a blue stain on the #5 cylinder head near the intake port that turned out to be coming from a crack in the casting.

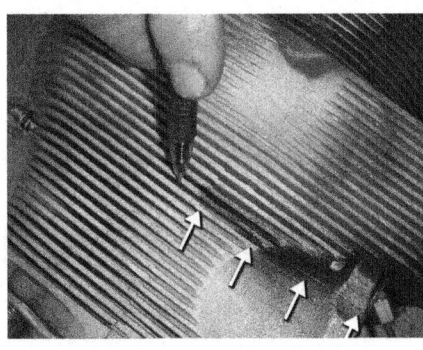

Crack near the intake port of the #5 cylinder head.

The mechanic contacted Sam, who authorized him to replace the cracked cylinder. Upon removing the jug, the mechanic noticed a considerable amount of rust pitting in the cylinder barrel and took advantage of the cylinder being off to take a closer look inside the crankcase. Shortly thereafter, Sam received a phone call from the mechanic and heard the words that every aircraft owner dreads: "Sam, I think you'd better stop by the shop and have a look at this." Uh oh!

Sam was shocked by what he saw when he got to the shop. The faces of the valve lifters were severely spalled. Several cam lobes were severely worn. Subsequent oil analysis showed the other engine was in similar trouble. Clearly, both engines were well down the road to self-destruction, and both would need to be torn down.

After consulting with me and others, Sam got quotes from several leading engine shops. The total tab for overhauling both engines, including removal and reinstallation, would be close to $80,000.

TOP: Severely spalled cam and lifters, the result of corrosion that occurred during 8 months of disuse.
MIDDLE: Closeup of the cam lobe, showing the extent of the damage.
LEFT: The large end of the connecting rod, showing extensive rust.

What happened?

Sam's Continental engines had a published TBO of 1,600 hours, but careful operators routinely get 1,800 to 2,000 hours from them, and high-utilization operators often take them to 2,400+ hours before overhaul. These engines sometimes have mid-TBO cylinder problems, but their "bottom ends" have a well-deserved reputation for being nearly bulletproof. It's certainly sad when these engines go bad well before their time, as did Sam's.

The culprit here is clear: corrosion due to eight months of disuse. Perhaps one can get away with this if the plane is based in Tucson or Albuquerque, but not in New Jersey (even in a hangar). Rust is the Number One reason why engines fail to make TBO.

The rust damage may have been further aggravated by the previous owner's choice of engine oil. I've long observed a correlation between corrosion damage and the use of multigrade oil in airplanes that fly sporadically (as many owner-flown airplanes do). I asked Sam to check the logbooks and learned that the previous owner had used Aeroshell 15W-50. In my experience, Aeroshell 15W-50 is an excellent oil for aircraft that fly regularly and often, but it's not a good choice for those that sit unflown for weeks or months at a time.

The perils of prebuy

But why wasn't this problem picked up during the prebuy? The key is that when the prebuy was performed, the engine (and oil) had only about three hours of operating time since the corrosion occurred. (That's how long it took to fly from the hangar in New Jersey to the shop in the Midwest.) Now, once the cam lobes and lifter faces develop corrosion pits, their destruction typically occurs quite quickly ... but not in three hours!

It's useful to think of a lifter or cam with corrosion pits as being like a brick wall with a few missing bricks. Each missing brick undermines several neighboring bricks which will ultimately work loose; each of those loose bricks then undermine more neighboring bricks, and before long the entire wall comes tumbling down.

In the case of a lifter face or cam lobe or cylinder wall, the "bricks" are the crystalline structure of case-hardened steel cam and the chilled cast iron lifters. Even microscopic corrosion pits disrupt that structure and inevitably lead to disintegration of the surfaces.

After just three hours of operation, the spalling of the lifters was probably still in its microscopic stages and hadn't progressed far enough to leave visible metal in the oil filter. ("Spalling" is a technical term that means the disintegration of a smooth surface by chipping or flaking.)

With the benefit of 20-20 hindsight, it's clear that it might have been a good idea to pull a few lifters during the prebuy and inspect them under a magnifier for corrosion pits. It would also have been a good idea to inspect the cam lobes with a borescope inserted through the vacant lifter bosses. In fairness to the shop, however, I must point out that this is by no means the sort of thing that is routinely done during a prebuy or even an annual inspection. (As a result of Sam's experience, my company has started pulling lifters and inspecting cam lobes during prebuy exams that it manages on Continental-powered aircraft that have a history of irregular use.)

> Rust is the Number One reason why engines fail to make TBO.

Why didn't the borescope inspection of the cylinders reveal rust damage to the cylinder bores? Once again, three hours of operation was probably just enough to scrape off most or all of the visible rust, and not nearly enough to allow the microscopic rust pits on the cylinder walls to progress to the point of visible spalling that could be seen with a borescope.

There were undoubtedly lots of rust particles in the oil filter, but rust is nearly impossible to see during filter inspection unless the filter media is inspected under a microscope. Had the engines been on oil analysis (which they weren't prior to Sam's acquisition), the report might have revealed higher-than-usual iron concentration, but this is expected after a period of disuse so the lab's recommendation would probably have been to "resample after 25 hours."

Preservation by pickling

There are a few important lessons we can learn from Sam's sad tale. The most important lesson for all aircraft owners is that a few months of disuse can wreak havoc on your expensive piston aircraft engine, even if the aircraft is hangared. Anytime you know that your aircraft won't be flown for a month or more—because you're taking a vacation in Europe, recovering from an illness, or temporarily lost your medical, for example—it's absolutely essential to "pickle" the engine to prevent corrosion damage.

Both Lycoming and Continental have service bulletins that spell out the recommended preservation procedure. Basically, it involves draining the oil and servicing the crankcase with special preservative oil, removing the top spark plugs, spraying preservative oil into each cylinder, and placing dessicant plugs into the spark plug holes and dessicant bags into the exhaust and induction pipes.

The pickling procedure is simple enough that even an owner can do it (legally) without requiring an A&P signoff. You can purchase a convenient Tanis engine preservation kit ("pickle kit") from suppliers like Aircraft Spruce. If you have your shop do it, it shouldn't cost more than a couple of hundred bucks, and it's probably the smartest money you could ever spend on maintenance.

Caveat emptor

For prospective buyers, there are several lessons to be learned here. One is that you're not likely to learn very much by inspecting an oil filter after just three or seven hours of operation. A prebuy examination of an aircraft known to have undergone a significant period of disuse should include a visual inspection of the cam and lifters if at all possible. This is not something that is routinely done unless you specifically ask for it, but it can be done without cylinder removal on engines with barrel-style lifters (i.e., all Continental 360-, 470-, 520-, and 550-series engines, as well as the Lycoming O-320-H2AD engine used in Cessna 172s built in the late 1970s).

Unfortunately, most Lycoming engines use mushroom-style lifters that cannot be removed without splitting the case, so there's no way to inspect the cam and lifters on these engines without pulling cylinders—something that's far too invasive and risky to be appropriate during a prebuy examination.

Another lesson for buyers is that the prebuy exam should always be done by a mechanic who has no prior connection with either the aircraft or the seller. In this case, the prebuy was done by a shop with a top-notch reputation, but it was unfortunately the same shop that had performed the last half-dozen annual inspections, including an annual just seven hours prior to the prebuy. The A&Ps at this shop naturally treated this airplane like an old friend and could not help but be predisposed to believe that it was in great mechanical shape. At a different shop—one that had never set eyes or hands on this aircraft before—the mechanics would have undoubtedly approached the aircraft with an appropriate attitude of skepticism. During a prebuy exam, skepticism is precisely what you want.

Finally, buying an airplane with a mid-time engine always involves a certain element of risk, since it's very hard to know exactly how the engine has been operated or maintained, and therefore to estimate how much useful time is left on the engine. For twins, the risk is doubled. That's one of the reasons I'm a big fan of buying airplanes with runout engines (at a suitably discounted price, of course); you buy the airplane knowing that you'll have to major the engine soon, and once you've done that you know precisely what you've got.

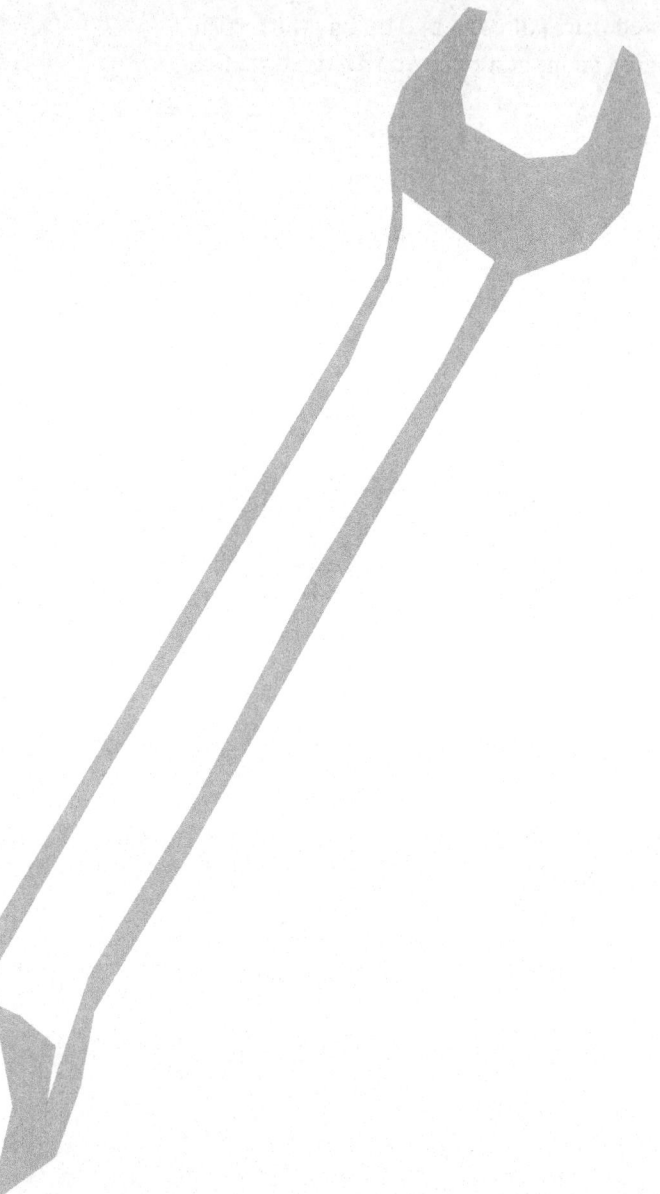

14
Cam Distress

When your engine finally needs to be overhauled, the most likely cause will be corrosion-induced distress to the cam lobes.

I once received an email from a Skylane owner in Miami, Florida who was understandably upset after receiving bad news from his IA, and looking for a second opinion:

> My 1980 Cessna 182Q went into annual, and I received a call from my IA a day later telling me that two cylinders had low compression (38/80). My IA pulled the cylinders, and just called me to say my engine is DEAD! I'm attaching some photos so you can see the reason for his pronouncement.
>
> The engine is at about 1,600 hours and 25 years SMOH. Published TBO for this engine is 1,500 hours. Please let me know what you think I should do. If I do need an overhaul, could you recommend an overhaul shop, or would you recommend a factory reman?

The owner attached some photographs of his Continental O-470-U engine showing a cam lobe in questionable condition and a severely spalled lifter.

When the cylinder came off and the IA looked at the cam lobe, he told the owner "your engine is dead!".

This lifter is severely spalled and clearly unairworthy. But on big-bore Continental engines, lifters are cheap and easy to replace. On Lycomings, the case must be split.

I added these photos to my large collection of similar photos from other engines, because this sort of thing is incredibly common in owner-flown piston aircraft, particularly those based in geographic regions with high corrosion risk. And with the possible exception of Hawaii and Puerto Rico, there's nowhere in the U.S. where the corrosion risk is higher than South Florida.

It doesn't take an engine expert to see that the lifter shown in the figure is in horrible shape. Fortunately, this is no big deal in Continental 360-, 470-, 520- or 550-series engines. Because these engines use "barrel-style" lifters that can be easily removed from the outside of the crankcase, replacing damaged lifters on these engines is easy and cheap. In fact, if you use PMA-approved lifters from Superior Air Parts, they only cost about $100 each. (If you buy them from Continental, the cost is almost twice as much.)

By comparison, had this been a Lycoming engine, a spalled tappet like this would be a real calamity. Almost all Lycoming engines use "mushroom-style" cam followers that cannot be removed from the outside, and therefore cannot be replaced without tearing down the engine. (The notable exception is the Lycoming O-320-H2AD engine in the Cessna 172N, which uses barrel-style lifters.)

Cam inspection per SID05-1

The real concern here is not the lifters but the cam. If the cam is unairworthy, then a teardown becomes unavoidable. When I zoomed in on the owner-supplied photo of the cam lobe to take a closer look at the questionable lobe it was obvious that it was exhibiting clear signs of distress.

But how much distress is too much? Is this cam lobe in bad enough shape to justify the IA's death sentence and to justify the owner spending $30,000+ to replace the engine?

Fortunately, Continental provides quite specific guidance on this subject in its Service Information Directive SID05-1B. This excellent document contains detailed instructions for performing cam inspections, criteria for assessing whether or not the cam is airworthy, and plenty of good photographs of cam lobes in various stages of distress.

This closeup of the questionable cam lobe suggested that the distress is still minor and the engine could remain in service with 100-hour lifter inspections.

A detailed review of SID05-1B makes it clear that Continental considers a certain amount of pitting of the cam lobe toe area to be acceptable. So long as the pits aren't deep enough to penetrate the case-hardened (carburized) outer layer of the cam lobe (which is about 15 thousandths of an inch thick), there is little risk of accelerated deterioration. The document goes on to say "for minor distress, the camshaft may be continued in service and re-examined upon the accumulation of 100 hours operation or 12 months, whichever occurs first."

I consider this to be very sound guidance, because in my experience cam and lifter distress is not a safety-of-flight issue, only a safety-of-wallet issue. In all the years I've been paying attention to such things, I've never once heard of a single case of an engine quitting or an airplane falling out of the sky due to a spalled lifter or cam lobe. In fact, I've seen dozens of cases of truly severe cam lobe damage—vastly worse than what this O-470-U exhibited—and in every such case, the pilot never noticed any sign of engine performance degradation.

Consequently, when in doubt, I see no risk in replacing the spalled lifter(s) and flying the engine another 100 hours. The worst that could happen is that the pitted cam lobe will tear up the new $100 lifter, and you'll wind up having bought an additional

100 hours before tearing down the engine. On the other hand, the new lifter might wind up surviving that 100 hours unscathed, in which case you dodged the bullet and may be able to extend the life of the engine for hundreds of additional hours.

Cracks vs. pits

On the other hand, SID05-1B indicates that Continental is very concerned about cracks (as opposed to pits) in the cam lobe surface: "If the visual cam lobe inspection reveals the presence of indentations or crack like features in the surface along the cam lobe apex, use a sharp pick or awl and lightly move its tip over the suspect surface area. If the suspect feature has any depth, the pick tip will repeatedly catch in the groove or pits. If the indentation or crack is determined to have depth, the cam must be examined by a Continental Motors service representative to determine any additional steps required. If the cam lobe inspection only reveals normal signatures… no further action is required."

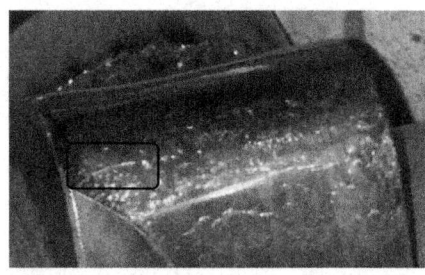

This closeup of the cam lobe at a slightly different angle reveals what might possibly be a surface crack…or perhaps just a piece of lint from a shop rag. The only way to know for sure is to probe it with a sharp pick.

This guidance also makes sense, because a crack in the cam lobe could result in the lobe coming apart fairly rapidly. When I studied a different photo of the questionable cam lobe under high magnification, I spotted a suspicious anomaly in the photograph that could have conceivably been a crack…or it might just have been a piece of lint from a shop rag. The only way to know for sure would have been to probe the lobe with a sharp pick and see whether the alleged crack had any significant depth to it. If it had turned out to have been an actual crack with significant depth, then I would have reluctantly agreed with the IA's doomsday prognosis.

Although SID05-1B doesn't mention this, I suggested to the owner that before performing the "sharp pick inspection" of the cam lobe, it would be a good idea to polish the cam lobe with some crocus cloth or very fine emery cloth to remove rough

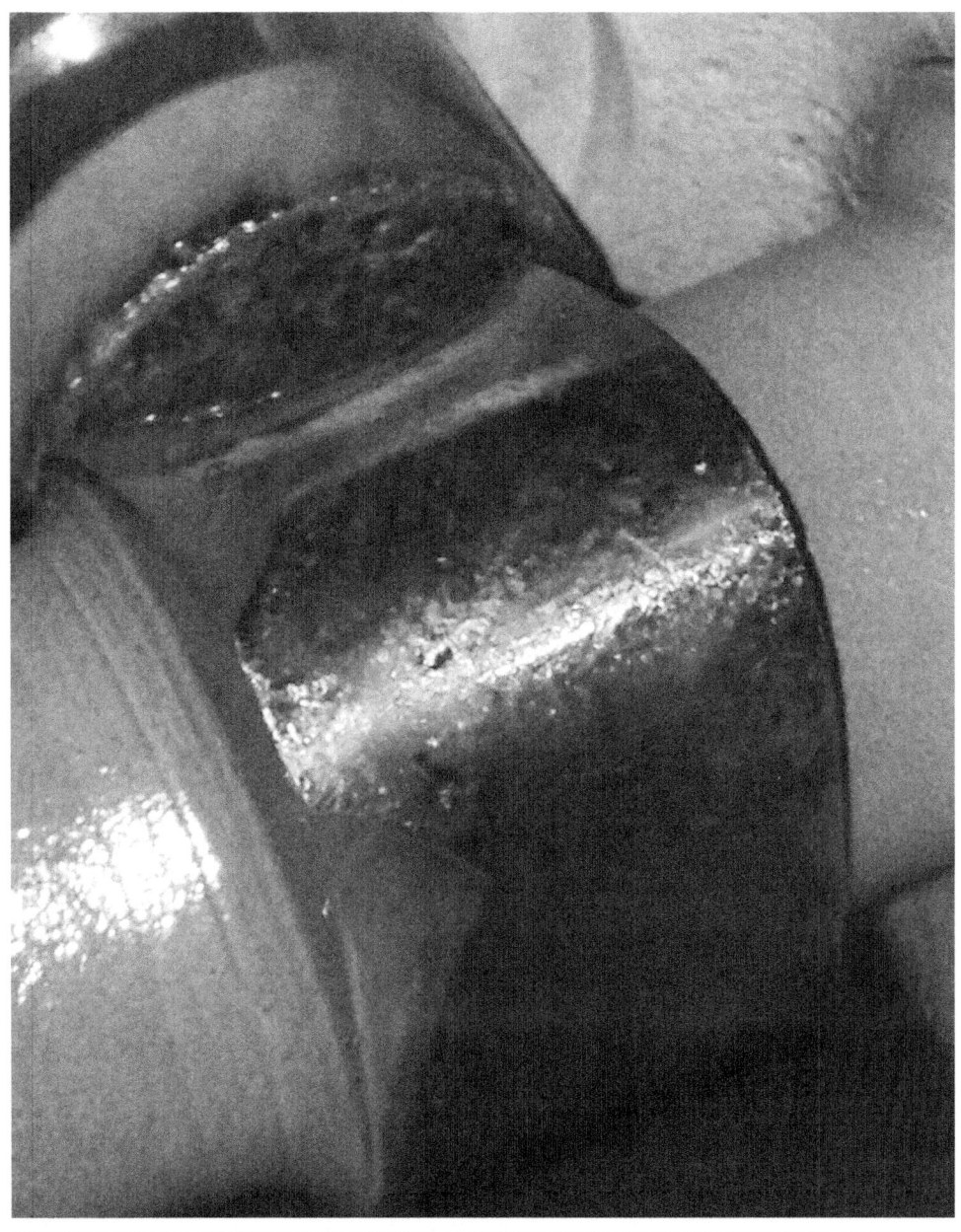

A higher-resolution image of the cam lobe, taken with a better camera and with better lighting, shows the cam lobe to be in worse shape than it originally appeared.

edges from the minor distress features. This is a trick that I learned from one of my mentors, the late Bob Moseley, who served as Continental's central region field representative for five years and possessed encyclopedic knowledge of these engines and their ailments. Again, this is a commonsense thing to do. We certainly wouldn't want to condemn an engine to euthanasia over a cam lobe flaw so insignificant that it could be polished away with crocus cloth.

The Skylane owner reported back that he'd printed out a copy of the SID and given it to his IA, together with my advice to give the cam lobe the benefit of any doubt and consider just replacing the bad lifter (together with any other spalled ones) and re-inspecting it after 100 hours.

The owner ultimately decided to install a factory rebuilt engine.

I genuinely hoped that the owner would be able to get another year or three out of his engine—I just hate to see an engine torn down before its time—but alas that was not to be. Closer inspection of the cam lobe revealed it to be in substantially worse shape than it initially appeared. At the end of the day—and after the IA brought in a colleague to inspect the cam lobe and render a second opinion—it became clear that this cam lobe had progressed beyond the point of "minor distress." The owner ultimately decided to bite the bullet and order a factory rebuilt engine.

Oh well… Sometimes you eat the bear, and sometimes the bear eats you.

15
Making Metal

How to ensure that nothing is coming apart inside your crankcase.

I'd been working with a Bonanza owner in Memphis for several weeks helping him chase down a problem with his Lycoming engine. That's not a typo: The aircraft was an A36 with a Machen conversion to a fire-breathing 350 hp Lycoming TIO-540-J2BD engine. The owner of this hot-rod Bonanza initially reported that the engine had exhibited several episodes of rough running after startup, but that it seemed to run smoothly once it warmed up.

The owner emailed me a data dump from his JPI engine monitor, which confirmed my suspicions that his "morning sickness" was caused by a couple of sticky exhaust valves in cylinders #4 and #5. Sticking exhaust valves is a fairly common malady in Lycomings, which is why Lycoming Service Bulletin 388C and Service Instruction 1425A call for doing a "valve wobble test" every 400 or 1000 hours (depending on what kind of exhaust valve guides are installed).

The owner wound up taking his sick engine to an excellent engine shop near Memphis. The shop pulled the rocker covers and found the #4 exhaust valve springs black with carbon from a badly leaking exhaust valve guide. #5 had the same problem, but not quite as bad.

But I've already discussed sticky exhaust valves in a previous chapter. What the shop found next was far more serious.

Flying nozzle

The engine shop decided to inspect the cam and make sure that it was not damaged by the valve-sticking episodes. In most Lycomings (unlike most Continentals), you can't remove the lifters from the outside of the engine, so the only way to inspect the cam is to pull a jug. The shop proceeded to pull cylinder #4, and it turned out to be a lucky thing they did. The owner emailed me:

> They pulled #4 cylinder and found evidence of damage from a screw hammering the bottom of the piston. They also found marks on the crankcase on one side of the cylinder base. The engine was removed and torn down. They found that the #1 cylinder oil spray nozzle and its helicoil had come out, bounced around inside the engine for some indeterminate period of time, managed to hit all six pistons, and scored two connecting rod end caps.

> What is strange is there was no indication of this in the oil analysis or any evidence when we cut open the oil filter at each oil change. However, when we checked the sump suction screen that should have been removed and inspected for metal at every oil change, we found the metal from the disintegrating oil spray nozzle and its helicoil. This is why the metal never made it to the filter.

> I checked with the shop that does my oil changes, and they admitted that they didn't know about the oil screen—they're mostly Continental dudes. After this, I will never forget it, and I'll make sure my A&Ps don't forget it.

> The inside of the engine, although marked by the flying nozzle, was extremely clean. The crankcase has to be repaired and certified as well as the camshaft. Little evidence of rust was detected on the lifters. All pistons and cylinders will be replaced. The turbocharger will also be overhauled. It looks like I'll be down for a couple of months. When I get the plane back, I'll need flying lessons again.

> I wonder how much longer it would have taken for this to cause a catastrophic engine failure? I believe monitoring the engine helped find this, but

clearly it would have been found much, much earlier had we been inspecting the pickup screen on a regular basis.

Monitoring for metal

The oil system of any piston aircraft engine provides two levels of filtration. There's a relatively coarse suction screen at the oil pickup tube whose job is to catch large chunks of metal (roughly 1/16 inch or larger) before they can get to the oil pump and possibly damage it. Then there's a fine screen or oil filter after the oil pump whose job is to catch tiny pieces of metal (roughly .001 inch or larger) before they can get to the engine's bearings and possibly contaminate them.

When implementing a condition-monitoring program, it's crucial to understand that there are three distinct sizes of metal particles that we're looking for:

- Large particles or flakes that get trapped by the suction screen;
- Tiny particles that are too small to be caught by the suction screen and get trapped in the oil filter; and
- Microscopic particles that are too small to be trapped by the oil filter.

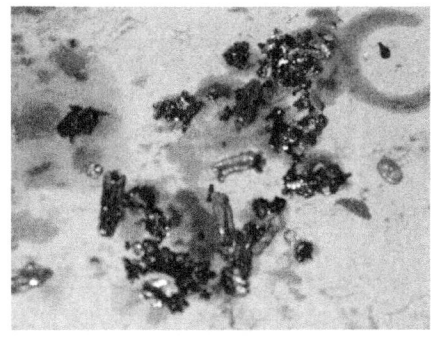

What was left of the Lycoming's flying oil nozzle. This shrapnel was too big to pass through the suction screen, so it was never spotted during oil filter inspections.

Lycoming piston damaged by the flying oil nozzle. All six pistons, two connecting rod caps, and the inside of the crankcase were damaged.

Therefore, our condition monitoring program must comprise three distinct elements.

Since **microscopic particles** are too small to be trapped by the oil filter (and too small to see even if some were trapped), we must place the engine on a spectrographic oil analysis program (SOAP) to detect abnormal wear events that throw off such

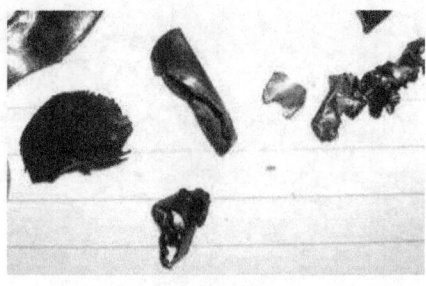

If you rely solely on oil filter inspection and oil analysis, you'll never know about stuff like this.

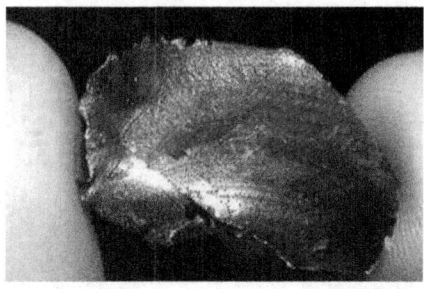

If there's something like this floating around inside your engine, you really want to know about it!

Running a mechanic's magnetic pickup tool around in the drained oil is a good idea for any engine, but particularly for Continental engines where the suction screen cannot be inspected.

microscopic metal particles. An oil sample should be captured at every oil change and sent to the lab for analysis.

To detect **tiny particles**, we must remove and cut open the oil filter at every oil change. Tiny particles can be hard to see, so it's essential to cut the filter media off its spool, spread it out flat, and carefully inspect each pleat under a bright light (and preferably with a magnifying glass). For engines that have only a fine oil screen instead of a spin-on full-flow oil filter, I strongly recommend adding a full-flow filter because it does a far better job of protecting the engine and provides a far better means for detecting problems before they cause a lot of damage.

To detect **large particles or flakes**, we cannot rely on filter inspection or oil analysis, because large stuff never makes it to the filter or into the sample jar. For Lycoming engines, we must remove and inspect the suction screen at every oil change. As we've seen, this step is often neglected and shockingly some A&Ps don't even know about it!

Unfortunately, Continental engines do not permit the suction screen to be removed and inspected. (To gain access to the suction screen on a Continental engine, you must drop the oil pan, something that usually can't be done while the engine is mounted in the aircraft.) So, for Continentals about the best we can do is to (1) drain the oil through a piece of window screen or cheesecloth and then inspect it for any large particles or flakes of metal, and (2) run a magnetic pickup tool around inside the oil drain bucket and see if it picks up any pieces of ferrous metal. (This isn't a bad idea for Lycomings, too.) Alas, very few A&Ps or aircraft owners perform these steps, either.

The result is that the worst engine problems—the ones that throw off large chunks or flakes of metal—often go undetected until it's too late. There's no excuse for this if we're doing our condition-monitoring job correctly.

If you do your own oil changes, make sure that you're inspecting the suction screen if your engine is a Lycoming, and that you're checking the drained oil for metal (using a screen and a magnet) if your engine is a Continental. If you have your oil changes done by a shop or mechanic, do not assume that they're doing this—check it out!

Lycoming Engines employ a suction screen at the oil pickup in the sump that should be inspected at every oil change. The largest metal particles are filtered by this screen. Unfortunately, these screens are not easily accessible on Continental engines.

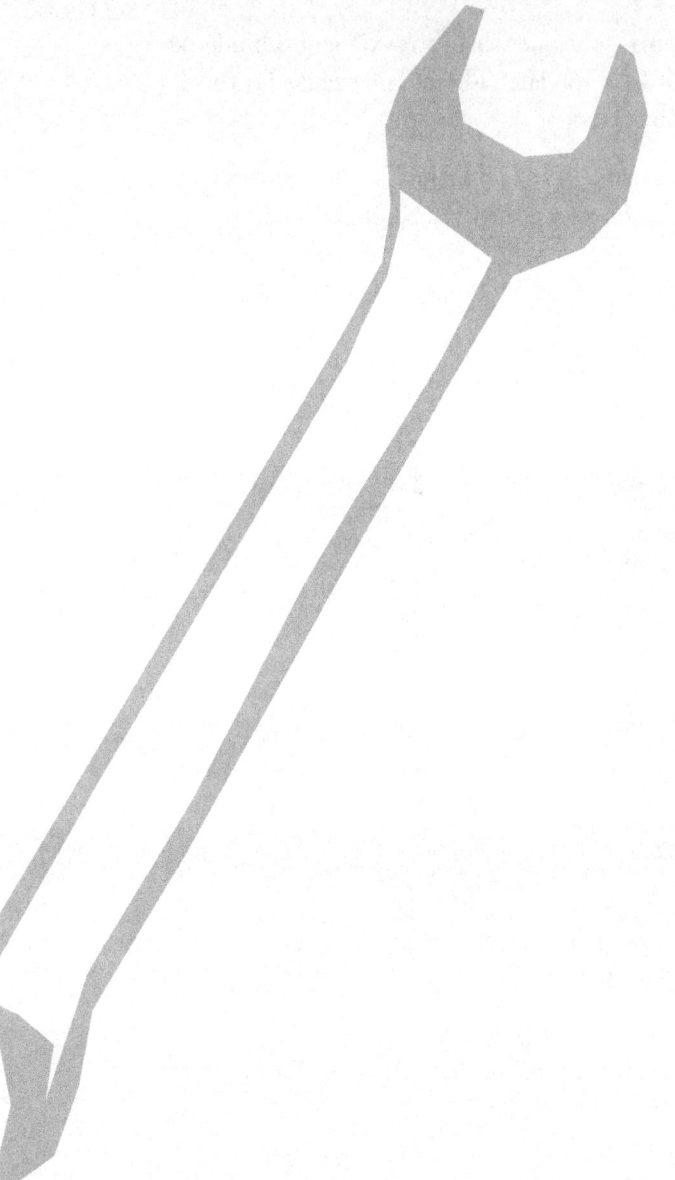

128 ENGINES

16
Teardown Dilemma

Owners often agonize over whether to do an engine teardown inspection after a prop strike, but the decision is really a no-brainer if you think it through logically.

Prop strikes come in all kinds of flavors. Some involve obvious sudden stoppages of the engine (for example, a gear-up landing). In others, the engine keeps running, perhaps with some short-term loss of RPM as the prop tips curl back. Still others occur when the engine is not running (such as when a car backs into your prop while the aircraft is parked).

Owners are often reluctant to remove their engine for a teardown inspection after a prop strike, especially if the engine was not running or was idling at the time. They don't want to incur the downtime. They worry about how the teardown inspection will look in their maintenance logs. They are reluctant even though insurance will invariably pay for such a teardown, no questions asked.

Here's an email I received from a Skylane owner who was obviously agonizing over the decision whether to do a teardown or not:

Continental Service Bulletin SB96-11B states that anytime the propeller sustains damage that requires it to be removed from the engine, an engine teardown inspection is mandatory.

My Cessna 182Q's prop sustained foreign object damage that didn't cause engine stoppage, or even loss of RPM, though it took several sizable gouges of aluminum off both blades. Definitely one and maybe both blades will need to be replaced, though most of the damage is on the back and trailing edge. If the replacement cost creeps up toward a new prop cost, that's not a hard decision, but the engine is another story altogether.

My IA says that since the engine didn't actually stop or lose RPM, and since the damage doesn't indicate any force applied against the prop rotation, he doesn't see the need to open the engine or the governor. I have read the non-mandatory Continental service bulletin that says any damage leading to prop removal dictates a teardown. The insurance company says it's up to me, but if I decide not to do a teardown inspection, any future claims traced back to damage from this incident would be disallowed. They are allowing for a teardown at my discretion.

My dilemma is that I really don't want a teardown on the logs, and it seems a real waste to open up an engine (IO-470-F) with 1,250 SMOH on the way to 1,500 TBO without doing an overhaul. The Continental factory cylinders are only about 350 hours since new from a recent top overhaul, and all mags and accessories have been replaced or overhauled in the last two years, so it seems like a complete waste to jump the gun on a major overhaul 3-5 years early, with a perfectly fine-running engine that may go past TBO due to the recent work.

One compromise would be to get the overhaul done with my own cylinders reworked and the accessories IRAN'd (inspected and repaired as necessary), but I am told that this type of overhaul is not looked on favorably at resale time, regardless of the shop doing it.

Then there's the safety issue. I guess it will be in the back of my mind (and I assume the mind of a potential buyer) that the prop was replaced or overhauled and the engine wasn't inspected, even with the IA's opinion.

I am really at a loss! Any guidance or points to consider would be greatly appreciated.

What the manufacturers say

Continental Service Bulletin SB96-11B offers the following guidance:

> A propeller strike is: (1) any incident, whether or not the engine is operating, that requires repair to the propeller other than minor dressing of the blades... or (2) any incident while the engine is operating in which the propeller makes contact with any object that results in a loss of engine RPM. Propeller strikes against the ground or any object, can cause engine and component damage even though the propeller may continue to rotate. This damage can result in catastrophic engine failure.
>
> Any time foreign object damage requires propeller removal for repairs other than minor dressing of the blades, the incident is considered a propeller strike.
>
> **Following any propeller strike, complete disassembly and inspection of all rotating engine components is mandatory and must be accomplished prior to further flight.**

It's clear that Continental considers a teardown inspection to be mandatory any time the prop sustains damage that requires its removal. Of course, even so-called "mandatory" service bulletins aren't truly mandatory for Part 91 operators, so an owner can decline to do a teardown if his A&P agrees.

Lycoming's guidance is often considerably fuzzier than Continental's, and prop strikes are no exception. Lycoming Service Bulletin 533C instructs that the safest procedure is to do a teardown inspection following any incident involving propeller blade damage. However, it includes the stipulation that the inspecting mechanic may overrule this guidance and may approve the engine for return to service without disassembly and inspection if he feels that it is prudent to do so.

AD 2004-10-14 further requires inspection of the crankshaft gear installation following any sudden stoppage or prop strike for most Lycoming engines. Since it's an AD, the owner really has no wiggle room on this.

Why are we typically so reluctant to do it?

The decision to tear down an engine is among the most agonizing maintenance decisions an aircraft owner can make. There are three main reasons for this:

1 **It's expensive.** It typically costs $15,000 for a minimal teardown inspection, or up to $40,000 for a major overhaul if that is deemed necessary.

2 **It's inconvenient.** The aircraft is typically down for 60 to 90 days.

3 **It's invasive and therefore risky.** NTSB data shows that most catastrophic engine-failure accidents occur in the first 200 hours or 24 months after overhaul. And introducing new parts into your engine introduces the possibility of an AD down the road…

Clearly, engine tear down is not a decision to be taken lightly.

Why NOT do it?

Legalities aside, the engine experts I've talked to all feel that doing a teardown after any flavor of prop strike is important for safety. And the insurance underwriters I've consulted all say that your insurance will pay for the teardown without hesitation. To my way of thinking, this makes the decision a no-brainer: Think of it as an opportunity, not a problem; just do it!

Here's how I replied to the agonizing Skylane owner to help him sort through his decision-making process:

> You should completely discount your feelings about 'I really don't want a teardown inspection on the logs.' A teardown inspection doesn't decrease the perceived value of the engine, and even if it did, your engine has essentially zero residual value at 1,250 SMOH anyway.
>
> On one hand, you are considering engine life-extension past TBO (something I highly recommend). On the other hand, you are worrying about what effect a teardown and bottom-end IRAN (inspect and repair as necessary) might do to the resale value of the aircraft. Clearly, you haven't decided whether you plan to keep the aircraft or sell it. So, let's think through both possibilities.
>
> If your objective is to keep the aircraft for the next several years, then you should set aside all thoughts of resale value and do what's most economically efficient to ensure safe operation. If the insurance company is willing to pay for an engine removal, teardown inspection and reinstallation, it strikes me as foolish not to take them up on that offer. While the engine is opened up, you can replace the main bearings and check out the cam and other bottom-end components, so that when the engine is back together and on the airplane, you can confidently fly it to 2,000 SMOH or more without worrying about the condition of the bottom end. (If you don't do the teardown, you'll worry…trust me on this.)
>
> If you plan to sell the aircraft soon, I still cannot see any good reason not to take the insurance company up on its offer of a teardown inspection. Although your IA is probably correct that there is no engine damage, the fact that the prop required major repairs but no teardown was done could (and should) raise questions in the mind of a prospective buyer. Continental SB96-11B is explicitly clear that they consider a teardown mandatory if the prop sustains damage.
>
> Continental says the teardown is mandatory. Your insurance company says they'll pay for it. Why not do it?"

Insurance will nearly always pay for an engine teardown following a prop strike.

A happy outcome

A couple of months later, the Skylane owner emailed me again:

> I thought I'd drop you a note for a final 'thank you.' My plane flew again (finally!) this weekend after we got a new prop and finished the engine teardown inspection and the aircraft annual inspection.
>
> Your advice on what to consider was spot-on. I worked directly with the engine shop to get the details on what the teardown revealed and they were very cooperative, especially when I started asking the right questions. On condition, I ended up replacing lifters, cam and bearings and reusing all existing cylinders that passed muster, then went with new engine mounts on the reinstall. They overhauled the mags as part of the teardown inspection (so insurance paid for that).
>
> All considered, given service times of cylinders and components, I feel I purchased half or two-thirds of a TBO for about $3,500 out-of-pocket, which is really a heck of a deal.
>
> Your response really helped me to clarify my situation, organize my thoughts, and undoubtedly saved me a lot of money and regret. Thanks again!"

I just love it when a plan comes together!

PART IV
Key Systems

17
Lubrication

What you need to know about piston aircraft engine oil.

This Chapter addresses the lubricating oil we use in our piston aircraft engines, and we'll be covering a lot of territory. We'll discuss the various types of engine oil—monograde versus multigrade, mineral oil versus synthetic—and the pros and cons of each. We'll talk about aftermarket oil additives and whether they're beneficial or just hype. We'll touch upon oil consumption, optimum oil level, and how often to change the oil.

Six key functions

If I asked you to explain the purpose of engine oil, I imagine you'd probably say something like "to lubricate moving parts and reduce friction and wear." Now that's certainly correct as far as it goes, but *lubrication* is only one of six key functions that oil must perform in your piston aircraft engine. In fact, the lubrication needs of our big displacement, slow-turning piston aircraft engines are quite modest, compared to the high-revving engines in our automobiles. Lubrication demands tend to vary with the square of RPM, so a car engine has vastly more demanding lubrication requirements than an airplane engine does.

In addition to lubrication, piston aircraft engine oil is called upon to cool, clean, seal, actuate, and preserve. Some of these functions (especially cooling, cleaning and preserving) are arguably just as important as lubrication.

Oil serves as a vital *coolant* for engine components that can't be air-cooled. Pistons, for example, are exposed to just as much heat of combustion as cylinders, but they don't have cooling fins or exposure to airflow. The only thing that keeps them from melting is the large quantity of oil that is splashed and squirted onto the bottom of the pistons to carry away the heat.

Another key function of oil is to *keep the engine clean*. Compared to car engines, piston aircraft engines are positively filthy creatures. They burn heavily leaded fuel and allow large quantities of lead salts, carbon, sulfur, water, raw fuel and other nasty combustion byproducts to blow by the rings and pollute the bottom end of the engine. The oil must be able to keep these contaminants dispersed and hold them in suspension so that they don't accumulate on internal engine parts in the form of sludge.

Oil also acts as a *sealant* to prevent the leakage of gases and liquids (including the oil itself) past piston rings, O-rings, gaskets, and various other kinds of seals.

If your airplane has a constant-speed propeller, oil serves as the *hydraulic actuating fluid* that is used to adjust the blade pitch. If it is turbocharged, the wastegate is probably hydraulically actuated by oil as well.

Last, but certainly not least, oil is required to *protect and preserve* expensive components like crankshafts, camshafts, lifters and cylinder barrels from rusting during periods when the airplane is not being flown. Because we tend to fly our airplanes less often and more irregularly than we drive our cars, the preservative requirements of our aircraft engines are far more demanding than for automotive engines.

Friction, wear, lubrication

The purpose of lubrication is to reduce friction and wear on the engine's moving parts. Friction occurs because even the smoothest surfaces have microscopic peaks and valleys. Whenever surfaces come in contact, these tiny peaks adhere to one another via tiny "micro-welds." If those surfaces are in relative motion, the micro-welds constantly fracture and re-form, resulting in friction and wear. Friction is the resistance to relative motion and wear is the loss of material. Both are due to the fracture of the micro-welds.

There are several different kinds of lubrication. The most effective kind is called "hydrodynamic" lubrication. It occurs when a fluid—most commonly a liquid like oil—is interposed between the moving parts. The relative motion of the parts creates sufficient pressure in the lubricant to keep the parts from touching.

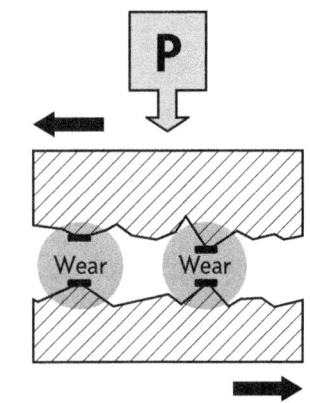

Friction and wear are caused by fracturing of "micro-welds" between moving surfaces.

Think of a water skier being supported by the skier's relative motion to the lake and the resulting pressure of the interposing water. The water pressure prevents the skier from sinking and contacting the lake bottom. Or, consider a car hydroplaning on a rain-slicked road, its locked tires being separated from the pavement by water pressure. That's hydrodynamic lubrication.

This kind of lubrication works well if the relative speed of the parts is high enough to overcome the load pushing them together. If the relative speed is not high enough, then there will not be enough lubricant pressure to keep the parts separated. (Think of the tow boat slowing down until the skier sinks.)

If hydrodynamic action cannot keep the parts separated, we must rely on another form of lubrication known as "boundary" lubrication. Boundary lubrication relies on a thin, soft, solid film deposited on the moving parts—typically by chemicals called "extreme pressure additives." That thin film reduces friction and wear by chemically interfering with micro-weld formation.

Rub your hands together vigorously and you'll feel the friction between your palms in the form of heat and resistance to movement. Believe it or not, you just created and then fractured a few zillion micro-welds on the surface of your palms! You can reduce the friction in several ways: for instance, by coating your palms with Vaseline, or by dusting them with talcum powder. Think of the Vaseline as hydrodynamic lubrication and the talc as boundary lubrication.

Pressure, splash, boundary

Most of the lubrication in our engines is hydrodynamic. It's accomplished in two different ways. Some moving parts—such as the main and rod bearings—are "pressure lubricated" by oil that is distributed directly to those parts by means of a series of drilled passages in the crankcase and crankshaft known as "oil galleries."

Many other moving parts—pistons, rings, cylinder barrels, cam lobes, lifter faces, gears—receive no direct lubrication from oil pressure. These critical components rely entirely on what is called "splash lubrication." When the engine is running, lots of oil extrudes from the crankshaft's pressure-lubricated main and rod bearings and is flung in all directions by the rapidly turning crankshaft, splashing oil on everything in the vicinity.

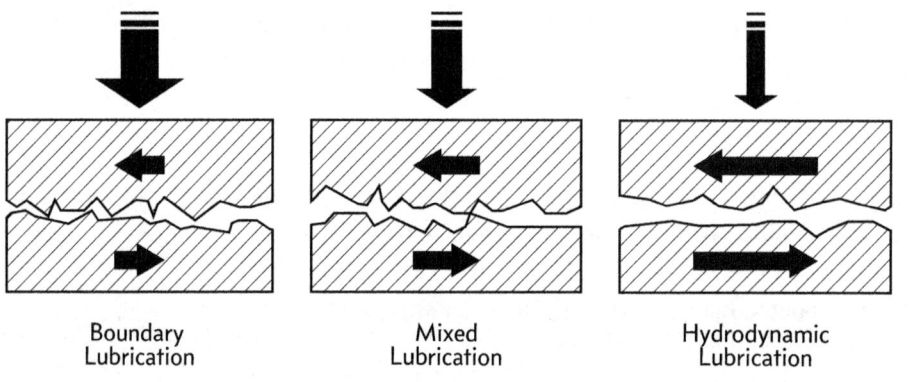

Hydrodynamic vs. boundary lubrication.

Other critical parts are too slow-moving and too heavily loaded to be separated hydrodynamically and rely mostly on boundary lubrication. One interesting example is the interface between the piston rings and cylinder barrel. Most of the time, the piston is moving rapidly inside the cylinder, and the ring-to-barrel interface is lubricated hydrodynamically so there's no metal-to-metal contact. At top-dead-center, however, the piston slows to a complete stop and then reverses direction. The slowing of the piston defeats any effective hydrodynamic lubrication in the critical ring-reversal area at the top of the stroke. For a brief but critical period during each crankshaft rotation, the rings stop moving, sink through the oil film—like a skier whose tow boat stopped—and contact the cylinder walls. The only thing that protects this "ring reversal area" from accelerated wear is boundary lubrication.

Boundary lubrication is also crucial during the first few seconds after the engine is started. That's especially true if the aircraft hasn't flown for a while and most of the oil has stripped off of the parts. Until oil pressure stabilizes and the crankcase fills with splash oil, dry engine parts have to rely on boundary lubrication alone.

Types of engine oil

Now let's talk about the different types of oil, and the pros and cons of each. Here are the six most widely-used piston aircraft engine oils in the U.S.:

Multigrade oils:

- Aeroshell 15W-50 (50% synthetic)
- Exxon Elite 20W-50 (25% synthetic)
- Phillips X/C 20W-50 (0% synthetic)

Monograde oils:

- Aeroshell 100/80 (non-AD break-in oil)
- Aeroshell W100/W80 (0% synthetic)
- Aeroshell W100 Plus/W80 Plus (0% synthetic)

Aeroshell 15W-50 is the most popular multigrade oil. It dominates the GA market, although it's my least-favorite oil (for reasons that I'll address shortly). It is referred to as a "semi-synthetic" oil, because it is a 50-50 hybrid of petroleum-based mineral oil and a synthetic oil called polyalphaolefin or PAO. To this half-and-half mixture of mineral and synthetic base stock, Shell adds chemicals called "viscosity index improvers" to give the oil its multigrade properties. It also has a complex additive package including ashless dispersants (AD) to help prevent sludge, corrosion inhibitors to help prevent rust, and an extreme-pressure additive called butylated

Some popular aircraft engine oils.

triphenyl phosphate (bTPP) to provide boundary lubrication.

Exxon Elite 20W-50 is a very similar formulation to Aeroshell 15W-50, except that its base stock is 25% synthetic PAO and 75% petroleum-based mineral oil. Its additive package is nearly identical to the one used in Aeroshell 15W-50.

Phillips X/C 20W-50 is an inexpensive, no-frills, all-petroleum multigrade. It's made strictly from dead dinosaurs, with no synthetics. Phillips adds viscosity index improvers to obtain multigrade properties, and ADs to prevent sludge.

Moving on to the monograde oils, Aeroshell "straight 100" is an SAE 50 mineral oil with minimal additives. It's used primarily as a break-in oil, and occasionally as an operating oil for radial engines.

Aeroshell W100 is by far the most popular monograde oil. The "100" means that its viscosity rating is SAE 50, and the "W" prefix means that it contains an AD additive package to help keep particulates in suspension and minimize sludge formation. (Note that this "W" prefix denoting AD additives should not be confused with the "W" suffix in "15W-50" which denotes that SAE 15 is the oil's "winter" rating.)

Aeroshell W80 is a less-viscous version of Aeroshell W100 used as a wintertime oil by operators who prefer monograde oil.

Aeroshell W100 Plus is simply Aeroshell W100 to which Shell has added the same package of anticorrosion and antiscuff additives that they use in 15W-50.

Monograde vs. multigrade

Monograde oil is simply mineral oil plus an additive package. It has viscosity, or thickness, that varies rather dramatically with temperature. At operating temperature—around 200 degrees Fahrenheit—it's quite thin and flows very freely. But at room temperature, it's thick and gooey. Get it cold enough and it won't pour at all.

Multigrade oils are much less thick and gooey at cool temperatures. They still get thicker as temperature decreases, just not nearly as much. To make a multigrade oil, you start with a thin monograde oil (something like SAE 10 or 15) and then add an artificial thickening agent called a "viscosity index improver" or VII. This is a man-made polymer that has the unusual property that it gets thicker and more viscous when heated—precisely the opposite of what mineral oil does. By combining monograde oil with a VII, the manufacturer can obtain pretty much any desired viscosity-to-temperature curve. Multigrade aircraft oils are roughly 90% base stock, with the remaining 10% composed of VIIs and other additives.

At operating temperature, Aeroshell W100 and 15W-50 have essentially the same viscosity. At room temperature or colder, the difference in viscosity is obvious and dramatic. The W100 pours like blackstrap molasses, while the 15W-50 pours like Aunt Jemima Lite. Or if you prefer, ketchup versus tomato juice.

At ambient temperature, multigrade oil is much thinner than monograde. (At operating temperature, the viscosities are the same.)

Polymers

Oil is made up of giant molecules called polymers. Some are natural like mineral oil, others man-made like PAO. Different polymers have different shapes. The molecules of mineral oil have a lot of side branches, while the molecules of synthetic oil are smoother and less "branchy."

Mineral oil gradually degrades the longer it remains in service. The little branches gradually shear off the molecules—a phenomenon known as "polymer shearing"—and that causes viscosity to decrease. Because synthetic oil is less branchy, it suffers far less from polymer shearing and

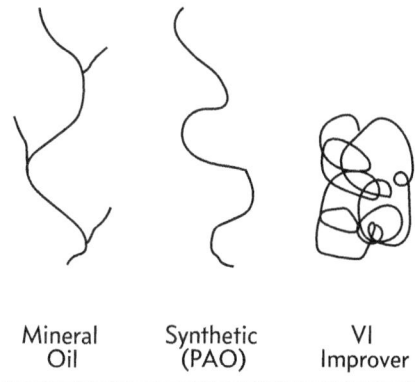

Polymers are giant molecules. Mineral oil molecules are "branchy," synthetic oil molecules are smooth, and viscosity index improvers get more viscous at high temperatures.

retains its viscosity longer. This means that synthetic oil can go a lot longer between oil changes, at least in automotive applications.

Viscosity index improvers are strange man-made molecules that roll up in a tiny ball when cool but unroll and become more linear as temperature increases. That's why they thicken the oil at high temperatures.

Pros and cons

The smooth, less branchy molecules of synthetic oils like PAO do offer some significant advantages over mineral oil. They provide improved lubricity, simply because they are smoother and more slippery. They also last longer, because they suffer less from polymer shearing and thermal breakdown.

But mineral oil also has its advantages. Its branchy molecules do a much better job of holding particulate contaminants in suspension so that those contaminants can be drained out at the next oil change instead of settling out as sludge. In fact, the full-synthetic aviation oil Mobil AV 1 was withdrawn from the market in the mid-90's (in a hail of litigation from aircraft owners) because so many engines were ruined by lead sludge deposits. To put it bluntly, synthetic oil simply can't deal with filth.

Mineral oil is also a better sealant, because its branchy molecules are less likely to sneak by O-rings and gaskets in the form of oil leaks.

The advantages of synthetic oils—which are quite compelling for automotive and turbine use—are far less significant for piston aircraft engines. The improved lubricity is less important simply because piston aircraft engines have such modest lubrication requirements. And the extended oil-change intervals that synthetics offer in automotive applications are of little use in piston aircraft engines, because the oil gets so filthy with blow-by that the oil must be drained every 50 hours or less due to contamination.

Obviously, the advantage of multigrade oil is that it doesn't thicken nearly as much at cold temperatures. This offers a significant benefit during cold weather if you must start without a preheat. With multigrade oil, the oil flows more quickly to the main and rod bearings, and splashes more quickly onto the cam, lifters, pistons and cylinders.

On the other hand, the fact that multigrade oil remains thin and pourable at cool temperatures also can mean that it drains off engine parts more quickly after the engine stops running. You can see this for yourself by checking how much longer it takes for the oil level on the dipstick to stabilize after shutdown. It takes many, many hours with a thick monograde oil like Aeroshell W100, less with multigrade like 15W-50. Consequently, multigrade doesn't provide as long-lasting a physical barrier against corrosive attack during extended periods of disuse. This is not important for "working airplanes" that fly every day or two but can be important for airplanes that fly irregularly and sometimes sit unflown for weeks at a time.

Corrosion is the #1 reason that piston aircraft engines fail to make TBO. We almost never wear these engines out, we rust them out. This is a big problem in the owner-flown fleet, because owner-flown airplanes tend to fly irregularly. Working airplanes that fly every day or two almost always make TBO without breaking a sweat.

For these reasons, I personally prefer mineral oils over semi-synthetics for piston aircraft engines that operate on leaded avgas. The advantages of synthetics simply don't benefit these engines the way they do automotive and turbine engines, and synthetic oil and leaded fuel just don't mix well. I also prefer monogrades (like Aeroshell W100) over multigrades (like Aeroshell 15W-50), except when multigrade is necessitated by exposure to unpreheated cold-starts in sub-freezing temperatures.

For aircraft operating is cold climates, I recommend using a mineral-based multigrade like Phillips X/C 20W-50 during the four coldest months of the year, switching to a monograde like Aeroshell W100 for the remaining eight months for maximum corrosion protection. Of course, if the airplane flies frequently and regularly so that corrosion risk is low, then there's nothing wrong with using a multigrade all year around.

When the time comes that we start using unleaded avgas instead of 100LL, I'll start looking favorably on semi-synthetics like Aeroshell 15W-50 and even full-synthetics like Mobil AV 1. Until then, I'll be sticking with petroleum lubricants.

Aftermarket additives

For as long as I can remember, there have been pitchmen promoting "miracle in a can" oil additives that claim to eliminate friction and wear, increase fuel economy, improve your landings, raise your IQ, and rescue your marriage.

The granddaddy of these is Marvel Mystery Oil. Folks have been pouring this red, sweet-smelling stuff into aircraft engines for more than 80 years. It was developed in 1923 by Burt Pierce, the inventor of the Marvel carburetor, and was intended as a fuel additive to clean carburetor jets. Why folks started using it as an oil additive escapes me. The name "mystery oil" came from the fact that Burt Pierce refused to divulge its formula.

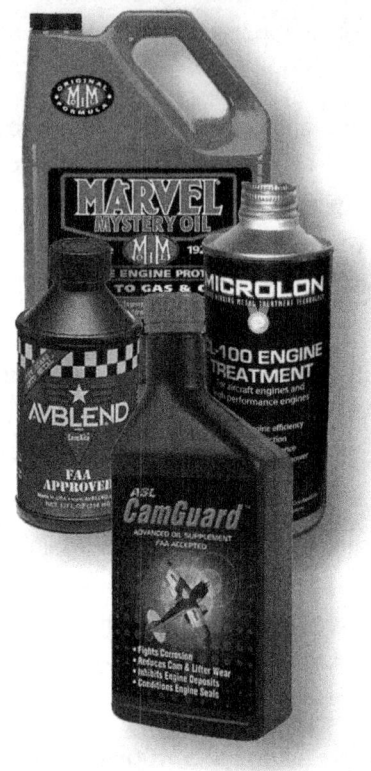

Aftermarket oil additives.

So far as I can tell, MMO doesn't do any harm if used in moderation, but it doesn't seem to do much good, either. Its formula is no longer a mystery since the Feds required the current manufacturer, Turtle Wax Inc., to publish the ingredients on a material safety data sheet. Turns out MMO has some interesting ingredients, including pig fat, perfume, and red food coloring in a base of petroleum distillate solvent. In case you were wondering, I don't use MMO in my engines.

Some aftermarket oil additives—notably Microlon and Slick 50—contain a slippery resin called PTFE, which is made and marketed by DuPont under the tradename Teflon. The Microlon and Slick 50 folks claim that the stuff bonds with metal surfaces in your engine and virtually eliminates friction. This is pure bunk. Manufacturers of non-stick cookware will tell you just how difficult it is to get Teflon to bond to anything. Users of non-stick cookware will tell you just how easy it is to ruin the Teflon coating because it's so fragile. NASA did a study of Teflon oil additives awhile back and concluded that not only are they ineffective as a friction reducer, but they can actually damage an engine by clogging oil filters and tiny oil passages in hydraulic lifters. DuPont specifically warns against using Teflon in internal combustion engines. Nevertheless, the promoters of Microlon somehow convinced the FAA to approve its use in aircraft engines. My advice is to just say no.

AvBlend has been around for two decades. It's a repackaged automobile oil additive called Lenkite that has been around for even longer. Back in the 90s, my friend Howard Fenton of Second OILpinion in Tulsa ran a test of AvBlend involving several piston twins that used AvBlend in one engine and not in the other. When we looked at the oil analysis results for these airplanes, we could not tell which engine had the AvBlend and which didn't. Like MMO, AvBlend appears to do no harm, but we can't find any evidence of benefit either.

Given this dismal history of miracles-in-a-can, when another one called ASL CamGuard hit the market, I was understandably skeptical. But in July 2007, I gave it a try and was very impressed with the results. In Chapter 28 I'll relate more about my experimentation with CamGuard and my conclusions from that experience.

Oil consumption

Take any group of aircraft owners, add a pitcher or two of beer, and it's not long before they're comparing notes on who's airspeed is the highest and whose oil consumption is the lowest. The subject of airspeed is beyond the scope of this article. But I'm here to tell you that low oil consumption is highly overrated.

Lots of factors affect oil consumption. Six-cylinder engines use more oil than four-cylinder engines. Big-displacement engines use more oil than smaller-displacement engines. Continental engines typically use more oil than Lycomings of similar horsepower. Chrome-plated cylinders use more oil than steel cylinders. Nickel-carbide cylinders use less oil than steel. Airplanes that fly a lot of short trips use more oil than ones that fly fewer long trips. And so forth.

Anything from a quart in 20 hours to a quart in four hours is in the normal range. I've seen engines that burned a quart in four hours throughout their entire life and made it past TBO without any problem. I've also seen engines that used hardly any oil and then wound up needing a top overhaul after 500 hours.

Continental doesn't consider oil consumption to be a cause for concern until it exceeds a quart in three hours and says that it isn't an airworthiness issue until it exceeds about a quart per hour. Now, I don't think any of us would allow our engines to get to the point of burning a quart per hour, simply because it's so darned embarrassing to run out of oil before you run out of fuel.

I do get concerned about any sudden increase in oil consumption. If an engine has been using a quart in 12 hours for most of its life and suddenly starts using a quart in six hours, that means something has changed, and to me that's a red flag. Until we figure out WHAT changed, we can't be sure whether it's serious or benign.

It also gets my attention if the oil starts to turn dark and opaque quickly after an oil change—say, within 10 hours or so. That indicates excessive blow by, and we'll want to do some testing and troubleshooting to determine why this is happening and which cylinder is the culprit.

Many owners install aftermarket air-oil separators to reduce oil consumption and keep the belly clean. I don't like air-oil separators, because they return all sorts of ugly, nasty stuff to the crankcase that would otherwise be expelled out the breather. Another of my mentors, engine guru John Schwaner of Sacramento Sky Ranch, said it best: "Air/oil separators are like hooking a line up to your anus and piping it back into your mouth. Excuse the crude analogy, but who wants the water, acids and other combustion residuals pumped back into one's engine?"

Dipstick level

Oil consumption can be exacerbated by filling the sump to the top mark on the dipstick. Many engines don't like that and will promptly toss a quart or two overboard until the oil level decreases to a level that the engine likes better. When an engine does that, it's trying to tell you something.

It's usually best not to keep the oil level at the top of the dipstick.

There's no reason to fill the oil to the brim. Before an aircraft engine can be certified, the FAA requires the manufacturer to demonstrate that the engine runs just fine in all normal flight attitudes with the oil sump filled to just one-half capacity. Most engines can function with even less oil than that. If you're curious how much less, just consult the engine's type certificate data sheet, available on the FAA website. For example, the Continental IO-550-N engine in a Cirrus SR22 has an 8-quart sump but will run fine down to 3½ quarts. Most Lycoming O-320s also have an 8-quart sump but need only two quarts.

As a rule of thumb, I recommend operating most engines at about 2/3 of the maximum sump capacity. That means running 8-quart engines at about 5 or 6 quarts, and 12-quart engines at about 8 quarts. If your engine uses a lot of oil, you might want to add one more quart to those figures for a little extra cushion.

Oil-change interval

Because piston aircraft engines put so much filth into the oil, regular oil and filter changes are an absolute must. Never go more than 50 hours or four months between changes, whichever comes first. Some experts recommend changing the oil every 25-30 hours or three months. If you have an older plane that has only an oil screen rather than a full-flow filter, you should not exceed 25 hours between oil changes.

If appreciable metal is found in the oil filter, or if the oil analysis report comes back with any red flags, we will normally put the engine on a reduced oil-change interval until the source of the metal is identified or the problem resolves itself.

If you're approaching an oil change and you know the airplane will go unflown for weeks or months, it's best to change the oil before the downtime rather than afterwards. Dirty oil tends to be corrosive, and you really don't want your expensive crankshaft and camshaft bathed in that stuff for any longer than necessary.

Never go more than 50 hours or four months between oil changes. (Some experts recommend 25-30 hours or three months.

Oil is also an excellent indicator of what's going on inside your engine. In later chapters we'll address oil filter inspection and spectrographic oil analysis, and how we use them to perform engine condition monitoring and to assess engine health.

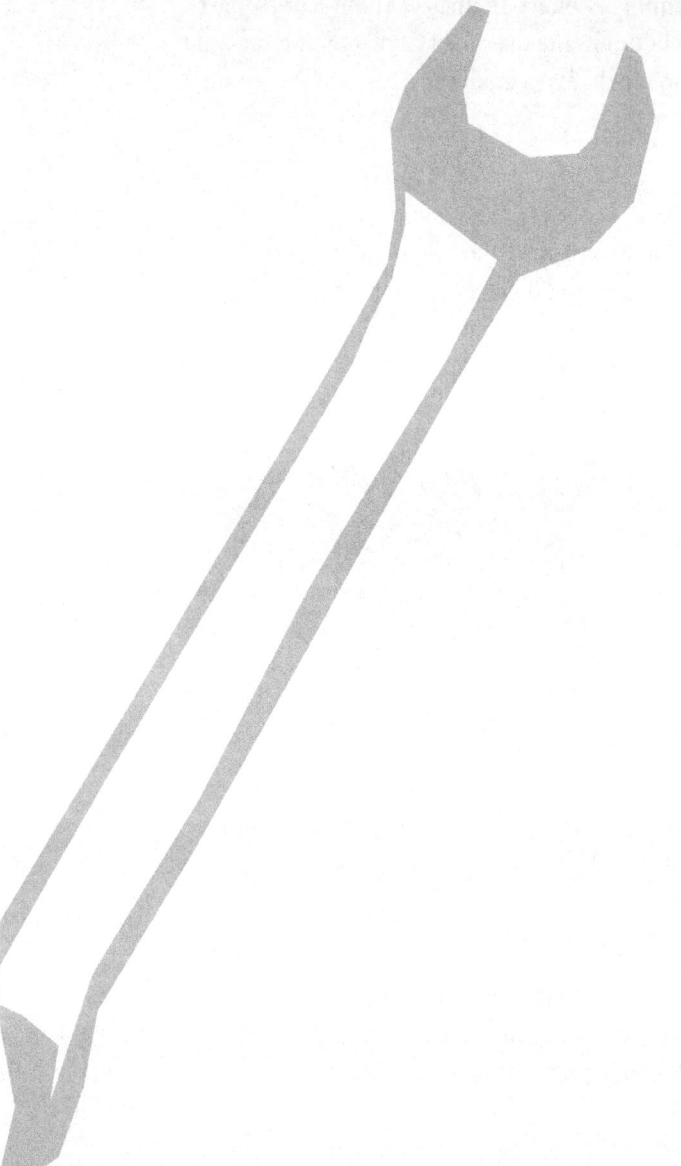

18
Ignition TLC

Mags and spark plugs don't need a lot of maintenance, but what they do need is often neglected.

Most piston aircraft engines still use magneto ignition. (Notable exceptions include the Rotax 912-series and various experimental engines that are fitted with modern electronic ignition systems.) Aircraft magnetos haven't changed much since they were first introduced near the beginning of World War I.

The name "magneto" comes from the fact that these devices use rotating permanent magnets and fixed coils to produce the high voltage required to generate a spark. This makes them independent of the aircraft's electrical system, so the engine will keep running even if the aircraft suffers a complete electrical failure. For additional reliability, most aircraft engines have two independent magnetos and two spark plugs in each cylinder, one energized by each magneto. This redundancy ensures that no single failure of a magneto, spark plug, or ignition lead can compromise engine operation. As a side benefit, the use of dual spark plugs in each cylinder provides more consistent and efficient combustion.

There are presently two manufacturers of magnetos for horizontally opposed piston aircraft engines: Champion (formerly Slick) and Continental (formerly Bendix). Both manufacturers call for minor mag maintenance every 100 hours, and major maintenance every 500 hours.

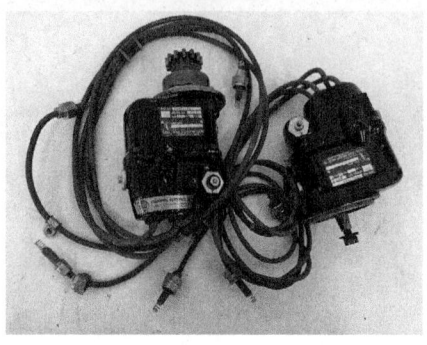

Champion (Slick) 6000-series magnetos.

Continental (Bendix) S-1200-series magnetos.

100-hour mag maintenance consists of removing the distributor cap, inspecting the breaker points for pitting and burning, checking the point gap and adjusting if necessary, cleaning the inside of the distributor cap, and checking the ignition timing and adjusting if necessary. All this can be accomplished without removing the mags from the engine, and normally doesn't require more than a half-hour per mag.

500-hour mag maintenance requires removing the mags from the engine and disassembling them. It entails a detailed internal inspection, lubricating the internal gears and bearings, and adjusting the internal mag timing ("e-gap") to ensure maximum output voltage. Figure a couple of hours per mag, plus parts.

My airplane uses Bendix (Continental) S-1200-series mags, and I normally perform the 500-hour maintenance on them myself, a task that typically takes me a couple of hours

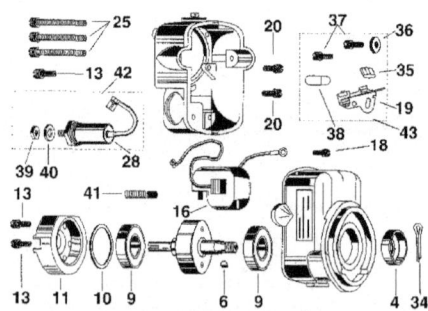

At 500-hour intervals, mags need to be completely disassembled, inspected, lubricated, and adjusted to correct internal timing ("e-gap").

per magneto (including removal and reinstallation). Although it's not required, I routinely replace the points, the distributor block, and the carbon brush assembly every 500 hours. The parts cost is roughly $100 per mag, and I consider it money well-spent.

Slick mags originally were manufactured by Slick Electro, renamed Unison Industries in 1980. For many years, Unison offered a very attractive trade-in program for factory rebuilt Slick mags, so most operators didn't bother to perform 500-hour maintenance on these mags, they simply exchanged them for rebuilt mags every 500 hours. Champion Aerospace acquired the Slick magneto product line in 2008, and increased the prices of rebuilt Slick mags rather dramatically. So, now it's generally more cost-effective to perform 500-hour maintenance on Slick mags than to exchange them, just as has always been the case with Bendix mags.

Spark plug maintenance

Spark plugs need regular maintenance, too. They need to be cleaned and gapped every 100 hours. For maximum life, they should also be reinstalled in rotated positions, with each plug alternated between odd- and even-numbered cylinders, and between top and bottom positions.

The reason for rotation is that mags generate high-voltage energy with alternating polarity. Consequently, the plugs in even-numbered cylinders wear differently from those in odd-numbered cylinders. (On six-cylinder Continental engines like mine, the plugs in even-numbered cylinders wear the center electrode, while those in odd-numbered cylinders wear the ground electrodes.) By rotating the plugs every 100 hours, the plugs will wear evenly and last longer.

Massive-electrode spark plugs like this one should last 400 to 500 hours, provided they are cleaned, gapped and rotated every 100 hours. Fine-wire iridium plugs last three times as long but cost three times as much.

Regular massive-electrode spark plugs will typically last 400 to 500 hours. When the center electrode—which starts out round and gradually wears into the shape of a football—has worn to half its original dimension (across the narrow axis), the plug should be scrapped and replaced. Champion sells a simple go/no-go gauge that takes the subjectivity out of determining whether plugs are worn out.

When it's time to replace the plugs, figure $32 street price for new Champion massive-electrode plugs—or a shade over $380 for a six-cylinder engine (2018 dollars). Tempest brand plugs are a bit cheaper—figure $26 each—and I think they're better than Champions. Fine-wire iridium-electrode plugs generally last three times as long, but cost three times as much, so the cost per hour is about the same as for massives.

For years, I cleaned and gapped my own spark plugs. But I fly a twin, and doing preventive maintenance on 24 plugs is a lot of work. So, I started sending my plugs off to Aircraft Spark Plug Service in Van Nuys, California. They'll clean, gap, bomb-test and recertify your plugs, and return them sealed in plastic with new copper gaskets, for about five bucks per plug, with turnaround typically a week or less. Given today's hourly shop rates, I consider this a real bargain.

Mag checks—done the right way

From your first days as a student pilot, you were undoubtedly taught to perform a "mag check" as part of each pre-takeoff runup. But do you know how to do it correctly, what to look for, and how to interpret the results? Surprisingly, many pilots don't.

To begin with, most Pilot Operating Handbooks (POHs) tell you to note the RPM drop when you switch from both mags to just one, and give some maximum acceptable RPM drop and sometimes some maximum acceptable RPM difference between the two mags.

For example, the POH for my Continental-powered Cessna 310 specifies that an RPM drop more than 150 RPM on either mag or a difference more than 50 RPM between the two mags is unacceptable. Many Lycoming-powered aircraft specify a maximum drop of 125 RPM and a maximum difference of 50 RPM.

In my view, however, this archaic RPM-drop method makes little sense for aircraft that are equipped with a modern digital engine monitor (as most are these days). EGT rise is a far more reliable and revealing indicator of proper ignition performance than RPM drop. Consequently, I recommend focusing primarily on the engine monitor, not the tachometer, when performing the mag check.

Look for all EGT bars rising and none falling when you switch from both mags to one mag. The EGT rise will typically be 50°F to 100°F, but the exact amount of rise is not critical. In fact, it's perfectly normal for the rise to be a bit different for odd- and even-numbered cylinders. Also look for smooth engine operation and stable EGT values when operating on each magneto individually. A falling or erratic EGT bar or rough engine constitutes a "bad mag check" and warrants troubleshooting the ignition system before flying.

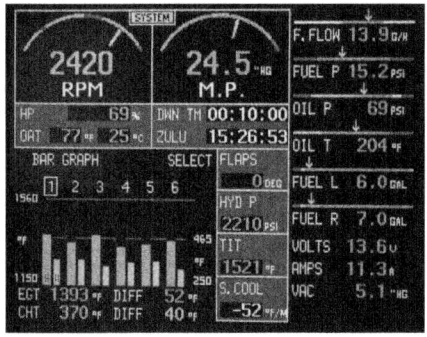

When performing a mag check, focus primarily on your digital engine monitor, not your tachometer.

Most engine monitors have a "normalize mode" that levels all the EGT bars at mid-scale and increases the sensitivity of the display. It's a good idea to use this mode during mag checks because doing so will make any ignition anomalies much more obvious. Bring the engine up to the POH-specified runup RPM (commonly 1700 for direct-drive Continental engines, 1800-2000 for Lycomings), place the engine monitor in normalize mode, perform the mag check (BOTH-LEFT-BOTH-RIGHT-BOTH), then return the engine monitor to its default mode.

In-flight mag check

The usual pre-flight mag check is a relatively non-demanding test and will only detect gross defects in the ignition system. To make sure your engine's ignition is in tip-top shape, I suggest performing an in-flight mag check at cruise power and a lean mixture—preferably a lean-of-peak (LOP) mixture.

An in-flight LOP mag check is a far more discriminating way to test your ignition system. That's because a lean mixture is much harder to ignite then a rich one. A marginal ignition system can pass the normal pre-flight mag check, but it takes one in excellent shape to pass an in-flight LOP check. That's why I like to call it an "ignition system stress test."

The in-flight mag check is performed at normal cruise power and a lean mixture (preferably LOP); the leaner, the better. Run the engine on each individual mag

An in-flight LOP mag check is the best way to make sure your ignition system is in tip-top shape.

for at least 15 or 20 seconds while watching the engine monitor in normalize mode. Ensure that all EGTs rise, that they are stable, and that the engine runs smoothly on each mag. If you have a constant-speed prop, don't expect any RPM drop. Focus primarily on the EGTs, and secondarily on any perceptible engine roughness when running on one mag.

If you see a falling or unstable EGT, write down which cylinder and which mag, so you or your mechanic will know which plug is the culprit. If you don't write it down, I guarantee you'll forget the details by the time you get back on the ground.

Bad mag checks

If you perform a mag check (ground or flight) and don't like what you see, then what? How can you tell what's wrong, and what should you do to correct it?

A non-firing spark plug affects only one cylinder, while a faulty mag affects all cylinders.

To begin with, the phrase "mag check" is a bit misleading. The overwhelming majority of "bad mag checks" are caused by spark plug issues, not magneto issues. (We really should call it an "ignition system check.") By watching EGTs, it's usually easy to tell whether a bad mag check is due to a spark plug problem or a magneto problem: A faulty spark plug affects only one cylinder (i.e., one EGT bar on your engine monitor), while a faulty magneto affects all cylinders (and all EGT bars).

If you detect a non-firing spark plug during your pre-takeoff runup, the most common cause is oil fouling. You can try to clear an oil-fouled spark plug by running the engine for 30 seconds or so at normal run-up RPM with the mixture leaned out to peak EGT. If that doesn't cure the problem, then the plug may be lead-fouled or damaged, and you'll want to have it inspected and cleaned or replaced before flight.

On the other hand, if you observe a non-firing plug during an in-flight mag check, there's usually no need to panic because the cylinder will not be damaged by running on only one plug. If the engine runs smoothly on both mags, simply proceed to your destination and deal with the problem when you get there.

Mag timing issues

During a pre-takeoff mag check, if you get an excessive RPM drop when you switch to one mag but all EGTs rise and the engine runs smooth, chances are that it's not a bad magneto but rather retarded ignition timing. This is sometimes caused by mechanic error in timing the mags during maintenance (especially annual inspections), but it can also be caused by excessive magneto cam follower wear (possibly due to inadequate cam lubrication) or some other internal mag problem. Retarded ignition timing also results in higher-than-usual EGT indications. Mildly retarded timing is not a serious problem, but it does cause some loss of performance so should be addressed as soon as practicable.

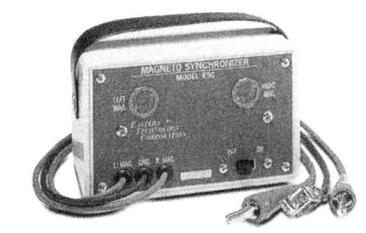

If you notice lower-than-usual EGTs and higher-than-usual CHTs after the airplane comes out of maintenance, have the ignition timing re-checked before further flight.

Conversely, advanced ignition timing results in lower-than-usual EGT indications, and higher-than-usual CHT indications. Advanced timing is a much more dangerous condition than retarded timing because it can lead to detonation, pre-ignition and serious engine damage. If you observe low EGTs and high CHTs after an aircraft comes out of maintenance, do not fly until you've had the ignition timing re-checked.

High-altitude misfire

As mentioned earlier, a lean mixture is more difficult to ignite than a rich one. In turbocharged airplanes, there's also another factor to consider: altitude. The higher a turbocharged airplane flies, the more difficult it is for the spark to jump the gap between the spark plug electrodes, and the more likely that the spark will instead "arc-over" inside the magneto itself.

Such "high-altitude misfire" is bad for two reasons. First, it can cause the engine to run rough—sometimes frighteningly, change-of-underwear rough. Second, it can damage the magneto internally, and in extreme cases cause the magneto to fail mechanically. Not good.

There are two fundamental strategies for preventing high-altitude misfire: make it easier for the spark to occur at the spark plug gap (where it belongs) or make it harder for it to arc-over inside the mag (where it doesn't).

To prevent high-altitude misfire, keep plugs gapped at the tight end of the allowable range.

The simplest way to make it easier for the spark to occur at the spark plug gap is to tighten up the gap. Most aviation plugs have specs calling for a gap of .016 - .019 inches. Keeping the gap at the tight end of the range (.016) makes high-altitude misfire less likely. Of course, the gap increases as the plug wears, so it's important to re-gap the plugs on a regular basis, typically every 100 hours or less for a turbocharged engine.

There are two ways to make it harder for arc-over to occur inside the magneto during high-altitude flight. One is to use a magneto that is physically large, which greatly reduces the likelihood of internal arc-over between widely-spaced components. For example, the huge Bendix S-1200 mags that I use on my Cessna T310 have distributor block electrodes spaced 1.2 inches apart, nearly twice as far as smaller mags like Bendix S-20s and Slick 6300s. However, the S-1200s are a good deal heavier and more expensive than their smaller brethren, and are too large to fit in some engine installations (such as top-induction Continental -550s).

The other way to inhibit arc-over at high altitudes is to pressurize the mags with upper-deck air. This works well, but it's something of a mixed blessing. Pressurized mags tend to have more problems and need more maintenance than unpressurized mags, because the pressurization pumps moist air through the magnetos (particularly when flying through clouds and precipitation), and often causes corrosion and contamination issues.

Magnetos and spark plugs are simple and reliable, but we cannot take them for granted. By adhering to manufacturer's recommended maintenance intervals and monitoring their condition frequently by performing ignition system checks, you can stay one step ahead of an ignition failure and count on safe and reliable operation.

160 ENGINES

19
Where Fuel and Air Meet

Evolution of piston aircraft engine fuel metering systems

If you fly a piston aircraft, chances are it has a spark-ignition (SI) engine that burns gasoline. There also exist compression-ignition (CI) engines—also called diesels—that burn kerosene, but in today's GA fleet they're still few and far between. While CI engines spray liquid fuel at high pressure directly into the combustion chambers, SI engines combine air and gasoline together to form a combustible air-fuel charge that is ingested into the combustion chambers and then ignited electrically by spark plugs.

To be combustible, the charge must have an air-fuel ratio between 8-to-1 and 18-to-1 by weight. Anything less than 8-to-1 is too rich to burn, and anything more than 18-to-1 is too lean to burn. The chemically perfect mixture—what a chemist would call "stoichiometric"—has an air-fuel ratio of about 15-to-1 by weight, and the mixtures we actually use when flying tend to be fairly close to that ratio.

The process of creating a combustible charge with the desired air-fuel ratio is called "metering" and it's not as easy as it sounds. The hard part is to keep the air-fuel ratio constant at all power settings and altitudes, and to do so in a fashion that doesn't impose unacceptable workload on the pilot.

Carburetors

The traditional way to accomplish this is with a carburetor. The word derives from the French verb *carburer* (to combine with carbon); more specifically, in the context of engines, it means to increase the carbon content of air by mixing it with a volatile hydrocarbon.

The earliest carburetors were developed in the 1860s and used drips, wicks, and various other mechanisms for combining fuel and air. These "evaporative carburetors" relied on evaporation of volatile fuel into the air passing through them. They didn't work very well unless the engine operated at a constant RPM and power setting. Notable among these was a design by German engineer Wilhelm Maybach that introduced the float-and-needle-valve mechanism most commonly used in contemporary carburetors. Maybach's carb was used by Karl Benz's 1897 "Patentwagen."

The early 20th century brought the development of "proportioning carburetors" capable of measuring the volume of air flowing into the engine and determining the proper amount of fuel required to create the desired air-fuel ratio across a wide range of engine speeds and loads. Numerous such designs were developed—Claudel-Hobson, Beardmore, Zenith, Bendix-Stromberg—but the one most commonly found in piston GA airplanes is the Marvel-Schebler aircraft carburetor, invented by George Schebler and Burt Pierce and originally used in Indy racing cars from 1911 to 1935. Their Marvel-Schebler Carburetor Company changed ownership numerous times—Borg-Warner, Facet Aerospace, Zenith Fuel Systems, Precision Airmotive, Volaré Carburetors—and finally emerged as Marvel-Schebler Aircraft Carburetors, LLC in 2011.

The Marvel carb

The basic design of the Marvel carb is simple. It has two main sections: a float chamber and a throttle body. The float chamber is a fuel reservoir that works a lot like the tank on a flush toilet. Fuel enters the float chamber through a filter and a needle valve. As the float chamber fills with fuel, the float rises and closes the needle valve, shutting off the fuel supply once the fuel level reaches the desired level. As the carb draws fuel from the float chamber, the float drops and opens the needle valve to admit more fuel into the float chamber.

The engine ingests air through the throttle body, which contains a controllable throttle butterfly plate that regulates airflow into the engine according to the position of the cockpit throttle control. Most Marvel carbs are MA-series "updraft" carburetors where air flows from bottom to top (designed to mount below the engine), but some are HA-series "sidedraft" units where airflow is horizontal (designed to mount behind it). Except for throttle body orientation, both series are quite similar.

The Marvel Schebler MA-4-5 updraft carburetor.

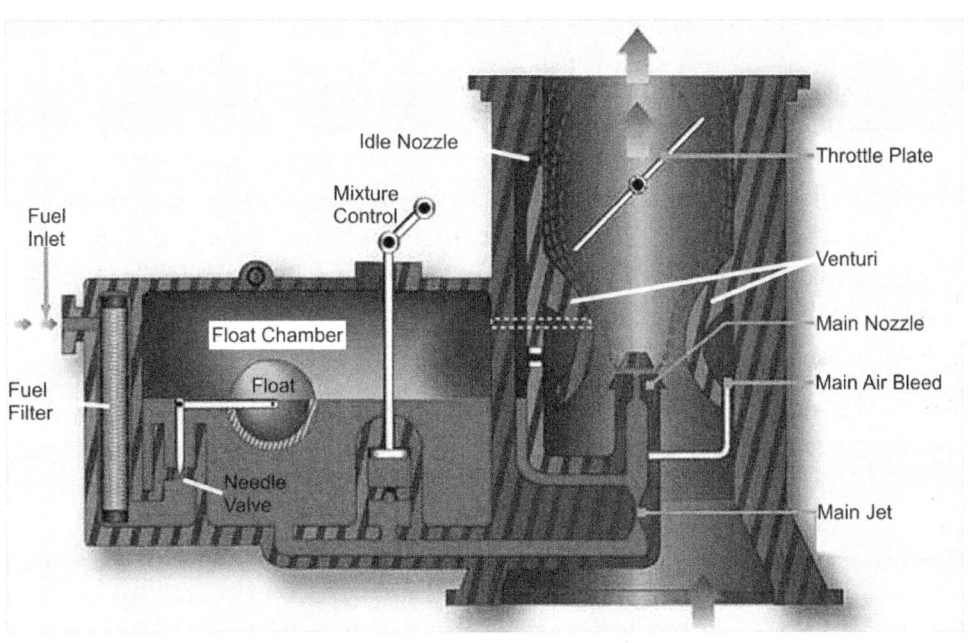

Diagram of float-type carburetor.

The throttle body contains a venturi through which the induction air passes before it reaches the throttle butterfly. Thanks to Bernoulli's principle, the venturi causes the air's velocity to increase and its pressure to decrease. The greater the volume of airflow through the throttle body, the more the air pressure decreases inside the venturi.

A fuel discharge nozzle fed from the float chamber has its tip strategically positioned in the center of the venturi. The pressure drop (suction) generated by the venturi causes fuel to be sprayed by this nozzle into the airflow, where it is atomized into a mist of tiny droplets. The greater the airflow, the greater the suction, and therefore the greater the fuel flow discharged by the nozzle. (This is what makes the carb "proportioning.")

A few complications

This simple scheme works fine at most power settings. However, when the throttle is closed and the engine is at idle, there's so little airflow and so little venturi suction that the main nozzle won't provide enough fuel to keep the engine running. Consequently, the carb has a special "idle circuit" that feeds extra fuel to a secondary nozzle located just downstream of the throttle butterfly and allows the idling engine to suck additional fuel into the airstream to provide smooth idling. This idle circuit has a needle valve adjustment that allows the idle mixture to be fine-tuned.

Another complication occurs when the throttle is advanced rapidly from idle to high power (e.g., at takeoff or go-around). The airflow increases immediately, but it takes awhile for the increased fuel flow to reach the cylinders, and that can cause the engine to hiccup at throttle-up. The solution is to add a throttle-operated "accelerator pump" to the carb to inject a bolus of extra fuel through the main nozzle to prevent the engine from stumbling. Some carbs also have a "power enrichment valve" or "economizer circuit" that provides an extra-rich mixture at wide-open throttle.

Yet another difficulty involves changes of altitude. The carb's venturi effect produces a fuel flow proportional to airflow volume, but the engine needs an optimum air-fuel ratio by weight, not volume. As the aircraft climbs higher and the density of air decreases, the weight of the induction air decreases even though its volume remains constant, causing the mixture to get progressively richer. The usual solution is to provide a cockpit mixture control that allows the pilot to lean the mixture as necessary as the aircraft climbs. (A few fancy carburetors—mostly on big radial engines—incorporate an aneroid-operated "auto-lean" mechanism to eliminate the need for manual leaning, but most don't.)

Throttle body injection

Float-type carburetors have some inherent limitations. One is that they don't work properly in inverted flight or other negative-G situations. Another is that they're quite susceptible to ice build-up in the venturi and on the throttle butterfly caused by the sharp temperature drop due to evaporation of the fuel. Also, it's tricky to get them to work with turbocharged engines.

One way to solve these problems is to reposition the main nozzle downstream of the throttle butterfly instead of upstream, and use carefully regulated fuel pressure (instead of venturi suction) to meter the fuel sprayed by the nozzle. Such a device is known as a "pressure carburetor" or alternatively "throttle body injection" (TBI).

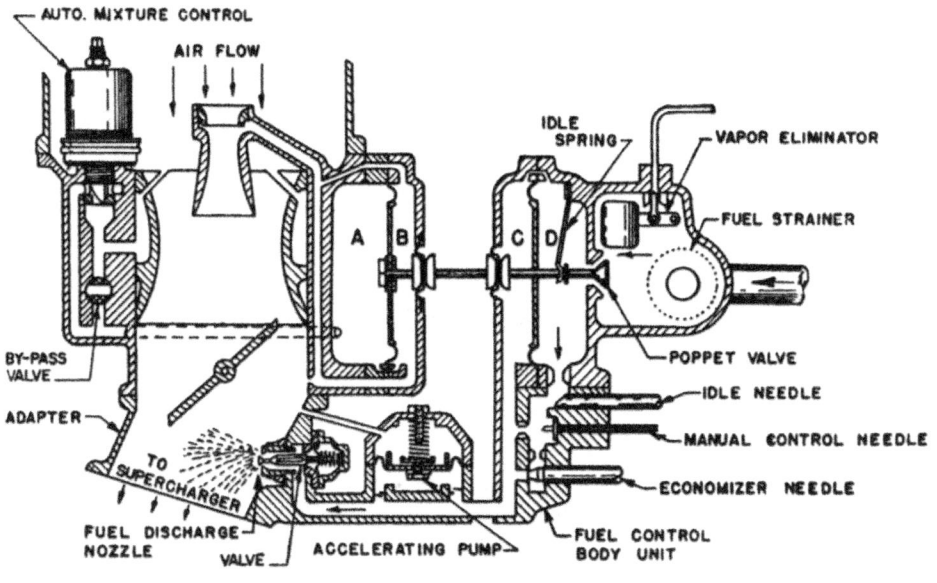

Diagram of Bendix-Stromberg pressure carburetor.

The Bendix-Stromberg pressure carburetor is mostly found on big radial recip engines. It uses a clever four-chamber regulator to meter the proper amount of fuel to its downstream discharge nozzle. Chambers A and B contain air, and are separated by a diaphragm. Chamber A contains pressurized air from "impact tubes" (essentially tiny pitot tubes) in the throttle body, while chamber B contains suction from the venturi. The pressure difference between A and B varies with airflow through the throttle body, and generates a force on a rod that controls fuel flow to the nozzle. Similarly, chambers C and D contain fuel, and are separated by another diaphragm. Chamber D contains pressurized fuel from the fuel pump, while chamber C contains lower-pressure fuel after it has passed through a metering orifice. The pressure difference between D and C varies with fuel flow to the nozzle, and generates an opposing force on the control rod. Equilibrium is reached when the airflow force and the fuel flow force are identical, and that occurs when the air-fuel mixture is optimal.

Rotec TBI.

Vastly simpler throttle body injection designs from Ellison and Rotec have become quite popular among owners of experimental amateur-built airplanes, particularly ones used for aerobatics where conventional float-type carbs don't do so well.

Intake port injection

Whether float-type or TBI, carburetors create an air-fuel mixture at the throttle body, and it then must travel through complex induction plumbing to reach the individual cylinders. Because air turns corners more readily than fuel droplets, a certain amount of centrifugal separation invariably occurs, resulting in different cylinders receiving slightly different air-fuel ratios. To achieve an even mixture distribution among the cylinders, it's better to inject metered fuel directly into the cylinders' intake ports. In today's GA fleet, the two most common systems for doing that are Bendix-RSA fuel injection and Continental Motors fuel injection.

The Bendix-RSA system—used on all fuel-injected Lycomings and a few Continentals—uses a throttle body and four-chamber fuel metering servo that's nearly identical to the one used by the Bendix-Stromberg pressure carburetor. The principal difference is that the throttle body doesn't have a fuel discharge nozzle. Instead, metered fuel from the servo is routed to a flow divider that divides the fuel into four or six equal fractions and routes them to fuel discharge nozzles threaded into the intake ports of the individual cylinders.

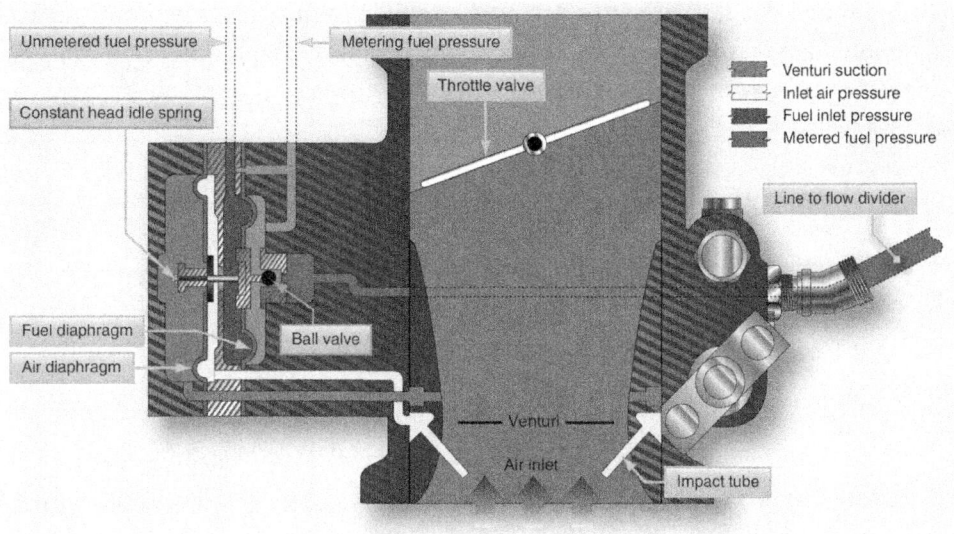

Diagram of Bendix-RSA fuel servo.

The Continental Motors system is considerably simpler, with no chambers or diaphragms. It consists of an engine-driven injection pump whose output pressure varies with engine RPM. Pressurized fuel from this pump is routed to a fuel control unit that contains two valves, one controlled by throttle position and the other by mixture control position. Metered fuel pressure from the fuel control unit is routed to a flow divider and then to individual nozzles in the cylinder intake ports, just as with the RSA system.

Both systems—RSA and Continental—are continuous-flow systems that steadily spray fuel into the intake ports, regardless of whether the intake valve is open or closed. Electronically controlled pulsed-flow injection systems of the sort used in automobiles for decades are just now starting to find their way into Full Authority Digital Engine Control (FADEC) piston aircraft engines.

Lycoming's RSA fuel injection system isn't field-adjustable (other than idle mixture and RPM). If full-power or mid-range fuel flow needs to be adjusted, the RSA fuel servo has to be removed and sent out to a fuel metering specialist who can test the servo on a calibrated flow bench and repair or overhaul it as necessary. The Continental fuel injection system, by contrast, is extensively field-adjustable to provide optimum fuel flow for all power settings from idle to full-power without the need for component removal or special equipment. The next Chapter describes the Continental system in more detail and explains how these important adjustments are accomplished.

20
Continental Fuel Injection

The Continental fuel injection system is simple and very reliable, but only if the system is properly adjusted and maintained.

There are presently two dominant fuel injection systems used in piston aircraft engines. Fuel-injected Lycoming engines (and a handful of Continentals) use the Bendix-RSA system which meters fuel to the engine by measuring the mass airflow into the engine (using blast tubes in the induction system), then converts this to an appropriate fuel flow by means of a complex arrangement of aneroids, diaphragms, springs and metering valves. The Bendix-RSA system is very elegant and sophisticated, and minimizes pilot workload by making most mixture adjustments automatically. This elegance and sophistication comes at a price, however—the system is maintenance-intensive, difficult to troubleshoot in the field, and most problems require that the fuel control unit be sent in to a specialized repair facility.

In contrast, the Continental fuel injection system is extremely simple, with a parts count that is less than most carburetors. This makes it very reliable, easy to troubleshoot, and easy to adjust in the field without a flow bench or other exotic equipment.

The Continental continuous-flow fuel injection system is used on most of their high-performance engines. It's very simple and reliable, and easy to troubleshoot and adjust.

The Continental system has four basic components: an engine-driven fuel pump, a fuel control unit, a manifold valve (also known as a distributor valve or flow divider), and fuel nozzles. Some installations also use an altitude-compensating pump, although most don't. I'll first describe these subsystems in detail as they appear on naturally aspirated big-bore Continental engines like the IO-470, IO-520 or IO-550, before delving into the procedure for adjusting the system for optimal performance.

Fuel pump

The fuel injection pump is an engine-driven, positive-displacement, carbon-vane pump, with output volume and pressure that varies with engine RPM. On big-bore Continental engines (IO-470, -520, and -550), the pump is mounted on the accessory case at the rear of the engine; on the IO-240 and IO-360 engines, it is mounted at the front of the engine.

The drive-end of the fuel pump assembly contains a mounting flange, a drive shaft, and a "dry bay" which incorporates an oil seal (to keep engine oil out of the pump) and a fuel seal (to keep fuel out of the crankcase). The dry bay has a drain fitting that deals with any small amounts of fuel or oil that make their way past the seals. If you notice any fuel or oil dripping from the dry bay drain tube, this tells you that the fuel seal or oil seal (respectively) are shot.

Just aft of the dry bay is the fuel pump itself. The pump uses a pair of floating carbon vanes mounted in a slotted hub and rotating within an eccentric chamber. If you've ever seen a dry vacuum pump disassembled, the fuel pump is of similar design, only much smaller. Unlike a dry vacuum pump, however, the fuel pump is continuously cooled, lubricated and cleaned by fuel, so catastrophic fuel pump failures are exceedingly rare.

Aft of the pump chamber is the vapor separator tower. Fuel from the aircraft fuel tank is plumbed to a fuel inlet about two-thirds of the way up the tower's cylindri-

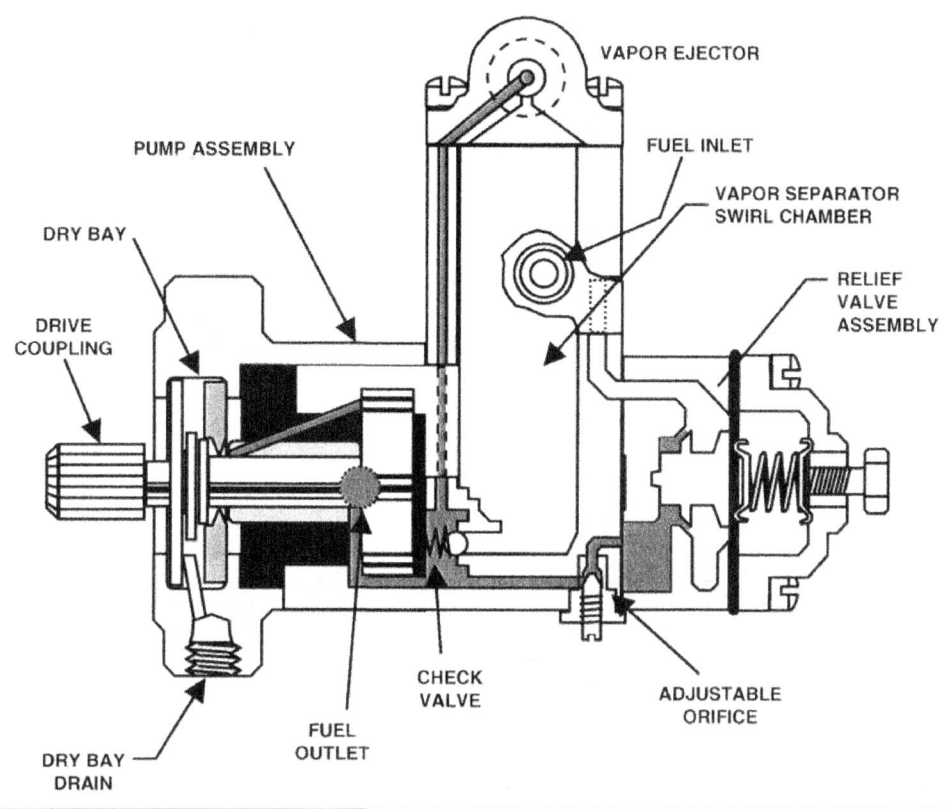

Cutaway of fuel pump used on naturally aspirated engines.

cal chamber, and swirls around the chamber to produce a centrifuging action which helps ensure that only liquid fuel can reach the fuel pump chamber. Any fuel vapors rise to the top of the tower, where a vapor ejector captures them. The vapor ejector contains a small orifice or "jet" through which a small quantity of pressurized fuel is diverted, creating suction that pumps vapors (and any excess liquid fuel) out the return line and back to the tank.

At the base of the vapor separator is a bypass check valve that is normally closed but opens any time the fuel pump inlet pressure is greater than the fuel pump outlet pressure. The purpose of this check valve is to allow pressurized fuel from the aircraft's electric boost pump to bypass the engine-driven vane pump when the vanes are not turning. This occurs routinely during pre-start priming and might also occur in the event of an (extremely rare) in-flight failure of the engine-driven pump.

Pressurized fuel from the pump chamber is routed three ways. A small amount is directed up a vertical passageway at the front of the vapor separator tower to pressurize the vapor ejector jet at the top of the tower. Some additional fuel recirculates through an orifice and relief valve (explained below) back to the fuel inlet port on the vapor separator tower. The remainder of the pump output—most of it—goes to the fuel pump outlet, where it passes through a fuel hose to the fuel control unit.

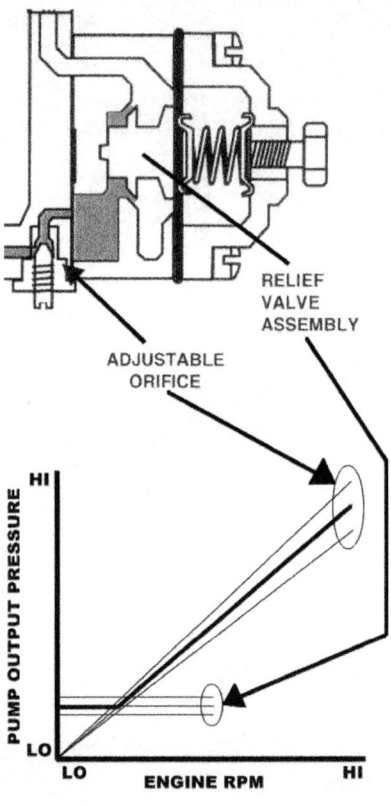

The metering portion of the naturally aspirated fuel pump contains two adjustments. The adjustable orifice is used to adjust the high-RPM output for full rated fuel pressure at takeoff power. The relief valve is used to adjust the idle-RPM fuel pressure.

The aft-most portion of the pump assembly is a metering unit that contains an adjustable orifice and an adjustable pressure relief valve. The adjustable orifice makes it possible to adjust the output pressure and volume of the fuel pump. Turning the orifice adjustment screw counterclockwise opens the orifice, allowing more fuel to recirculate back to the inlet, and therefore decreasing the output, allowing less fuel to recirculate and therefore increasing the output pressure of the pump. (As we'll see later when we discuss fuel system adjustments, this is how you adjust the system for full rated fuel pressure at takeoff power.)

The output pressure of the fuel pump varies with engine RPM. As RPM is reduced to minimum idle, output pressure would normally fall to a very low value insufficient to keep the engine running. To solve this problem, an adjustable spring-loaded relief valve is included in the fuel recirculation path in series with the adjustable orifice. The relief valve is wide open during normal engine operation, so pump output pressure is governed solely by the adjustable orifice. At idle-range RPM, however, spring tension closes the relief valve to maintain adequate pump output pressure. The graph on the previous page shows a simplified illustration of how pump output pressure varies with engine RPM, and how these two adjustments affect output pressure.

Altitude-compensating pump

For most naturally aspirated Continental engines (IO-240, IO-360, IO-470 and IO-520), mixture management is decidedly not a "set it and forget it" situation. As the aircraft climbs, the pilot must manually lean the mixture to compensate for the reduction in ambient air pressure. This is because the naturally aspirated fuel injection system has no means to sense changes in ambient pressure.

In certain IO-550-series engines, however, Continental changed to an altitude-compensating fuel pump that does provide "auto-lean" capability by sensing ambient pressure and adjusting fuel pump output pressure accordingly.

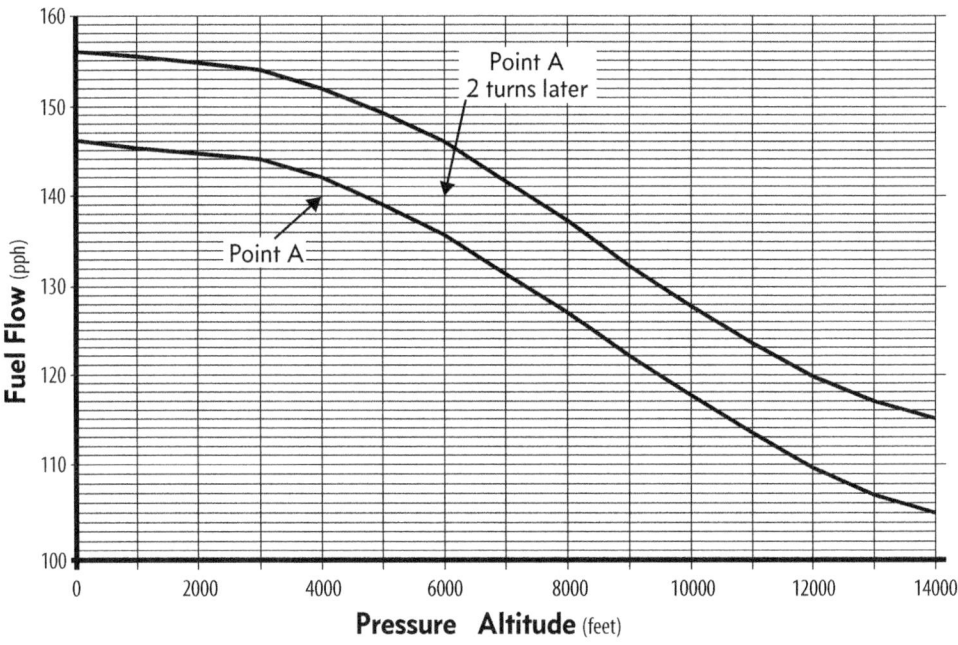

The altitude-compensating fuel pump installed on certain IO-550-series engines provides automatic leaning with altitude. The altitude-compensating aneroid must be flight tested and then adjusted (one turn per 1,000 feet) until it meets the specified leaning schedule.

The new altitude-compensating pump has the same screwdriver-adjustable fixed orifice and low-pressure relief valve as the standard naturally-aspirated pump—but it has an additional bypass orifice controlled by an altitude-compensating aneroid referenced to outside ambient pressure. Thus, the new pump has three adjustments, rather than the two adjustments found on other Continental fuel pumps. The first two adjustments (fixed orifice and low-pressure relief valve) can be adjusted on the ground as usual, but the third (altitude-compensation) requires flight testing. This makes the new system somewhat more difficult to set up.

The new altitude-compensating fuel pump was originally developed by Continental at the request of Beech who wanted a "set it and forget it" auto-lean capability in its new Bonanzas and Barons. Unfortunately, the system proved to be problematic, and a good number of new IO-550 engines experienced premature top-end failures due to excessively lean mixtures during climb. Part of the problem was that the system is harder to adjust properly because of the extra fuel pump adjustment and the need for flight testing.

In addition, there have been some installation-related problems. The altitude-compensating aneroid gets its ambient pressure reference from the dry bay drain port. In some installations, the drain line was placed in a low-pressure area of the engine compartment, causing the aneroid to sense deceptively low pressure and respond by leaning the mixture too much. Such problems can be easily resolved by repositioning the drain, but lots of cylinders were toasted before the nature of this problem was fully understood.

Ironically, Beech endured so many warranty claims related to this system that they asked Continental to go back to the old non-compensating pump. This may well have been an overreaction, since the system works quite well when it is installed and adjusted properly.

Fuel control unit

Fuel from the engine-driven pump—also known as "unmetered fuel"—travels through a fuel line to the fuel control unit (FCU), located next to the throttle butterfly on all the IO-470, IO-520 and IO-550 engines. The FCU incorporates three components: a fuel screen, a mixture control valve, and a throttle control valve.

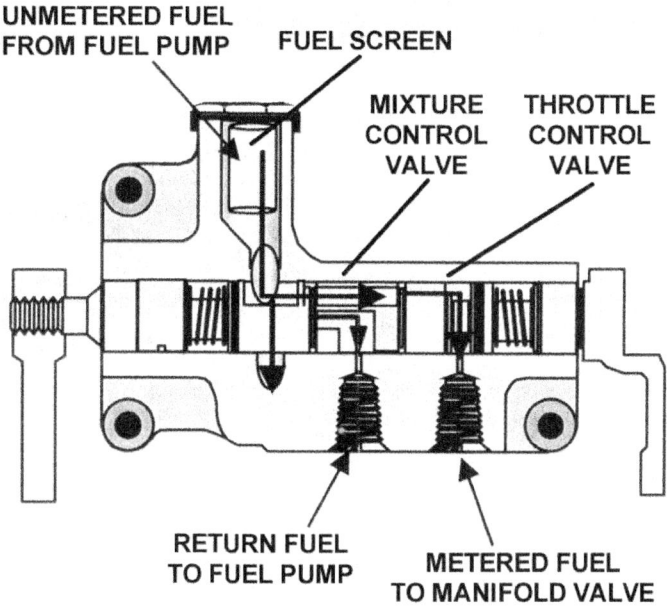

The fuel control unit (FCU) consists of a fuel screen, a mixture control valve, and a throttle control valve. It takes unmetered fuel in from the fuel pump, sends metered fuel to the manifold valve (in accordance with mixture control and throttle), and returns any excess fuel back to the pump.

The internal construction of the FCU is shown above. Upon arriving at the FCU, the fuel first passes through the fuel screen where it is filtered. (This screen should be removed and cleaned at every annual and 100-hour inspection but is often overlooked.) The filtered fuel then passes to the mixture control valve.

The mixture control valve is controlled by the cockpit mixture control and determines how much of the unmetered fuel goes to the engine and how much returns back to the fuel pump. With the mixture control in "idle cutoff" position, 100% of the unmetered fuel will return to the fuel pump. With the mixture control at "full rich," most of the fuel can proceed to the throttle control valve, although a small amount still is returned to the pump.

The throttle control valve in the FCU is linked to the throttle butterfly so that both the valve and the butterfly are operated together by the cockpit throttle control shown in the figure below. The FCU's throttle control valve is designed to restrict fuel flow to the engine in the same proportion as the throttle butterfly restricts airflow to the engine, so that the mixture ratio remains constant throughout the range of the throttle.

This FCU arrangement is used in the IO-470, IO-520 and most IO-550 engines. In certain other engines (notably the IO-240 and IO-360), the mixture control valve is incorporated into the fuel pump itself, leaving only the throttle control valve at the throttle body, and eliminating the need for the fuel return line from the FCU to the pump. In this configuration, the throttle control valve is integrated into the throttle butterfly assembly and referred to by Continental as the "fuel metering unit" (FMU) rather than the "fuel control unit" (FCU).

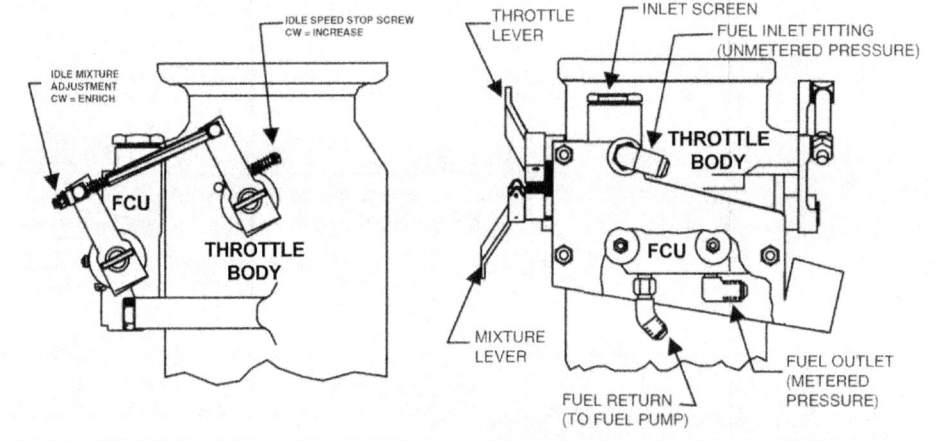

The FCU is mounted to the induction throttle body, and the FCU's throttle control valve is linked to the throttle butterfly so that both move together with the cockpit throttle control.

Manifold valve

Metered fuel from the FCU passes through a fuel line to the manifold valve (sometimes called the "distributor valve" or "flow divider" or colloquially "spider") mounted on top of the engine. The manifold valve performs two distinct functions: it divides the metered fuel equally between the fuel nozzles, and it provides a positive fuel cut-off during engine shutdown.

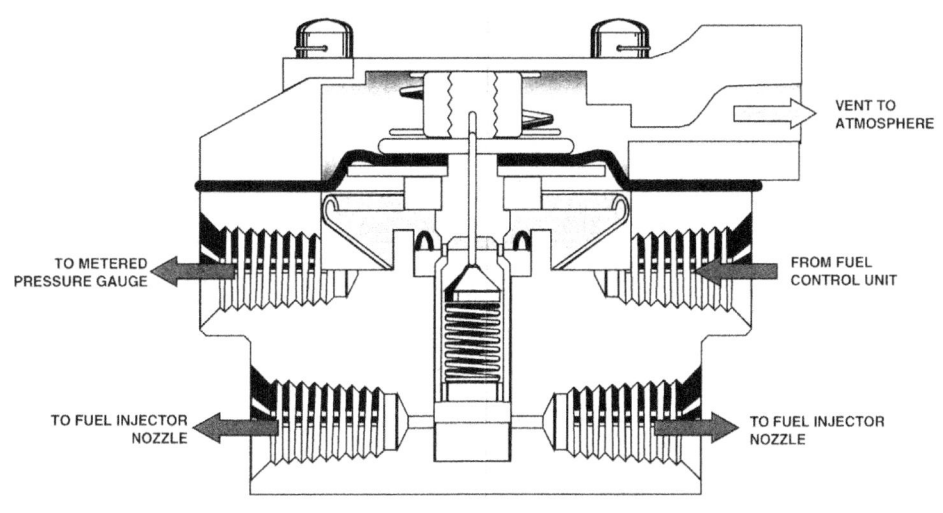

The manifold valve divides metered fuel into equal parts for each nozzle, and provides a clean fuel cut-off at engine shutdown.

The bottom of the manifold valve contains one calibrated orifice per cylinder. The outlet ports are connected to the nozzles by small fuel lines. The conventional wisdom used to be that all the injector lines had to be of equal length, but recent testing showed that this was unnecessary. Consequently, in some of Continental's newer top-induction-system engine designs, the manifold valve is located well off-center and the various injector lines are of unequal length. (One more old wives' tale down the drain.)

The manifold valve also provides a clean cut-off of fuel to the nozzles when metered fuel pressure drops below a threshold value at shutdown. Most Continental manifold valves are set to a 4 PSI threshold (although there is also a 2 PSI version that is used on certain engine models). As metered fuel pressure rises above 3.5 PSI, it pushes up on the diaphragm at the top of the manifold valve and compresses the spring above the diaphragm. This opens the poppet valve and allows fuel to enter the lower chamber. As the fuel pressure rises to 4.0 PSI, it lifts the plunger in the lower chamber and permits fuel to flow out of the distributor ports to the nozzles.

At engine shutdown, as metered fuel pressure falls below 4.0 PSI, the closing spring pushes the plunger down into the bore, closing off the distributor ports at the bottom. As pressure continues to fall below 3.5 PSI, the upper chamber spring pushes the diaphragm down, closing the poppet valve. This provides a double seal and ensures a positive fuel cutoff to the nozzles.

The manifold valve air vent must be open to ambient air pressure and face away from ram air entering the cowling. Since a cracked diaphragm could result in fuel dribbling out the vent, it's best to connect the vent to an overboard line that would route any such fuel out of harm's way.

Fuel injection nozzles

The basic function of the fuel injection nozzle is to take metered fuel from the manifold valve, pass it through a calibrated orifice that establishes a precise fuel flow into each cylinder, and atomize the fuel into the intake port of the cylinder (just outside the intake valve) where it is vaporized by cylinder head heat and mixed with induction air to provide the necessary fuel-air mixture during the cylinder's intake stroke.

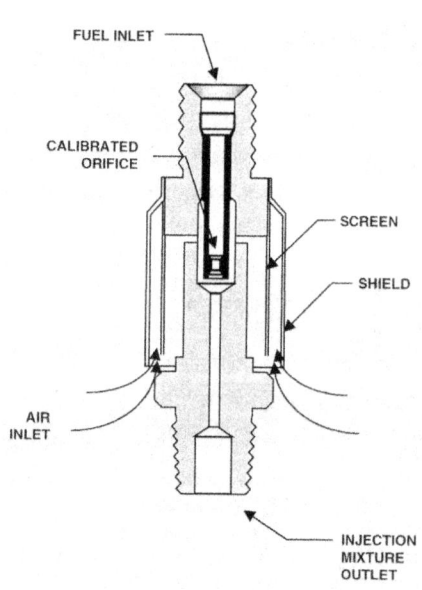

Fuel injection nozzles. Naturally aspirated nozzles (shown here) use ambient air to help atomize the fuel at reduced throttle settings, while turbocharged nozzles use upper-deck air pressure.

At high power settings, the high fuel pressure flowing to the nozzle provides good atomization of the fuel, and the high cylinder head temperature provides excellent vaporization. At reduced throttle, however—and especially at idle—the fuel flow to the nozzle is greatly reduced, and CHT is also much less. Consequently, the nozzle is designed to mix air with the fuel and break the fuel stream up into discrete droplets under reduced-throttle conditions.

Here's how this works. The nozzle is manufactured with a series of circumferential air openings that permit ambient air to be mixed with the metered fuel just prior to the calibrated orifice. The nozzle is fitted with a screen and shield to keep airborne contaminants from plugging the tiny air openings. At wide-open throttle, induction manifold pressure is relatively close to outside ambient pressure, so very little air enters the nozzle. At reduced throttle settings, however, ambient pressure is much greater than manifold pressure, causing substantial air flow into the nozzle air openings and improved atomization of the fuel.

The nozzle's air openings are essential for another reason as well. If these air bleeds didn't exist, then at reduced throttle settings the high manifold vacuum would cause fuel to be "sucked" from the nozzle at a higher-than-desired rate, resulting in over-rich operation (especially at idle). The nozzle's air bleeds prevent this from happening. In fact, if the air bleeds or air filter screen become plugged, the telltale symptom is rough idle operation (due to over-rich idle mixture).

On turbocharged engines, ambient air cannot be used to supply the nozzle air bleeds. If it were, then any time manifold pressure was boosted to above ambient, fuel would spray out of the air bleeds instead of air being sucked in! Consequently, nozzles on turbocharged engines are furnished with a sleeve that permits the nozzle air bleeds to be supplied by upper-deck air pressure from the turbocharger's compressor. Since upper-deck pressure is always greater than manifold pressure, nozzle bleed airflow is maintained just as in the naturally aspirated case.

Putting the pieces together

Now that we've examined the principal components of the Continental continuous-flow fuel injection system, let's step back and review how the entire system is connected.

Fuel from the aircraft fuel tank enters the inlet port of the engine-driven fuel pump, where the vapor separator centrifuges out any vapor and returns it to the tank. The remaining liquid fuel is pressurized by the vane type pump, producing unmetered pump output fuel pressure that varies with engine RPM.

The unmetered fuel is sent to the fuel control unit, where it passes through a filter screen, then through the mixture control valve (operated by the cockpit mixture control), and finally through the throttle control valve (linked to the throttle air butterfly and operated by the cockpit throttle control). The resulting metered fuel is sent to the manifold valve, while any excess unmetered fuel is returned from the FCU to the fuel pump.

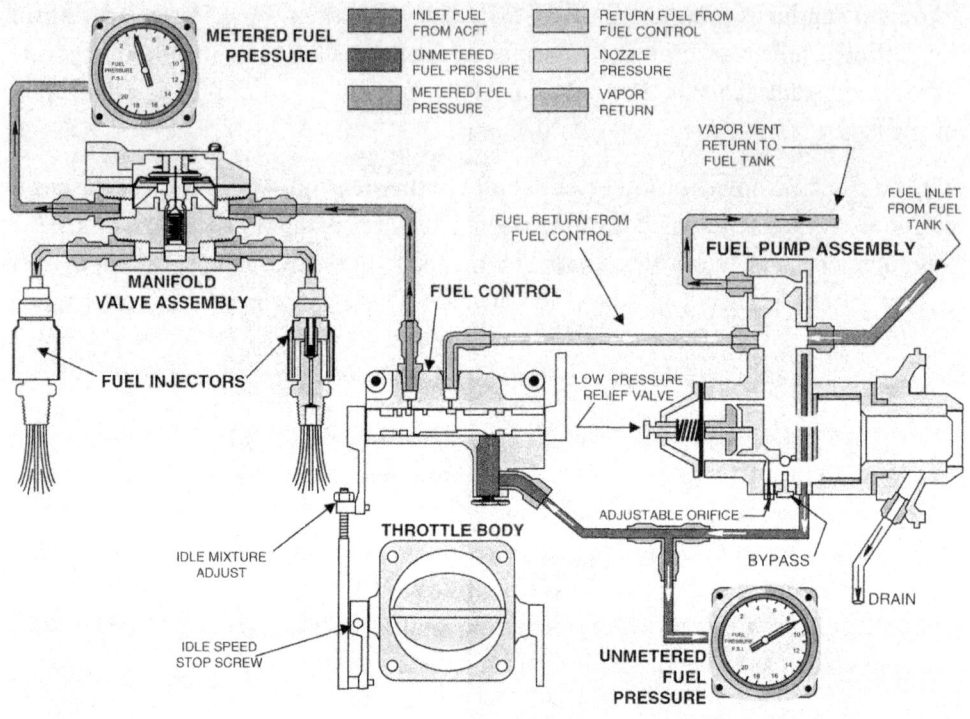

The entire Continental continuous-flow fuel injection system used on normally aspirated engines.

The manifold valve divides the metered fuel into equal parts and distributes it to the nozzle for each cylinder. It also provides for a clean fuel cutoff during engine shutdown when metered fuel pressure drops below 4.0 PSI.

The nozzles pass the fuel through a calibrated orifice and atomize it into the intake port of the cylinder. At reduced throttle settings, the nozzles also mix air into the fuel stream to improve atomization.

Setting up the system

The Continental continuous-flow fuel injection system is marvelously simple and generally works very well. When it doesn't, the most likely cause is that the system is out of adjustment. Although all factory new and rebuilt engines, and many field overhauled ones, have their fuel systems set up during a test cell run before delivery, the fuel pressures are virtually guaranteed to change when the engine is installed on the aircraft, due to differences in induction system, fuel supply system, and operating environment between the test cell and the aircraft.

Thus, it is always necessary to perform the full fuel injection system adjustment procedure whenever a new, rebuilt or overhauled engine is installed on the aircraft. Unfortunately, this is often not done—or is done incorrectly—resulting in poor engine operation and sometimes even engine damage.

Any time a fuel system component is replaced, the fuel injection system adjustments should be rechecked. The same is true any time the engine is modified—by adding an aftermarket intercooler or GAMIjectors, for example. It's also a good idea to recheck the adjustments from time to time—every 500 hours or so—since they can gradually drift out of tolerance as the vanes and springs in the fuel pump get older.

Detailed adjustment procedures and specifications for all Continental fuel-injected engines traditionally appeared in Continental Service Information Directive SID97-3, and now have been incorporated into Continental Manual M-0 ("Standard Practices Maintenance Manual"). This guidance provides step-by-step procedures for adjusting the fuel injection system, as well as tables that provide the proper fuel pressures for each individual Continental engine model. This guidance is the "bible" for fuel system setup, and no adjustments should be made without having this document in hand and understanding it thoroughly.

> Any time a fuel system component is replaced, the fuel injection system adjustments should be rechecked.

To adjust the fuel injection system, it is first necessary to plumb two external pressure gauges into the system. One gauge is hooked up to measure unmetered fuel pressure at the output of the fuel pump, while the other gauge is connected to the manifold valve to measure metered fuel pressure. Idle RPM adjustments are made to a specified unmetered fuel pressure, while full-power adjustments are made to a specified metered fuel pressure. For turbocharged engines, the second gauge must be a differential pressure gauge, with the reference port connected to upper-deck pressure.

Continental recommends the use of the Model 20 ATM-C Porta Test Unit made by Aero Test Inc. of Romulus, Mich. This is an expensive unit, however, and unless you adjust a lot of fuel injection systems, you may want to make due with a low-cost homebrew test rig instead. A couple of 0-60 PSI gauges plus some hoses and tee-fittings will do just fine, so long as the gauges are accurate. If you're adjusting a turbocharged engine, one of the gauges must be a differential gauge.

Step-by-step procedure

Once the external gauges have been teed into the fuel system and all air has been bled from the fuel lines, the basic adjustment procedure involves the following steps:

Step 1: Start the engine and warm it up until engine temperatures and pressures have stabilized in the normal operating range.

Step 2: With the mixture control full rich, reduce the throttle to the idle RPM specified in Continental's guidance. For most engines, this is 600 RPM. Continental recommends that the idle speed adjustment screw at the throttle body be backed off a couple of turns to assure that the RPM can be reduced to the specified value.

Step 3: Adjust the idle RPM unmetered fuel pressure by adjusting the low-pressure relief valve adjustment screw on the fuel pump. Turn the screw clockwise to increase pressure, or counterclockwise to decrease pressure. Continental's guidance specifies a maximum and minimum idle unmetered pressure, but it is desirable to set the pressure to the minimum limit in order to provide a slightly richer mixture at partial throttle settings.

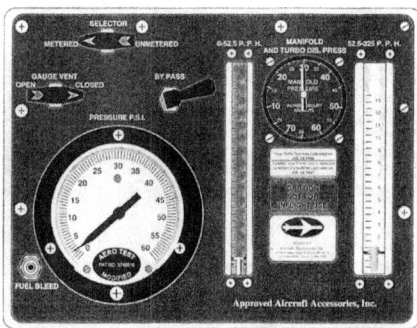

Continental recommends this Model 20 ATM-C Porta Test Unit for checking and adjusting fuel injection system pressures. However, a homebrew pressure gauge setup can also be used.

Step 4: With the engine at specified idle RPM (normally 600), slowly move the mixture control from full rich toward idle cutoff and note the maximum RPM rise. A rise of 25 to 50 RPM indicates proper idle mixture. A rise of more than 50 RPM means the idle mixture is too rich, while a rise of less than 25 RPM means it is too lean. If the RPM rise is out of tolerance, adjust the idle mixture adjustment at the fuel control unit until the specified RPM rise is obtained. Then re-check the idle RPM unmetered pressure to make sure it's still within limits (and preferably at or near the lower limit).

Step 5: Advance the throttle to full power and check the metered fuel pressure to see whether it is within the limits specified in Continental's guidance. If it isn't, adjust the fuel pump's adjustable orifice (for naturally aspirated engines) or altitude-compensating aneroid adjustment (for turbocharged engines) until the full-power metered fuel pressure is within limits. It is desirable to adjust this pressure to the maximum limit. (For turbocharged engines equipped with a fuel pressure regulator, the regulator must be disconnected and the full power metered fuel pressure adjusted to 5% higher than the maximum specified limit. Then the regulator is reconnected and adjusted to the specified limit.)

Step 6: If the full power metered fuel pressure adjustment was changed during Step 5, then idle RPM unmetered fuel pressure and idle mixture must be rechecked and readjusted as necessary (per Steps 3 and 4). This is because the full power adjustment interacts with both idle adjustments.

This completes the adjustment procedure for most engines, and the external pressure gauges can now be disconnected.

In the case of an IO-550 with an altitude-compensating fuel pump, however, an additional procedure is required. This involves a flight test in which the aircraft is climbed at full throttle and full rich mixture while fuel flow is recorded at each 1,000-foot altitude. After the test flight, the fuel flow readings are checked against the specified fuel-flow schedule as specified in Continental's guidance. If the fuel flows are not within the specified ranges at all altitudes, the altitude-compensating aneroid must be adjusted to bring the altitude compensation within limits. Each turn of the adjustment screw increases or decreases the auto-leaning schedule by approximately 1,000 feet. Adjustment of the altitude-compensating aneroid interacts with the idle and full power fuel pressure adjustments, so they must be rechecked and readjusted as necessary.

Just do it!

If your fuel-injected Continental engine won't idle smoothly below 1,000 RPM, or if your takeoff fuel flow is higher or lower than what the POH says, it's a near certainty that your fuel injection system is out of adjustment. Don't try to compensate for such misadjustment with the cockpit mixture control. Get your system adjusted properly.

If it's been awhile since you had your fuel system pressures checked—especially if you're not sure that they were adjusted properly when the engine was installed on the aircraft or you've made any engine modifications since then—you'd be well advised to have it done next time your bird is in the shop. It's no big deal—figure two to four hours of labor for most engines, perhaps a bit more if you have the altitude-compensating pump. Trust me, it's worth it. You simply won't believe how much better your engine will run once the fuel system has been adjusted to specs.

21
Fuel Gone Wrong

When your fuel system hiccups, it can really get your attention.

My first indication that something might be awry came on takeoff from New Haven (Conn.) Tweed Airport, headed for Hanscom Field near Boston. All the way across the country from my home base in California, the takeoff fuel flows for the left and right engines, as displayed on my Shadin digital fuel flow instrument, had been almost perfectly matched—differing at most by one or two pounds per hour. But on takeoff from New Haven, the right engine was showing 10 pounds per hour less than the left. The right EGT was indicating higher than the left, too, although not alarmingly so. But I made a mental note that something seemed... well...different.

My month-long annual-from-hell was over at last. With two freshly rebuilt turbochargers, two brand new tailpipes, 24 new spark plugs, all new wheel bearings, a new main gear tire, a new battery, newly-upholstered seats, and a whole bunch of other stuff, my airplane was finally back in service.

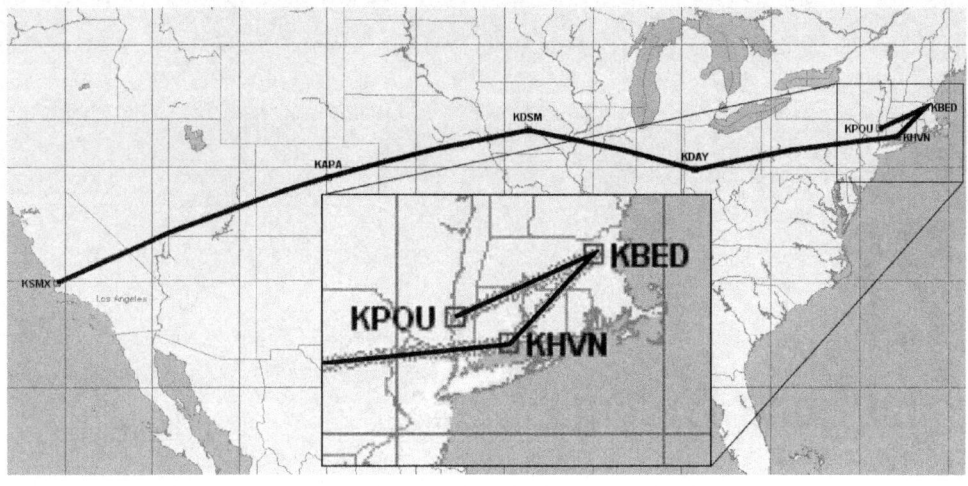

Route of flight from Santa Maria (Calif.) to Denver, Des Moines, Dayton, New Haven, Boston (Bedford), and Poughkeepsie (N.Y.).

After coming through its post-maintenance test flight with flying colors, I'd departed on this long-anticipated trip to the East Coast. Having gotten a late start departing from Santa Maria, Calif., I'd overnighted in Denver, stopped for lunch in Des Moines, spent two nights in Dayton before finally flying to New Haven to spend a few days at Yale University.

I'd put a bit over 10 hours on the airplane since the annual, and so far it had been running like a top. I'm always a little nervous right after doing major maintenance, and this year's annual certainly qualified as major. But everything seemed to be working perfectly, and I was starting to relax, stop worrying, and enjoy the flying.

Still, the fuel flow split on takeoff from New Haven had gotten my attention. It didn't seem serious enough to warrant scrubbing the flight. But it wasn't quite right, and it was the first glitch of any sort I'd seen since the annual. Certainly worth keeping an eye on, I told myself.

It's only about 100 nautical miles from New Haven to Boston—just over a half-hour in the Cessna 310. I'd filed for 3,000 feet, hoping to enjoy the New England scenery at close range, but was assigned 5,000 by ATC. That turned out to be okay, since the forecasted clearing hadn't happened and 5,000 put me just on top of a broken layer. The air on top was dead smooth, and "George" was flying the airplane.

What the !@#$% was that?

Without warning, the airplane shuddered. Instinctively, my eyes focused on the engine gauges. It was obvious that the right engine had experienced some sort of transient roughness—I caught a momentary drop in both EGT and manifold pressure of the right engine. Everything seemed to return to normal, but 10 or 15 seconds later, the same thing happened again.

I'd had the engines leaned very aggressively for cruise, as is my habit. Knowing that lean-of-peak operation tends to magnify the effect of any transient combustion event, I enrichened a bit and continued to watch.

Nothing happened. The roughness didn't return. Maybe some water in the fuel, I thought. But I knew that was unlikely, since it hadn't been raining and I'd drained the tanks and gascolators thoroughly at pre-flight and there wasn't a trace of anything but blue avgas.

My train of thought was interrupted by ATC, who handed me off to Boston Approach, who in turn gave me a descent clearance and told me to expect a visual approach. I landed at Hanscom Field in a stiff wind, with no further signs of engine trouble. After taxiing to the ramp, I did a quick runup, which was entirely normal. After parking and unloading the plane, I drained the gascolators again and found no signs of fuel contamination.

Transient anomalies are the most frustrating kinds of problems to deal with, because they're so difficult to troubleshoot. My approach to dealing with such things is to patiently watch to see what develops. In most cases, one of two things happens: Either the problem goes away by itself (in which case I ignore it), or the problem gets worse (in which case it's a lot easier to troubleshoot). In this case, I decided to take the "watchful waiting" approach.

After a great visit with friends and relatives in the Boston area, it was time for me to move on to my next planned stop, which was Poughkeepsie, N.Y. My family owned a little country cottage about 30 minutes' drive north of Poughkeepsie, and I planned to do a few days of serious sightseeing in the beautiful and historic Hudson River Valley before starting westward for California.

The takeoff from Hanscom field seemed normal. The right engine fuel flow was a bit lower than the left, just as it had been during the previous takeoff from New Haven, and the right EGT was a bit higher. But there was no sign of roughness during the takeoff or climb-out to my 8,000-foot assigned cruising altitude. Leveling off at 8,000 feet, I reduced to 65% power and leaned the engines for cruise.

At which point, things went to hell.

The right engine started running as rough as a stucco bathtub. It surged repeatedly, as the EGT and manifold pressure needles danced all over the dials. This was clearly one mighty unhappy powerplant! This anomaly was no longer "transient." I started richening the mixture on the right engine, and ultimately found a mixture at which the roughness seemed to go away—but it was a godawful rich mixture. Any attempt to lean to something more reasonable restarted the surging. Something was clearly wrong, but was it an emergency? Poughkeepsie was only about 30 minutes away, and I had lots of fuel on board, so I made the decision to continue (with the right mixture nearly full-rich) and try to sort things out on the ground there.

Preliminary diagnosis

This "felt" to me like a fuel system problem. The symptoms were consistent with a partially clogged fuel nozzle in one cylinder. A partially clogged nozzle would result in a reduced fuel flow indication on takeoff—remember, I have a Shadin digital fuel flow gauge, not a fuel pressure gauge. A partially clogged nozzle would also cause engine roughness when the mixture was leaned aggressively for cruise, because the affected cylinder would have an ultra-lean mixture that produced very little power (or possibly no power at all if it was too lean to support combustion). That theory was reinforced when I tried an in-flight mag check and found that the roughness did not get perceptibly worse with either magneto switched off. This pretty much ruled out ignition as a cause of the roughness.

At this moment, it would have been nice to have a probe-per-cylinder engine analyzer. Such instrumentation would have provided absolute confirmation of my diagnosis and told me exactly which cylinder had the clogged nozzle. I'd been meaning to install a digital engine monitor in the airplane for years but had not yet gotten around to it. So, all I had in the way of EGT instrumentation was the brain-damaged factory-installed single-probe gauge. I knew the EGT probe was installed in the lefthand exhaust cluster, and "saw" the combined exhaust from cylinders #2, #4 and #6. Based on the erratic EGT indications I was seeing when the engine was running rough, I theorized that one of the left-bank fuel nozzles was most likely clogged.

With its mixture control jammed forward, the right engine continued to run relatively smoothly. I completed the flight to Poughkeepsie and taxied to the FBO, where I found a technician willing to help. Pete turned out to be a taciturn northeasterner, a fellow who used words sparingly as if they cost $100 each. But it quickly became clear that he worked mostly on Lears, Falcons and Gulfstreams—the FBO's primary business was managing corporate jets for Fortune 500 companies—and would be perfectly happy to give me access to his toolbox rather than get involved with a piston engine. Which, as it happened, was just fine with me.

The JP Instruments EDM-760 twin-engine analyzer I installed in my Cessna 310 not long after this episode. It sure would have been nice to have one of these then!

Easy fix

The next day, I returned to the airport to have a look at the engine. After pulling the top cowling from the right engine, I performed a compression check, just to make sure that the rough-running episode hadn't eaten an exhaust valve or holed a piston. I got decent readings on all six cylinders, reassuring that no serious damage had been done.

After replacing the spark plugs, I removed the #2, #4 and #6 fuel nozzles from the cylinder heads, taking care not to get them mixed up (since they're GAMIjectors and not interchangable). I held each nozzle up to a bright light and looked through the tiny orifice. #2 and #6 were clear, but #4 was clearly obstructed. Bingo!

I blew some air through the #4 nozzle and then held it up to the light again. This time, the orifice appeared clear. I never was able to figure out what had caused the clog, but it appeared now to be gone. Just to be on the safe side, however, I gave it a good soak in MEK to make sure it was as clean as clean could be. I then squirted some aerosol no-residue contact cleaner through all three nozzles, and all three produced a nice fluid stream with no sign of clogging.

After reinstalling the fuel nozzles, buttoning up the cowling and making sure that all the tools were back in their assigned drawers, I found Pete and thanked him for the

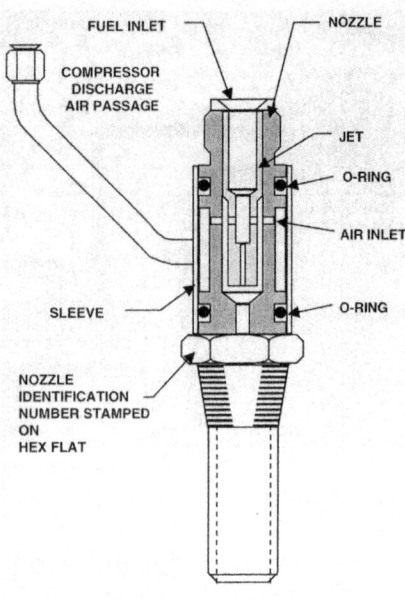

Fuel nozzles on turbocharged engines are referenced to upper-deck air by means of a sleeve sealed by O-rings. Any contaminants in the upper-deck air can clog the secondary orifice at the bottom of the nozzle.

use of his hangar and toolbox. I asked him to write up an invoice to cover his time and trouble, but he smiled and shook his head. Mighty nice fellow, that Pete.

A few days later, I departed Poughkeepsie for the West Coast. On takeoff, the fuel flows were perfectly matched. In cruise, both engines ran smooth as silk, even when aggressively LOP. The airplane ran like the proverbial top all the way back to California. Color me a happy camper.

Aftermath

Though no harm was done this clogged-nozzle incident bothered me. What caused it? How could it have been prevented? Do I need to re-think my maintenance practices? (I tend to overthink these things.)

Many years ago, when I first started swinging wrenches on my airplane, I habitually removed the fuel nozzles at each annual inspection and cleaned them in an ultrasonic cleaner. But I never saw the slightest evidence that the nozzles were getting dirty in service. It wasn't clear to me that there was anything to be gained by performing periodic cleaning of nozzles that spent their working life flowing solvent (avgas), and protected from contamination by at least two very fine screens (one in the gascolator, and a second in the fuel control unit). So for the past decade or so, I'd not been in the habit of cleaning my fuel nozzles on any regular basis. And in all that time I'd never had a clogged nozzle. Until now.

I related this whole story in an email to my friend George Braly, founder and chief engineer of GAMI and arguably the world's foremost expert on Continental fuel nozzles. Here's part of what George emailed me back on the subject:

If a six-probe EGT is installed, I see no reason whatsoever to engage in prophylactic cleaning of fuel nozzles. I think it's fair to say that we see more nozzles with crud in them in the first five hours after routine "prophylactic cleaning" than we do in the next 200 hours!

Further, if one performs a will-it-run-lean-of-peak test on every flight, then one will instantly know if a nozzle is plugged, because the engine will operate unusually rough.

I strongly believe that a lot of cylinder problems are directly attributable to a combination of (1) recent maintenance during which the nozzles were removed and reinstalled, (2) rich-of-peak operation (which allowed a clogged nozzle to go undetected), and (3) lack of six-probe EGT instrumentation (which did the same).

So my advice is: Clean nozzles only on condition if you have a six-probe EGT. If you don't, it's a harder call, but I guess I'd still prefer not to clean them routinely, even at annual inspections, so long as aggressive in-flight leaning does not reveal any unusual engine roughness.

However, nozzles left in place for a long time tend to seize in the threads, so they really need to be installed with proper anti-seize thread lube and minimal torque.

For turbocharged engines, the O-rings and rubber compression seals should be replaced periodically—at least every few years. Turbo nozzles are a bit more prone to clogging because they're supplied with upper-deck air that is filtered only by the engine's induction air filter. If the filter gets dirty, or if alternate air is used, all sorts of dirt can go through the upper-deck air lines into the nozzle air inlet holes, and thereby plug the secondary orifice (downstream of the fuel-metering orifice).

As for what caused my #4 nozzle to clog, George may well have hit the nail on the head with his last comment. Although the nozzles were not removed during the "annual-from-hell" and there was no significant work done to the fuel system, there was significant maintenance done to the induction air system. I replaced the turbocharger and several flexible induction couplings at the annual, and that could easily have introduced some small contamination into the induction air system.

In any case, it seemed clear that prophylactic cleaning of the nozzles during the annual inspection wouldn't have helped one bit. The problem occurred 10 hours after the aircraft got out of maintenance, and that corresponds exactly with George's observations about when most nozzle clogging problems occur (i.e., shortly after maintenance).

Two other things also seemed clear: First, the clog might have gone undetected far longer had I not been in the habit of leaning aggressively. Second, I'd probably be pressing my luck if I let another year go by without installing analyzer good engine monitor—so I installed a JPI EDM-760 and it has paid for itself many times over since then.

Misfueled!

On March 2, 2008, a Cirrus SR22 turbo was destroyed when it crashed on takeoff from Rio de Janiero, Brazil, killing all four people aboard. Shortly after the aircraft departed the runway, the airplane's engine lost power, and the aircraft hit a building and exploded. Further investigation revealed that the aircraft had been refueled with Jet A instead of 100LL.

This report reminded me of an incident many years earlier during which my own airplane was misfueled with Jet A at a busy GA airport just south of San Francisco CA. Fortunately, I caught the (mis)fueler in the act, red handed. Had I not been lucky enough to do that, I very well might not be writing this column.

Normally, I either fuel my aircraft myself (at a self-serve pump) or watch it being fueled (when avgas is supplied by truck). On this occasion, I'd radioed for the fuel truck and waited patiently for it to arrive. After 10 minutes of waiting, Mother Nature intervened and compelled me to walk into the terminal building in rather urgent search of a loo. By the time I took care of my pressing business and returned to the ramp, there was a fuel truck parked by my airplane and a lineperson pumping fuel into my right main tank. As I approached the aircraft, I observed to my horror that the truck was labeled "JET A."

Theoretically impossible

At first, I was not too worried, because I believed that misfueling my airplane with Jet A was physically impossible. That's because in 1987 all turbocharged twin Cessnas became subject to AD 87-21-02 which mandated installation of restrictor ports on all fuel filler openings. The restrictor ports were designed to make it impossible to insert an industry standard Jet A nozzle, while accommodating the smaller diameter avgas nozzle.

The FAA also mandated that jet fuel trucks install a wide spade-shaped fuel nozzle, and that vulnerable airplanes have restrictor ports installed into which the wide jet fuel nozzle would not fit. They also changed their certification requirements so that all new-production avgas-fueled aircraft be factory-equipped with restrictor ports. This theoretically made misfueling of piston aircraft with jet fuel impossible.

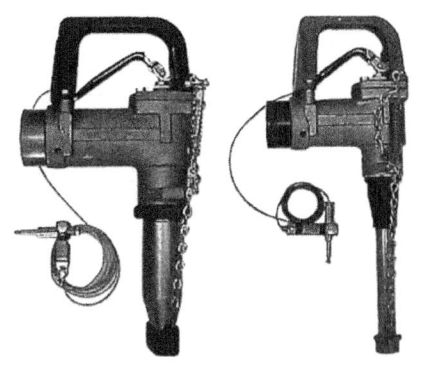

Jet A fuel trucks have been mandated to install a nozzle that should not fit into your avgas fill port.

But as I arrived at my airplane, I discovered that indeed my left main tank had been topped with Jet A. How could this be? A subsequent investigation by the San Francisco FSDO revealed that this particular Jet A fuel truck had not been fitted with the correct spade-type nozzle. (I suspect they got in trouble for that.)

Undoing the damage

I spent literally hours trying to find an A&P on the field that would assist me in purging the fuel system of its witches' brew of 100LL and Jet A. That turned out to be surprisingly difficult because no one wanted to go near my contaminated airplane, clearly afraid of the potential liability exposure. Finally, I persuaded one maintenance manager to help me out after writing and signing an omnibus waiver absolving the shop and its mechanics of any liability in connection with their work on my aircraft.

The purging process itself was quite an eye opener. We drained the tanks and the fuel line going to the engine driven fuel pump as completely as possible, putting the noxious effluent into a 55-gallon drum. Next, we pumped 100LL through the system using the electric boost pump and drained the disconnected fuel line into a 5-gallon bucket. The fuel in the bucket was tested for Jet A contamination using the paper-towel test: A few drops are placed on a paper towel and allowed to evaporate.

Pure 100LL will evaporate completely without leaving any residue on the towel, but even small amounts of Jet A contamination will leave an obvious oily ring. It took five purges with 100LL before we were satisfied that the system was essentially kerosine-free.

We reconnected the fuel line, cowled up the engine, the fueling company then topped off the airplane with 100LL (gratis), and I was finally good to go…fully six hours after the misfueling incident.

Lessons learned

I learned some important lessons that day. Perhaps the most important is that it's impossible to distinguish pure avgas and a mixture of avgas and Jet A solely by looking at it in a clear plastic fuel tester or GATS jar. My main tanks had been about half-full of avgas, so after the misfueling they contained roughly a 50-50 mix. If you take a jar full of pure 100LL and another jar full of a 50-50 mix of 100LL and avgas, I defy you to see any difference in color or clarity between the two. You can't. Trust me on this.

I had always been taught that you sump the tanks and observe the color—100LL is blue and Jet A is straw color. What I was not taught is that a mixture of 100LL and Jet A is also blue and that you simply can't tell the difference visually. In retrospect, I shudder to think what would have happened had I not caught that Jet A truck in front of my airplane.

I was also taught that since Jet A is significantly heavier than avgas (6.7 lbs/gal versus 5.85 lbs/gal), the Jet A and 100LL will separate just like oil and water, with the Jet A at the bottom (where the sump drain is) and the 100LL at the top. The truth is that 100LL and Jet A are highly miscible—they mix quite well—and the mixture takes nearly forever to separate.

There are at least two good ways to distinguish pure 100LL from kerosine-contaminated 100LL. One is by smell: Jet A has a very distinctive odor that is detectable even in small concentrations. The other (and probably best) is by using the paper-towel test: Pour a sample on a paper towel (or even a sheet of white paper), let it evaporate, and see if it leaves an oily ring.

Nasty stuff

What effect does Jet A contamination have on a piston engine? Enough to ruin your day.

Think of Jet A as being avgas with a zero-octane rating. Any piston engine that tries to run on pure Jet A will either fail to start or, if it does start, will go into instant destructive detonation. However, in real life, we almost never encounter that situation because the main tanks used for landing and takeoff are almost never completely dry when the aircraft is misfueled.

Therefore, the real-world problem is not running on pure Jet A, but running on a mixture of 100LL and Jet A. Depending on the mixture ratio of the two fuels, the effective octane rating can be anything between 0 and 100. A mixture with a lot of Jet A and just a little 100LL might be detectable during runup. A 50-50 mix might not start to detonate until full power is applied, and the engine might fail destructively 30 seconds or three minutes after takeoff. Just a little Jet A contamination might produce only moderate detonation that might not be noticed for hours or even weeks. Like so many other things in aviation, "it depends."

The Cirrus SR22 accident in Rio reminds us that the problem of misfueling is still with us, despite all the efforts of the FAA to eradicate it. We need to be vigilant. Always watch your airplane being fueled if you possibly can. Make sure its fuel filler ports are equipped with restrictor rings, if warranted. Don't just look at the fuel you drain from your sumps—sniff it, and when in doubt, pour it on a paper towel.

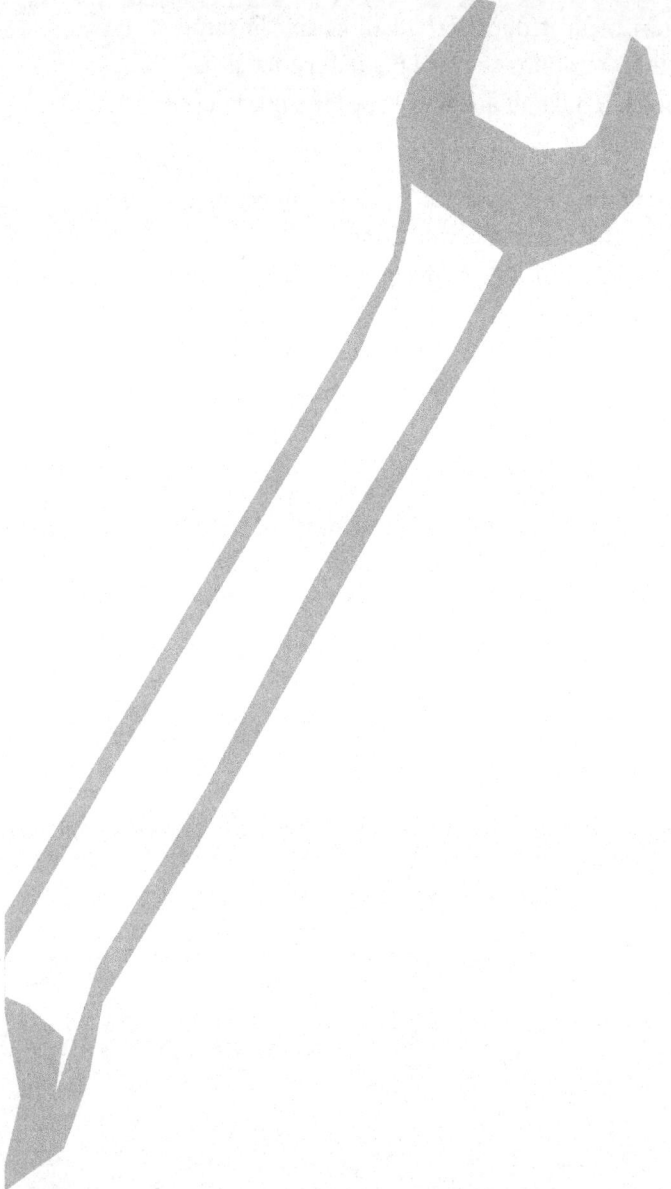

22
Turbosystems

Understanding turbocharging systems is the first step towards logical troubleshooting strategies.

The basic principles of turbocharging are quite simple. The turbocharger itself consists of exhaust-driven turbine wheel mounted on one end of a shaft, and a centrifugal compressor impeller mounted on the other end. Engine exhaust gases, which would otherwise simply be wasted energy, are used to spin the turbine at very high speed (typically 50,000 to 100,000 RPM). This drives the compressor, which is used to boost the pressure of the engine's induction air and therefore increase the engine's power output.

Turbocharging can be employed in two ways. One, known as *turbonormalizing*, is used to maintain sea level manifold pressure (roughly 30 inches of hg) at altitude, thereby eliminating the progressive horsepower reduction that occurs with normally-aspirated engines as the aircraft climbs. The other, known as *turboboosting*, boosts manifold pressure to a value significantly higher than sea level ambient (usually 35 to 45 inches of hg) to provide increased sea level horsepower. Boosted engines normally employ some means to provide adequate detonation margins, such as reduced compression ratio and intercooling.

In either case, the turbocharging system needs to include a means of controlling the turbocharger's compressor output pressure. Without such a control system, a turbocharged engine would be fundamentally unstable.

For example, a small increase in engine power would result in a small increase in exhaust volume. This would cause the turbocharger to spin faster, which would increase the compressor speed and therefore the manifold pressure. This would result in an additional increase in engine power, producing more exhaust volume, faster turbocharger speed, higher manifold pressure, etc. In other words, the system would "run away" and very possibly exceed maximum engine operating limits.

Likewise, a small decrease in engine power would cause a reduction in exhaust volume, a decrease in turbocharger speed, a reduction in manifold pressure, a further decrease in engine power, and so forth. In short, the engine would be nearly impossible to control.

The way turbocharger output is regulated is by means of a butterfly valve called a "wastegate" which allows a certain amount of exhaust gas to be vented overboard without going through the turbocharger. If the wastegate is fully open, almost all of the exhaust bypasses the turbocharger; if it is fully closed, virtually all of the exhaust must go through the turbocharger.

Automatic control

Most turbocharged aircraft employ an automatic wastegate control system to regulate the turbocharger. (The alternative is a manually-controlled wastegate, where the pilot controls the turbocharger from the cockpit, but this tends to cause high pilot workload and unstable engine power.) The automatic system employs a *hydraulic wastegate actuator* and a *pressure controller* to maintain turbocharger output at the desired pressure.

Turbocharger "hot section" components: turbine wheel (bottom) and housing (top).

The wastegate butterfly is normally held in the full-open position by a strong spring, allowing exhaust gas to bypass the turbocharger. Engine oil pressure applied to the wastegate actuator causes the wastegate butterfly to close, forcing exhaust gas to go through the turbocharger. The more oil pressure is applied to the wastegate actuator, the more the wastegate closes until, at about 50 PSI, the butterfly is fully closed.

The pressure controller monitors the output of the turbocharger's compressor (also known as "upper deck pressure" or UDP), and regulates the oil pressure to the wastegate actuator to hold the turbocharger output constant. The controller is a simple device that consists of an aneroid and a poppet valve. Here's how it works.

If the turbocharger output is less than the *set-point* of the controller, the aneroid expands and closes the poppet valve, increasing the oil pressure in the wastegate actuator and causing the wastegate to close. This causes more exhaust to pass through the turbocharger, spinning it faster, and increasing the compressor output.

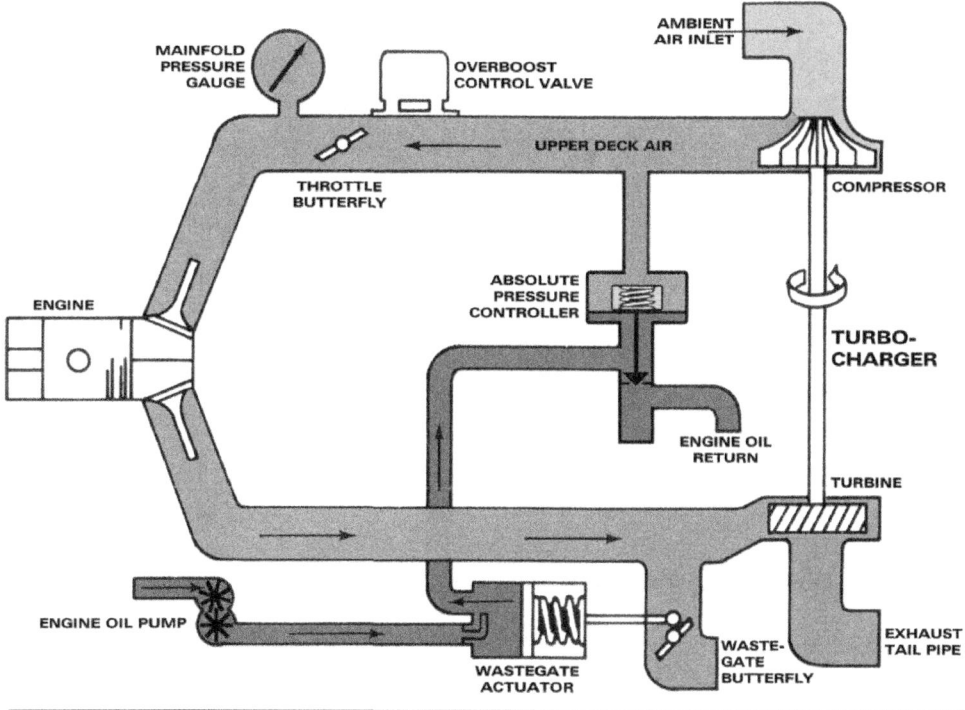

Flow diagram of automatically-controlled turbocharging system.

If, on the other hand, turbocharger output rises above the set-point of the controller, the aneroid contracts and opens the poppet valve, decreasing the oil pressure in the wastegate actuator and allowing the wastegate to open a bit. This lets more exhaust gas bypass the turbocharger, slowing it down and decreasing compressor output. Thus, equilibrium is quickly reached whereby the turbocharger output stays right at the set-point of the controller and the system remains stable.

Most unpressurized aircraft use an *absolute pressure controller* (APC) set to maintain turbocharger output at a few inches over engine red-line. The controller set-point is easily adjustable by screwing the controller's poppet valve seat in or out, and should be adjusted so that application of full throttle produces the proper red-line manifold pressure for takeoff.

> To understand how the system works, let's follow it through a flight and see what it does.

Most pressurized aircraft use a **variable** *absolute pressure controller* (VAPC). The difference between an APC and VAPC is that the VAPC's set-point is varied by means of a cam connected to the throttle control. At full-throttle, the VAPC works just like an APC (and is adjusted to produce proper full-throttle manifold pressure in the same fashion). But at partial throttle settings, the VAPC set-point is reduced so that the turbocharger doesn't have to "work so hard" when the pilot throttles back to reduced manifold pressure.

Turbocontrollers also come in a few other varieties. A "slope controller" varies its set-point pneumatically according to the pressure drop across the throttle body, eliminating the need for the VAPC's mechanical throttle linkage. A "pressure ratio controller" (PRC) also varies its set-point pneumatically, but based on the ratio between compressor input and output pressure. Sometimes these controller varieties are used in combination; e.g., an APC for low altitudes and a PRC for high altitudes.

What really happens

To gain a better understanding of how the system works, let's follow it through an actual flight and see what it does. To make things simple, let's suppose we're flying an unpressurized airplane like my Cessna Turbo 310 that uses a simple absolute pressure controller. (The differences when flying a pressurized airplane with a VAPC are minor and not really significant for purposes of this discussion.)

You've probably noticed that when flying a normally-aspirated airplane, full-throttle manifold pressure at takeoff never quite reaches sea level ambient (around 30 inches), but tops out at a few inches less than that due to unavoidable pressure losses in the induction system. Likewise, for a turbocharged airplane to achieve rated red-line manifold pressure on takeoff, the APC set-point must be a few inches higher to compensate for induction system losses. My Cessna Turbo 310's manifold pressure red-line is at 32 inches, and the APC set-point is adjusted to about 3 inches higher (about 35 inches) to produce red-line manifold pressure on takeoff.

Let's start the engines and taxi to the runup area. The APC sees that the turbocharger output is less than its set-point of 35 inches so it closes its poppet valve to call for wastegate to close. As soon as engine oil pressure comes up, the wastegate (which is spring-loaded to the full-open position) will close all the way. However, since the engine is at idle, there's not enough exhaust flow to spin up the turbocharger enough to produce 35 inches of UDP, so the wastegate remains fully closed throughout the taxi and probably even during the runup.

Now we taxi onto the runway and slowly apply full throttle for takeoff. As the engine develops more and more power, the exhaust flow increases dramatically and spins the turbocharger faster and faster, causing UDP to increase until it reaches the controller set-point of 35 inches. At that point, the controller opens its poppet valve to relieve the oil pressure to the wastegate actuator, allowing the wastegate to open as necessary to stop the turbocharger from spinning up any faster and thereby holding UDP right at 35 inches (and indicated manifold pressure right at the 32 inches red-line).

Climbing out of 1000 feet AGL, we reduce to cruise-climb power; I usually use full-throttle (32 inches of manifold pressure) and pull the prop control back from 2700 to 2450 RPM. This reduces exhaust flow and causes the turbocharger to start slowing down, but the controller immediately notices the resulting decay of UDP and closes its poppet valve to command the wastegate to close and force more exhaust through the turbocharger, causing the turbo to spin back up to the point where UDP is steady at 35 inches. This all happens so quickly that we're never aware that it's going on.

Climb

As we climb on up to the Flight Levels, outside ambient pressure decreases by about 1 inch per 1,000 feet of climb. This decreased pressure would normally cause a corresponding decrease in UDP (and therefore manifold pressure), but once again the controller compensates for this decay by gradually closing the wastegate more and more as we continue to climb, forcing more and more exhaust through the turbocharger and spinning it up faster and faster as required to maintain UDP at a constant 35 inches. In the cockpit, we notice that manifold pressure is staying more-or-less right where we set it (at 32 inches), without the inch-per-thousand-feet drop-off that we'd expect in a normally-aspirated airplane.

Of course, this can't go on forever. If we were to keep climbing higher and higher, and the controller were to keep closing the wastegate more and more to compensate for the decreased ambient air pressure, eventually we'd reach a point where the wastegate was fully closed and the controller was no longer able to maintain 35 inches of UDP. In my T310R at 75% cruise-climb power, this occurs at around FL220, while in most other turbocharged Cessnas, it occurs somewhat higher (FL250 or more). At this point, when the wastegate is fully closed and the automatic control system is no longer able to maintain constant UDP, the engine is said to be "bootstrapping" because the system is unregulated (and therefore unstable) and large manifold pressure variations may be observed.

Cruise

But we don't want to go that high today. Let's suppose we level off at FL180 and let the airplane accelerate to cruise speed. As the airspeed increases, the ram air effect causes a small increase in induction air pressure and a corresponding (somewhat larger) increase in UDP. Again, the controller notices this happening, and commands the wastegate to open a bit in order to slow down the turbocharger and hold UDP right at 35 inches.

Now that we're trimmed for level cruise at FL180, we slowly pull back on the prop controls to reduce RPM from 2450 to 2300 RPM. As we reduce engine RPM, exhaust volume is also reduced, causing the turbocharger to spin slower and reducing UDP. The controller reacts by commanding the wastegate to close in order to spin the turbo back up and restore 35 inches UDP.

Suppose we continue to reduce RPM gradually from 2300 to 2100 RPM, which is the bottom of the green arc on my T310R. As we do this, the controller closes the wastegate further and further in order to compensate for the reduced exhaust flow and maintain 35 inches UDP. But at some point around 2150 RPM, the wastegate will reach the fully closed position and any further reduction in RPM will cause the engine to bootstrap (indicated both by loss of MAP and instability of MAP readings). Upon observing the onset of bootstrapping, we increase RPM by 50 or so and see that the bootstrapping stops.

Descent

Okay, we've had enough fun, and it's time to head back to the barn. We switch off altitude hold on the autopilot, and roll in enough nose-down pitch trim to start a 1,000 FPM descent out of FL180.

As our indicated airspeed rises from its cruise value of 160 KIAS to around 200 KIAS, increased ram air tries to increase UDP above 35 inches, but the controller sees this and opens the wastegate enough to hold UDP steady. As we descend, outside ambient increases by about 1 inch per 1,000 feet, so the controller must continually open the wastegate more and more to prevent UDP from rising. In the cockpit, all we see is that MAP remains rock steady at 32 inches, right where we set it.

By the time we get down to pattern altitude, the wastegate is most of the way open. It stays there until we throttle back for our final descent and landing. When we do that, the reduction in engine power causes exhaust volume to fall, and the controller has to close the wastegate to make up for it and maintain 35 inches UDP. Eventually, as we close the throttle all the way prior to touchdown, even full-closed wastegate is not enough to maintain 35 inches UDP because the idling engine is hardly putting out any exhaust volume at all. The wastegate remains fully closed as we turn off the runway and taxi in to the ramp. It remains fully closed until we pull the mixtures to idle cutoff, at which point engine oil pressure goes away and the wastegate returns to its spring-loaded full-open position.

> When the turbosystem can't maintain constant pressure, large manifold pressure variations may occur and the engine is said to be "bootstrapping."

Fuel injection for turbos

Except for a few details, the fuel injection system installed on turbocharged engines is identical to the naturally aspirated system. Perhaps the most obvious difference between the turbo and naturally aspirated systems is the use of different nozzles.

As we saw in Chapter 20, naturally aspirated nozzles use ambient nacelle air for the tiny air bleeds that help atomize the fuel at reduced throttle settings. This works for naturally aspirated engines because induction manifold pressure is always less than ambient nacelle pressure, so the airflow through the bleeds is always from outside to inside.

On turbocharged engines, however, manifold pressure is often greater than ambient, so pressurized upper-deck air must be furnished to the nozzle air bleeds. This is accomplished with a sleeve that slides over the top of each nozzle and is plumbed to upper-deck air. The turbo-style nozzles are fitted with O-ring grooves to provide an airtight seal between the nozzle bodies and sleeves.

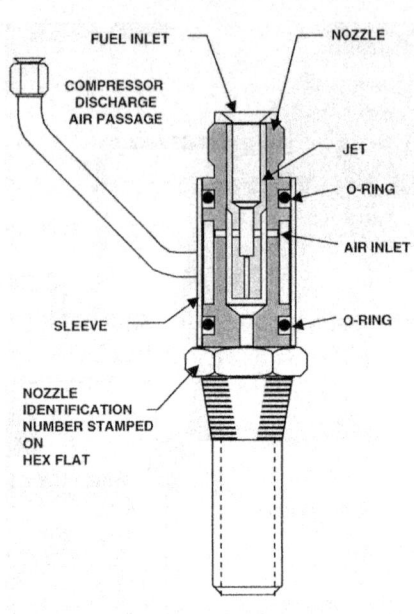

On turbocharged engines, fuel injection nozzles are referenced to upper-deck air pressure.

Another important difference on Continental engines is found in the fuel pump. The fuel pump used on naturally aspirated Continentals uses a simple screwdriver-adjustable fixed orifice to set the fuel pressure output of the pump at high engine RPM. On turbocharged engines, however, this orifice is regulated by an altitude-compensating aneroid that is referenced to upper-deck pressure.

The reason this aneroid is required is that upper-deck pressure in a turbocharged engine can change quite rapidly, especially at high altitudes when the system is bootstrapping. It would be quite difficult for a pilot to manually maintain constant mixture in the face of rapidly changing upper-deck pressure. In installations that utilize a VAPC, upper-deck pressure also varies with throttle position. The fuel pump aneroid maintains constant mixture automatically in the face of varying upper-deck pressure, minimizing pilot workload. For the most part, mixture control is a "set it and forget it" affair with this system.

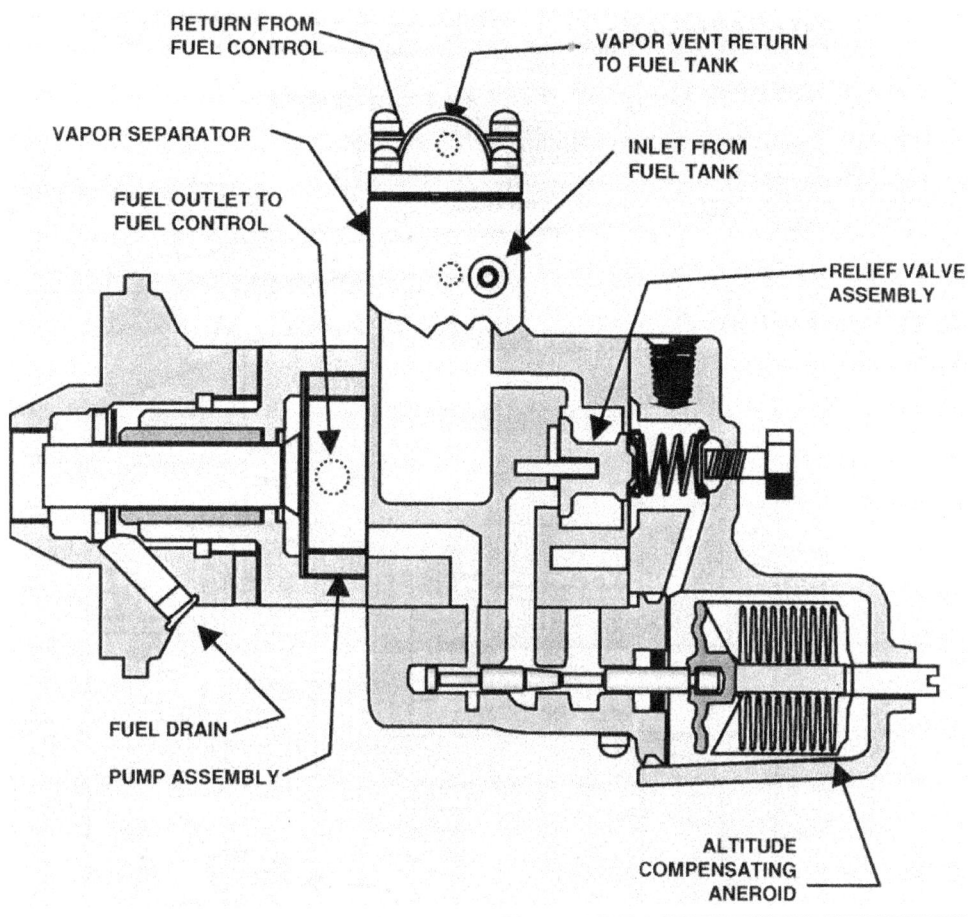

On turbocharged Continental engines, the fuel pump's adjustable orifice is controlled by an altitude-compensating aneroid referenced to upper-deck air pressure. This automatically maintains constant mixture in the face of changing upper-deck pressure (as occurs during bootstrapping).

A final wrinkle that appears in some but not all turbocharged Continental engines (mainly highly-boosted ones) is the addition of a fuel pressure regulator. The purpose of this regulator is to provide additional enrichment at high power settings for extra detonation margin. The idea is that the fuel pump is adjusted to produce 5% more fuel pressure than needed at full takeoff power, and then the regulator is adjusted to limit the fuel pressure to the desired value. This ensures that full rated fuel flow is available even at slightly less than red-line engine RPM and high OAT.

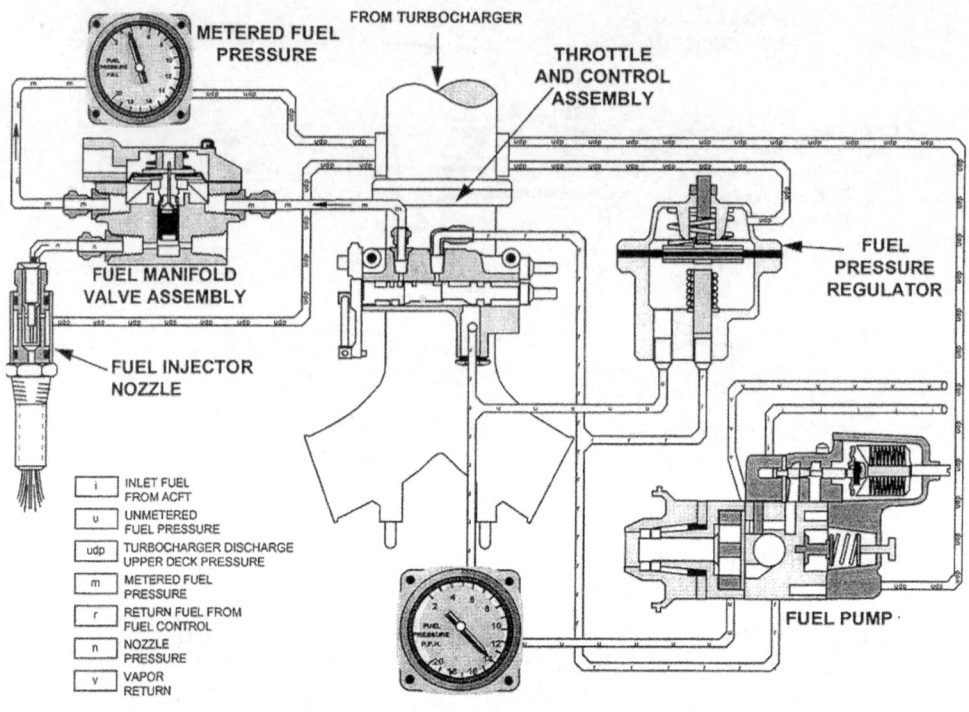

Typical schematic of the fuel injection system used on turbocharged Continental engines. Notice that the fuel pump aneroid and fuel nozzle air bleeds are referenced to upper-deck pressure. Note also the fuel pressure regulator, which is used on some but not all turbocharged engines.

A turbosystem does add some cost and complexity to an aircraft installation, compared to a normally aspirated engine. In my experience, however, if the turbo components are inspected regularly and given a little TLC, a turbocharged engine can operate trouble-free and deliver performance that allows the aircraft to cruise higher and faster, often taking advantage of favorable winds/weather, and benefiting from the improved range and fuel efficiency that high-altitude operation offers.

PART V
Powerplant Management

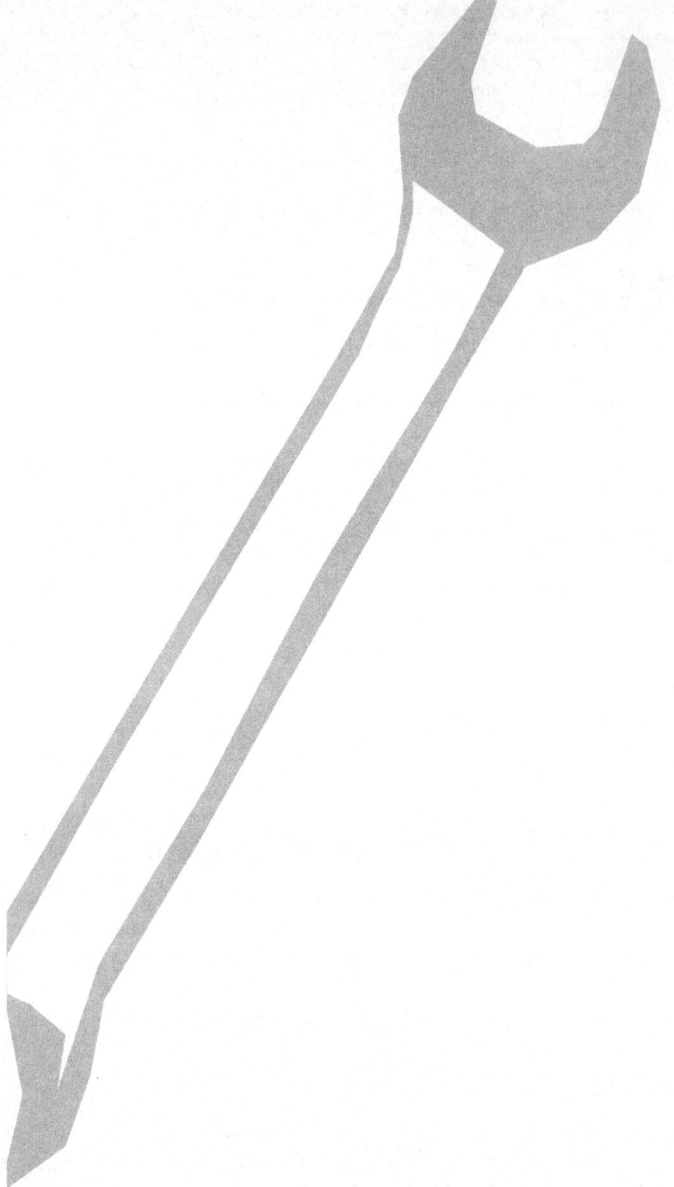

23
Leaning Basics

Why does leaning seem to be such a confusing subject for pilots?

Of the many tasks that we must perform as pilots, leaning the engine is one of the simplest. Leaning is vastly easier than shooting a circling approach in low IMC, or picking the smoothest route through a cold front, or deciding when to overhaul the engine. Yet no subject I know seems to trigger more discussion and debate among pilots, nor to provide more misinformation and bad advice.

Although I usually focus on maintenance-related topics, aircraft owners seem to ask me more questions about leaning procedures than just about any other subject. It's obvious to me that despite the simplicity of this subject, it remains poorly understood by a lot of aviators.

The best source I know for in-depth information about optimal powerplant management is the Advanced Pilot Seminars course developed by my good friends George Braly, Walter Atkinson and John Deakin. This outstanding seminar is offered both as a "live" course in Ada, Oklahoma and occasionally elsewhere, and is also available in a home-study on-line version. I've taken both the live and on-line versions, and both are excellent.

There's no need to be afraid of the red knob. Learning is easy if you learn to do it the right way.

The objective of the APS course is to offer pilots an in-depth understanding of powerplant management, both theory and practice. It offers a huge amount of information on the subject, and most APS graduates liken the experience to drinking from a firehose. But many pilots are reluctant to invest the time, money and neurons into gaining that level of understanding of powerplant management. Many are just looking for a simple, cookbook-like approach to leaning that doesn't require rocket science to master.

Forget the POH!

Most Pilot Operating Handbooks (POHs) provide precisely such simple, cookbook-style guidance. Here's an example, quoted verbatim from a Beech POH:

> For level flight at 75% power or less, the EGT unit should be used in the following manner:
>
> 1. Lean the mixture and note the point on the indicator at which the temperature peaks and starts to fall.
>
>   a) **Cruise (lean) mixture:** Enrich mixture (push mixture control forward) until EGT indicator shows a drop of 25°F to 50°F on the rich side of peak.
>
>   b) **Best power mixture:** Enrich mixture (push mixture control forward) until EGT indicator shows a drop of 75°F to 100°F on rich side of peak.
>
>   CAUTION: Do not continue to lean mixture beyond the point necessary to establish peak temperature. Continuous operation is recommended at 25°F or below peak EGT only on rich side of peak.
>
> 2. Changes in altitude and power setting require EGT to be rechecked and mixture reset.
>
> 3. A mixture resulting in an EGT 25°F on the rich side of peak should also result in fuel flow and TAS values approximately equal to those presented in the Cruise Power Settings tables in the PERFORMANCE Section. If not, the values derived from the Range, Endurance, and Cruise Speeds charts

must be revised accordingly. In very cold weather, EGTs 25°F rich of peak may not be obtainable.

This Beech POH calls for operating at "Cruise (Lean) Mixture" defined as leaning to peak EGT and then richening until EGT drops by 25° to 50° Fahrenheit. (Or in shorthand, "25°F to 50°F ROP.") Some POHs also authorize operating at "best economy mixture" (defined as peak EGT) for power settings less than 55% to 65% power.

Unfortunately, this POH guidance leaves a lot to be desired. **50°F ROP is almost precisely the WORST possible mixture setting from the standpoint of engine longevity.** The maximum cylinder head temperature (CHT) and peak internal cylinder pressure (ICP) occurs almost precisely at 50°F ROP. So, using Beech's "Cruise (Lean) Mixture" assures that your engine operates at the hottest, most stressful corner of its operating envelope.

"Best economy mixture" (peak EGT) is only slightly better—providing a bit cooler CHTs and a bit less internal stress on the engine, but not by much. Furthermore, peak EGT is certainly NOT the best economy mixture—minimum brake specific fuel consumption (BSFC) occurs at a substantially leaner mixture than that, well lean of peak EGT (LOP).

Why would so many aircraft manufacturers publish such bad advice in their POHs? Well for one thing, back in the 1960s and 1970s when many of the POHs were written, the relationships between EGT, CHT and ICP were not as well understood as they are today. The conventional wisdom at that time was that richer mixtures were better for the engine, and leaner mixtures were worse. A culture of fear evolved, promulgated by the flight instructors of the day: If you lean too aggressively, you'd blow up your engine.

With today's sophisticated instrumentation, we now know that this isn't true. The hottest, most stressful mixture is about 50°F ROP, and mixtures that are richer OR leaner are better for the engine. At 75% cruise power, you want to stay well away from that worst-case mixture setting, either by operating *at least* 100°F ROP (preferably richer) or *at least* 20°F LOP (preferably leaner), take your pick.

Given the choice between operating ROP or LOP, LOP operation has some compelling advantages. It's cleaner, cooler, less stressful on the engine, and uses a lot less fuel. Or as one APS mantra goes: "Leaner is greener."™

Also, many aircraft engines in the 1960s and 1970s typically would run unacceptably rough if you tried to lean them beyond peak EGT. Today, with tuned fuel nozzles and digital engine monitors, we can operate these engines deep in the LOP regime without roughness. Even most carbureted engines can be operated at least somewhat LOP if the pilot knows what he's doing.

The Beech POH's "Cruise (Lean) Mixture" (25°F to 50°F ROP) does offer a reasonable compromise between best power (100+°F ROP) and best economy (well LOP). What "Cruise (Lean) Mixture" does NOT provide is good engine longevity, which is something that the manufacturers might not care much about, but owners certainly do. (Premature cylinder replacement is a major expense item for an aircraft owner, but a revenue item for the manufacturer.)

CHT is the best proxy we have in the cockpit for internal cylinder pressure (ICP). Peak ICP and peak CHT occur at the same mixture setting. This is the mixture that's hardest on the engine because it creates the greatest stresses. Except at low power settings—say 60% power or less—it's a good place to avoid if you care about engine longevity.

So, while many pilots still follow the antediluvian POH guidance, we can do a lot better. Note that the leaning recommendations in the POH are NOT limitations; they are mere suggestions (and often not very good ones). A pilot is under no regulatory obligation to follow them, and that's a good thing.

How I lean

Over the past decade, I've evolved a dead-simple approach to leaning that has worked very well for me in my twin. My Continental TSIO-520 engines obviously love it—they both made it well past 200% of TBO with only minimal maintenance along the way. With minor variations, my approach should work for just about any piston-powered airplane.

Perhaps the most controversial aspect of my technique is that I don't use EGT as a leaning reference for cruise flight. EGT is extremely useful for troubleshooting engine problems, but as a leaning reference it leaves quite a bit to be desired. Absolute EGT values are meaningless; what matters is the relative value of EGT with respect to peak (e.g., 100°F ROP, 50°F ROP). So to use EGT as a leaning reference, you first have to determine what peak EGT is by leaning very slowly until you find the point where indicated EGT stops rising and starts falling. The process of doing this puts the engine in the area of maximum stress and ICP. It's really not a very kind thing to put your engine through every time you fly.

I find it a lot easier to lean in cruise by reference to CHT and fuel flow, an approach that's dead simple yet still keeps my mixture settings right in the ballpark and obviously has made my engines live long and prosper.

Here's how I do it. First, I decide upon my objective: Do I want to go fast (i.e., achieve maximum airspeed) or do I want to go far (i.e., achieve minimum fuel consumption)?

If my objective is to go fast, then I lean so that the CHT of my hottest-running cylinder does not exceed a pre-established target value. That target depends on the aircraft and to some extent the OAT, but for most legacy airplanes like my Cessna 310, a target of about 380°F for Continentals and 400°F for Lycomings works well. For modern airplanes like the Cessna TTx or Cirrus SR22 or Diamond DA40 with their superior engine cooling systems, those target values should be reduced by about 20°F. In cold OATs (below ISA), the CHT targets should also be lowered a bit.

For guidance in leaning, I recommend focusing on CHTs rather than on EGTs.

If the CHT of the hottest-running cylinder exceeds the target value, then I richen a bit more (if ROP) or lean a bit more (if LOP) to bring the CHT down to the target. Conversely, if the hottest CHT is lower than the target value, I can gain a bit more fuel economy by leaning a bit more (if ROP) or a bit more speed by richening a bit more (if LOP).

Personally, I always cruise LOP for all the reasons cited earlier (cooler, cleaner, cheaper, greener), but your mileage may vary.

If my objective is to go far, then I lean so that my GPS-coupled fuel totalizer system shows forecast fuel remaining at my destination to be not less than my target minimum fuel reserve (which for me is one hour of fuel at cruise fuel flow). If the totalizer forecasts that I will arrive at my destination with less fuel than this, then I lean further until the totalizer does show enough reserve fuel. If I find that I cannot lean enough to achieve the necessary fuel reserve figure without experiencing engine roughness, then I know I'll need to make a fuel stop.

If you choose to cruise ROP, then you also must make sure that you don't lean so far as to exceed your target CHT. If you can't find a mixture that simultaneously yields the required fuel reserve and doesn't exceed the target EGT, then you'll either have to reduce power, switch to LOP operation, or make a fuel stop.

If you don't have a GPS-coupled fuel totalizer, then you can calculate your reserves manually from fuel quantity, fuel flow and GPS-derived time-to-destination, but that's a lot more work. For anyone who flies a lot of long-distance fuel-critical missions (like I do), a GPS-coupled fuel totalizer is probably #3 on the "things you just gotta have" list, right behind a digital engine monitor and real-time satellite weather-in-the-cockpit.

Frequently-asked questions

Q: When I operate LOP, my EGTs are noticeably higher than when I operate ROP. Won't those higher EGTs harm my engine?

A: Indeed, if you run 20°LOP instead of 100°ROP, your EGTs will be higher—80°F higher to be exact. This is nothing to worry about. At cruise power, your engine is not capable of producing EGTs high enough to harm anything. High EGTs are not damaging to your engine. It's high CHTs that are damaging.

Q: If I operate at peak EGT or LOP, don't I risk burning my exhaust valves?

A: This question belies a common misconception that burned exhaust valves are caused by high EGTs. They aren't. Most burned exhaust valves are caused by poor sealing between the exhaust valve and seat, most often due to excessive valve guide wear. The best way to avoid burned valves is (1) to keep CHTs down, and (2) to run a lean mixture to minimize build-up of combustion byproducts on the valve stem. The leaner you operate (while keeping CHTs at prudent levels), the happier your exhaust valves will be.

Q: My engine monitor uses a spark plug gasket probe on cylinder #2 because the threaded boss on that cylinder is already occupied by the factory CHT probe. Is that why my #2 CHT always seems to run hot?

A: Yes, it is. A spark plug gasket probe often results in a CHT reading that's about 40°F hotter than a normal threaded probe on the same cylinder. To avoid this problem, you can purchase a "piggyback" probe for your engine monitor that will screw into the threaded boss on the cylinder, and that will allow the factory probe to be piggybacked on top of it. The piggyback probe sometimes reads slightly lower than the regular probe, but it's a whole lot closer than the spark plug gasket probe.

Q: Why do you recommend keeping CHTs at or below 380°F or 400°F, while Continental sets its CHT red line at 460°F and Lycoming sets it at 500°F? Aren't you being excessively conservative?

A: Both Continental and Lycoming specify CHT limits (460°F and 500°F, respectively) that should be considered emergency limits, not operational limits.

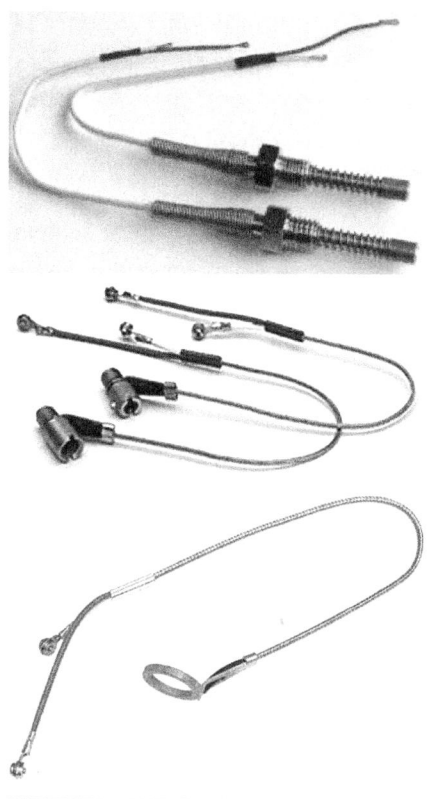

CHT probes come in three types: threaded (standard), piggyback, and spark plug gasket. A spark plug gasket probe generally results in a higher CHT reading than a normal threaded probe.

Allowing your CHT to get anywhere close to those red-line values for significant periods of time will most likely result in cylinder longevity issues such as burned exhaust valves and cracked heads. I get uncomfortable when I see CHT much above 400°F—which is the temperature at which aluminum cylinder heads lose one-half their tensile strength. I recommend a maximum target CHT of about 380°F for Continentals and 400°F for Lycomings. For modern designs like the Cessna TTx and Cirrus SR22 and Diamond DA40, reduce those CHTs by 20°F or so.

Maximum CHT (and maximum abuse) occurs at about 50°F ROP. The higher the power setting, the further away from that abusive mixture setting you need to stay to keep CHT at or below the target. As power decreases, this "zone to avoid" around 50°F ROP becomes narrower and narrower. When power gets down to about 60%, the avoidance zone disappears, and you can run the mixture pretty much anywhere you please without overtemping or overstressing anything. (The APS folks refer to this zone-to-avoid as "the Red Box", and I'll talk more about it in the next Chapter.)

In my view, the best way to manage our piston engines is the same way we manage turbine engines: by limiting temperature, specifically CHT (which is the best proxy we have for ICP). If the CHT is too high, you can reduce it by richening a lot (ROP) or leaning a little (LOP). Which method you choose is less important from the standpoint of longevity.

Q: All this LOP stuff may be fine for you fuel-injected guys, but I fly a C-35 Bonanza with a carbureted E-225 engine. I've been told that LOP operation is a bad idea for carbureted engines. Do you agree?

A: LOP operation is fine for any engine that can run smoothly in that configuration. However, LOP operation requires even mixture distribution among the cylinders. That's sometimes difficult to achieve in a carbureted engine, particularly carbureted Continentals which are famous for poor mixture distribution. Most carbureted Lycomings have pretty good mixture distribution and can operate at least mildly LOP without much effort.

There are a couple of techniques you can use to improve the mixture distribution of your carbureted engine and thereby enable the engine to be leaned more aggressively before it starts to run rough. One is to use a touch of carb heat during cruise (particularly in low OATs). The other is to avoid full-throttle operation, backing off the throttle until you can just see the slightest drop in MAP. The warm induction air and the slightly cocked throttle plate both improve fuel atomization and mixture distribution in your engine, and will enable you to lean more aggressively before the engine starts running rough.

You should feel quite comfortable experimenting with these techniques to see if you are able to operate LOP without creating uncomfortable engine roughness. Contrary to popular belief, you can't hurt anything by operating LOP. If you get your engine to run smoothly LOP, I suggest you try it (and you'll probably like it). If you can't, then you'll have to be content with ROP operation.

Q: You caution against excessive CHTs, but it is possible for CHTs to be too cold?

A: Yes, it's possible to run CHTs so cold that the tetraethyl lead (TEL) in the 100LL is not properly scavenged and starts creating metallic lead deposits in the combustion chamber and lead-fouling the spark plugs. However, in most engines, it takes very cool CHTs (down in the mid-200s °F or lower) for an extended period of time (hours) for this to cause a problem. We usually see this problem in airplanes used for fish spotting, pipeline patrol, search and rescue, and other "loiter mode" operations. Unless you fly at very low power settings (e.g., 50%) and/or at very high altitudes and very cold OATs (e.g., FL240 and -30°C), it's not usually a problem.

Q: I fly a Beech Musketeer with a Lycoming O-360-A4J and no CHT or EGT or fuel flow instrumentation. How should I lean my engine?

A: After stabilizing in cruise and reducing power to the desired cruise RPM, slowly lean the mixture until you feel the onset of perceptible engine roughness. Then slowly richen just to the point that the roughness goes away. With your limited instrumentation, that's the best you can do—and it's not a bad technique.

Having said that, I would strongly recommend that you consider installing a digital engine monitor in your airplane. To my way of thinking, having an engine monitor is even more important in a four-cylinder single-engine airplane than it is in six-cylinder single or a twin. If you fly a four-cylinder single and you lose a cylinder in flight, you don't have much left.

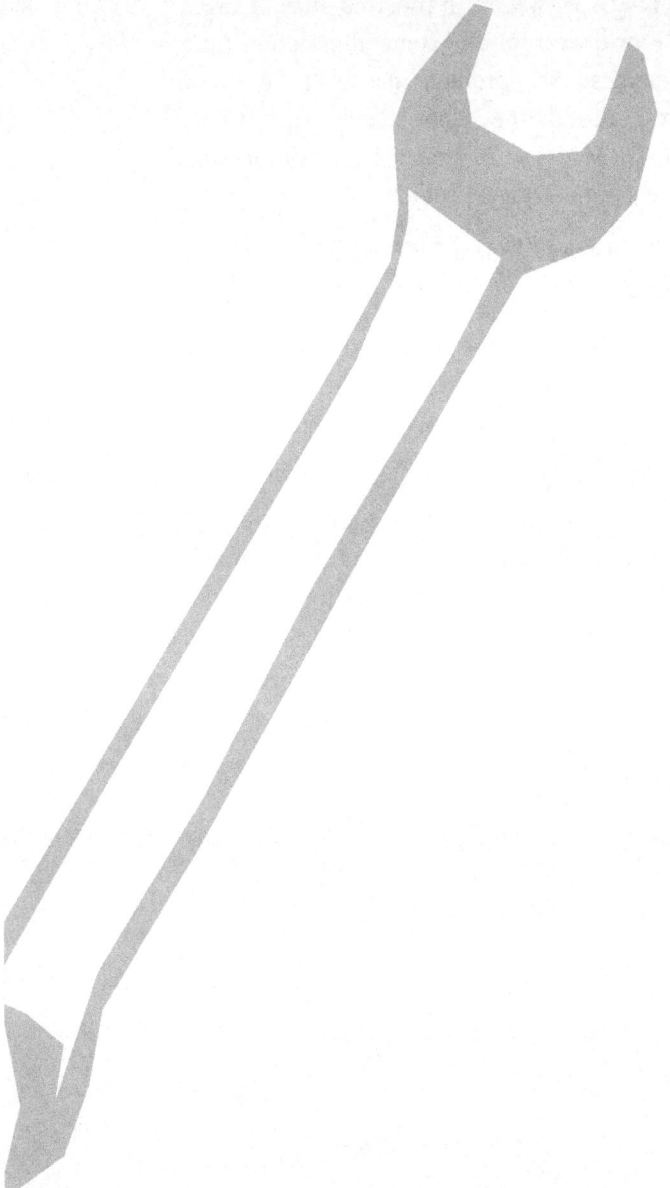

24
Advanced Leaning

Using the red box and red fin to better understand how NOT to lean your engine.

At my many July pilgrimages to EAA AirVenture in Oshkosh, I have had the opportunity to speak to thousands of pilots and aircraft owners on a wide variety of subjects ranging from reliability-centered maintenance to TBO busting to corrosion, and to conduct a half-dozen informal hour-long Q&A sessions addressing whatever maintenance-related issues were on their minds. In those sessions, I received more questions about one topic than all others combined: leaning.

Some of the questions focused on Old Wives' Tales about lean-of-peak (LOP) operation:

Q: Won't operating LOP hurt my engine, burn my exhaust valves, etc.?

A: It's a lot easier to damage your engine ROP, much less likely LOP.

Q: Can my carbureted engine be operated LOP?

A: Most can. Using carb heat helps. Only way to know is to try it. You can't hurt anything by experimenting with LOP operation.

Q: Can my injected engine be operated LOP without GAMIjectors?

A: Some can, some can't. Only way to know is to try it. You can't hurt anything by experimenting with LOP operation.

Q: Can my engine be operated LOP without an engine monitor?

A: Sure. I operated LOP for a decade before I installed my engine monitor. Now, I think it's really important to install an engine monitor, but that's true regardless of whether you run ROP or LOP.

Q: I've experimented with LOP, but I find that my EGTs are much higher when I run LOP than when I run ROP.

A: That's true. Why does that concern you? High EGTs are not damaging to your engine. It's high CHTs that are damaging to your engine. And LOP operation almost always results in lower CHTs.

Other questions focused on "the right way to lean" and sought cookbook answers:

Q: How many degrees LOP should I operate my engine?

A: That depends on many variables: power setting, altitude, temperature, etc. The answer might be anywhere from 0°F LOP to 100°F LOP.

Q: How many degrees LOP do you operate your own airplane?

A: I don't have a clue. I never use EGT as a leaning reference, so I don't know how many degrees LOP I operate. All I know is that it varies all over the place depending on various conditions, and it's not a particularly interesting number so I don't worry about it.

The problem with questions like these is that they are based on the misconception that there's "a right way" to lean an engine. In fact, there are lots of different "right ways" to lean an engine, and I employ them all from time to time.

In my turbocharged Cessna 310 I mostly climb very ROP, but occasionally I climb LOP when that's appropriate. I mostly cruise LOP, but it varies from slightly LOP to profoundly LOP depending on cruise altitude, OAT, and whether my objective is speed or fuel economy. I have thousands of hours flying Cessna 182s, and most of that time was spent neither ROP or LOP but rather right at peak EGT (and at appropriately reduced power). When I fly a Super Cub, I lean to the onset of engine roughness and I haven't a clue whether I'm ROP or LOP. All these ways of leaning are "right ways."

The key to leaning is not doing it "the right way" because there are so many different right ways to lean. Rather, the important thing is to avoid doing it "the wrong way" by avoiding situations that are potentially damaging or abusive to the engine.

The red box

My friends George Braly, John Deakin and Walter Atkinson of *Advanced Pilot Seminars* fame developed an important conceptual tool for conveying this idea. They call it "the red box" because it's generally depicted as a red-tinted rectangle superimposed over a graph of various engine landmark parameters (EGT, CHT, ICP, HP, BSFC) plotted as mixture is varied from full-rich to extremely LOP. The red box depicts the range of mixture settings that result in excessive internal cylinder pressures and therefore should be avoided. Mixture settings outside of the red box—whether on the rich side or the lean side—are all fair game.

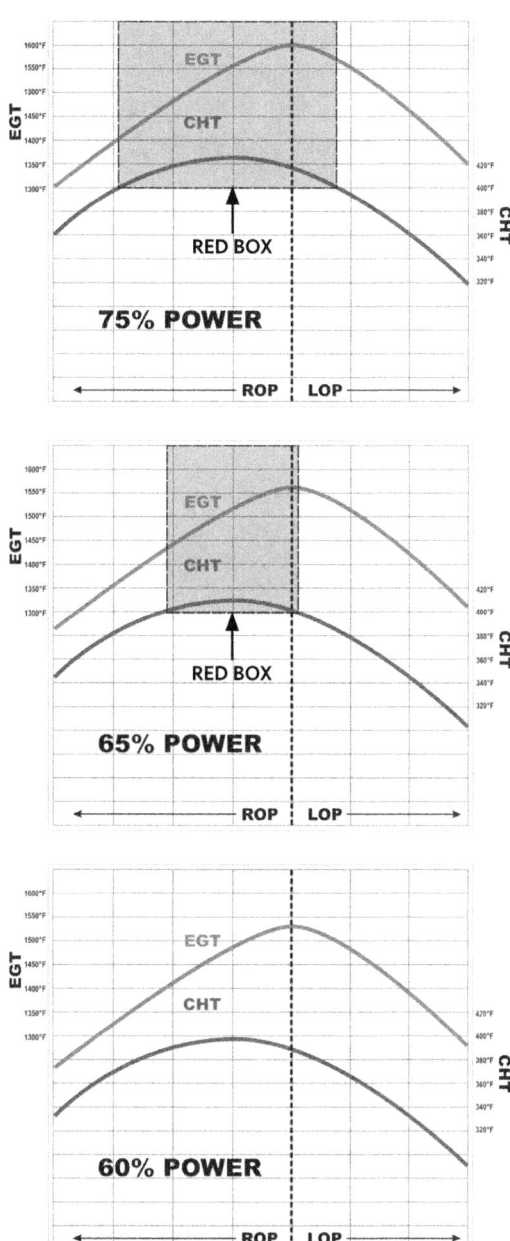

The lower the power, the narrower "the red box" becomes. Somewhere between 60% and 65% for most engines, it disappears completely.

The width of the red box varies with power. The lower the power, the narrower the red box becomes. At sufficiently low power (generally somewhere between 60% and 65% for most engines), the red box disappears completely, and you can run the engine at any mixture you like without abusing anything.

One practical problem with the red box concept is that it's based on limiting ICP; unfortunately we don't have an ICP gauge in our cockpits. It sure would be nice if we did, because it would make leaning pretty much of a no-brainer. In the GAMI test cell in Ada, Oklahoma, they instrument ICP by installing special "tricked out" spark plugs that contain pressure transducers capable of measuring instantaneous combustion chamber pressure. Sadly, we don't have these in our aircraft because the transducers are godawful expensive and the tricked-out spark plugs aren't certified.

In the absence of an ICP gauge, the best proxy for ICP we have in the cockpit is CHT. The good news is that the ICP and CHT curves have the same shape and peak at the same mixture. The bad news is that CHT is affected not only by ICP but also by several other factors that don't vary with mixture (notably OAT, IAS, density altitude and cooling system efficiency).

The figure on the previous page depicts the red box as encompassing all mixtures that result in CHTs above 400°F, and that's probably appropriate for most legacy aircraft when the OAT is at standard temperature (ISA) or greater. But if the OAT is colder than ISA or if the aircraft has a particularly efficient cooling system design (e.g., Cessna TTx, Cirrus SR22, Diamond DA40), the maximum acceptable CHT is lower.

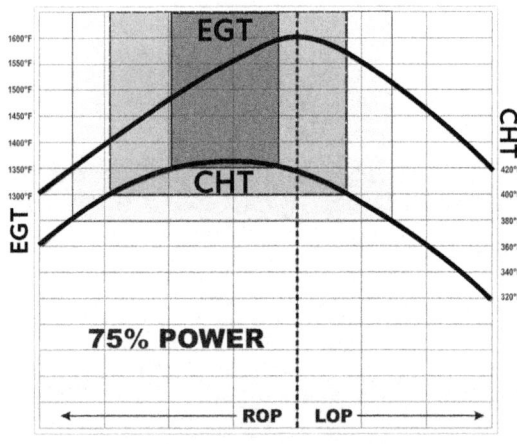

Modified "red box" chart, depicting a cautionary buffer zone in light grey and a highly abusive zone in dark grey.

Another problem with the red box concept is that it suggests that all mixture settings inside the red box are equally bad. That's obviously not true; the higher the ICP (and CHT), the more abusive the mixture. For this reason, I think it's useful to think of the red box has having a purple zone in the center depicting the mixtures that are ultra-abusive and to be avoided at all costs, and a yellow cautionary zone around the edges depicting a cautionary buffer zone to be avoided when possible for maximum engine TLC.

The red fin

Perhaps an even more useful variant of the red-box concept is one that has been popularized in the Cirrus community by my friend Gordon Feingold, but is relatively unknown in non-Cirrus circles. It is called "the red fin" and emphasizes that the width of the red box varies dramatically with power, and disappears altogether when power is reduced sufficiently.

Like the red box, the red fin depicts mixture settings that are abusive to the engine.

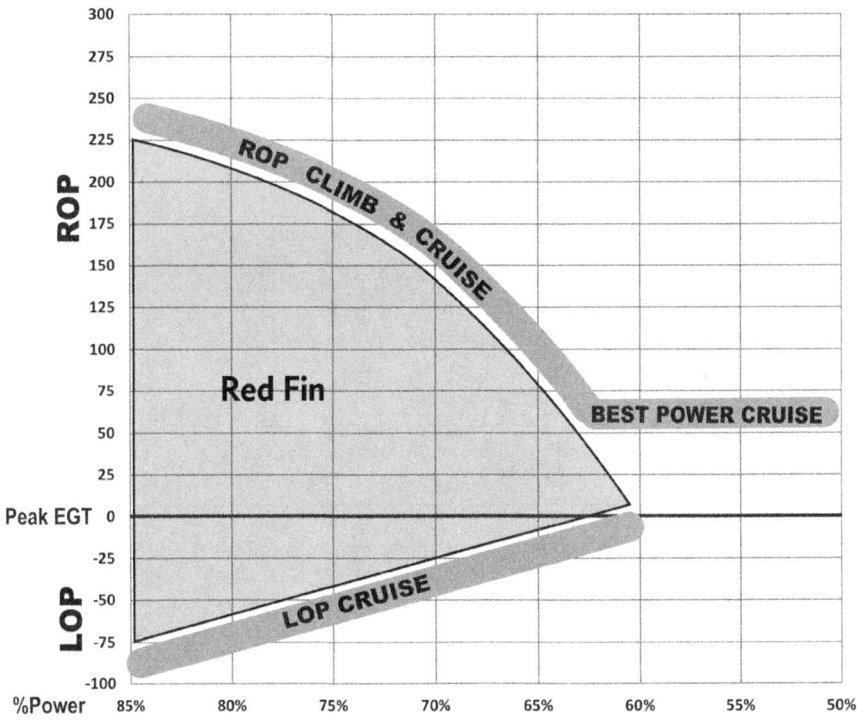

The "red fin" (shown as the large grey triangular shape) is an alternative depiction of the red box concept, and emphasizes that the width of the red box varies dramatically with power. The three most useful outside-the-red-fin zones for climb and cruise are shown on the outside of the red fin zone.

Settings outside the red fin—whether ROP or LOP—are fair game. The figure on the previous page depicts the three most useful outside-the-red-fin zones for climb and cruise. ROP mixtures are above the fin, and LOP mixtures are below it. At low power settings where the fin disappears, best-power mixture occurs at roughly 75°F ROP.

As with the red box, the red fin suggests that all mixtures inside the fin are equally abusive, but that's obviously not true.

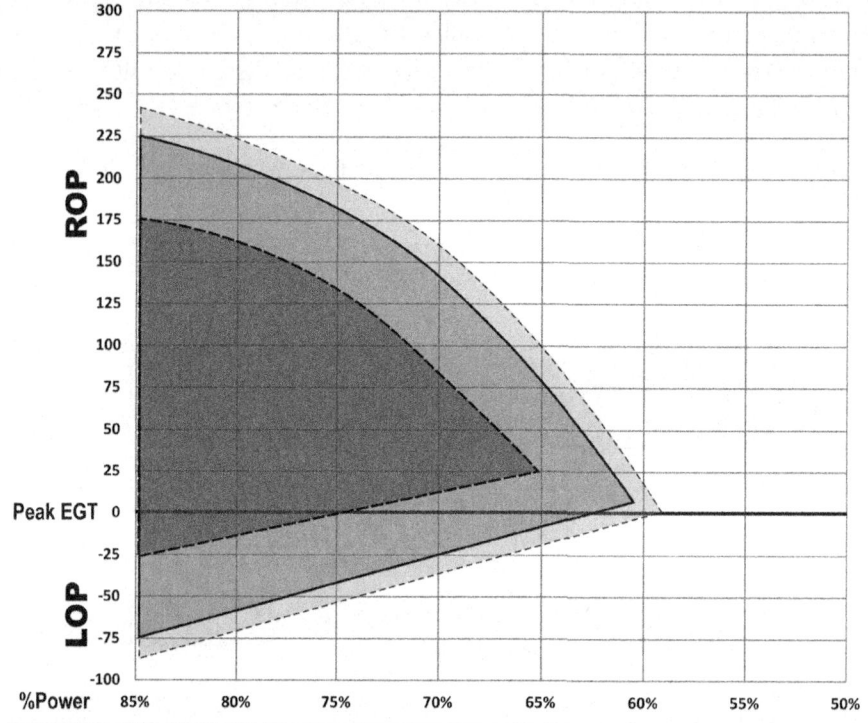

Modified "red fin" chart with the lightest gray depicting a cautionary buffer zone and the darkest gray indicating the highly abusive zone.

Flying the fin

Let's look at how we can use the red fin concept as a guide to mixture management throughout all phases of flight. The accompanying graphic depicts one method of managing the mixture, but certainly not the only method. (Remember, any mixture that lies outside the red fin is fair game.) It also assumes a normally-aspirated engine

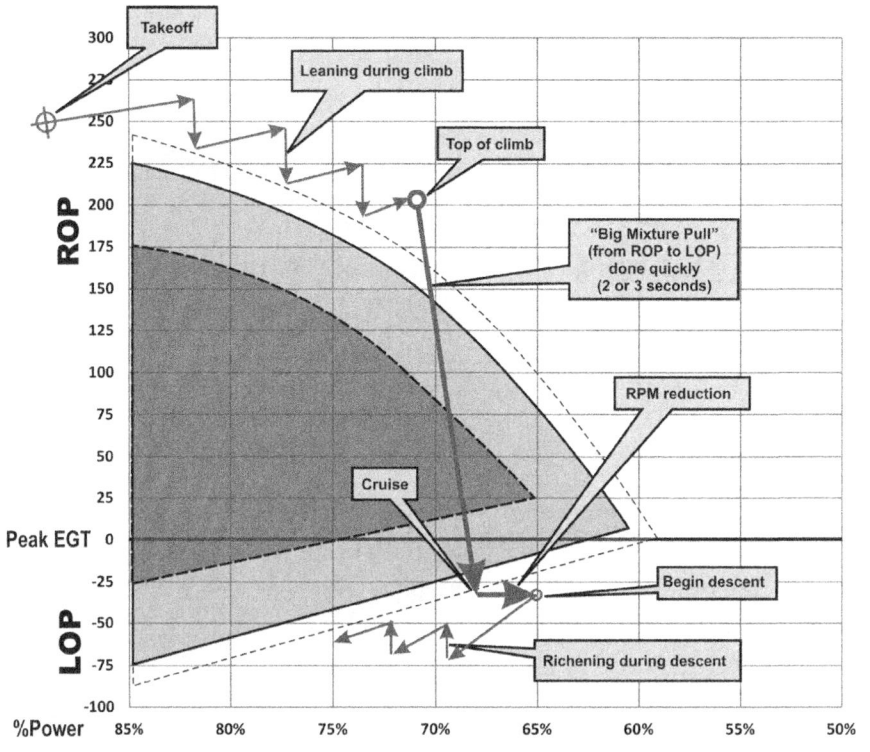

One way of managing the mixture during a flight with reference to the red fin. This assumes a normally aspirated engine with a conventional non-altitude-compensating fuel system.

with a conventional non-altitude-compensating fuel system. (Turbocharged engines and engines with an altitude-compensating system are a bit simpler to manage because you don't need to adjust the mixture during climbs and descents.)

The flight starts when takeoff power is applied at full-rich mixture (which is typically at least 250°F ROP for most properly-adjusted engines). We remain at wide-open throttle and let Mother Nature take care of reducing manifold pressure as we climb. With most engines, this results in a mixture that gets progressively richer with increasing altitude and decreasing manifold pressure, so from time to time we lean the mixture manually to keep it "in the zone" on the rich side of the red fin. (In my turbocharged airplane, I don't need to do this because manifold pressure doesn't decrease as I climb so there's no need to touch the mixture.)

When we reach top-of-climb, level off, and commence the cruise phase of the flight, we perform a "big mixture pull" (BMP) to transition from ROP to LOP. This should be done quickly to minimize the amount of time spent inside the red fin (and especially the ultra-abusive purple zone). About 2 or 3 seconds is about right for the BMP. Note that we lose a bit of power as we transition from ROP to LOP; that's normal and expected, and will be reflected by a small loss of airspeed.

I recommend NOT using the "lean-find mode" of your engine monitor when doing this, because this requires you to lean very slowly to locate peak EGT. That results in spending a considerable amount time inside the red fin (and the dreaded purple zone), which is exactly what you don't want to do. If you feel compelled to locate peak EGT, it's much better to perform a quick BMP to get into the LOP zone below the fin, and then slowly richen to locate peak EGT from the lean side.

Personally, I don't care about locating peak EGT, so I skip this step altogether. I just do a quick BMP to a known-safe LOP fuel flow—or until I hear and feel a small power loss that tells me I'm safely LOP below the fin—and then fine-tune the mixture using either CHT or my fuel totalizer as a primary reference.

As we begin the descent phase, we remain LOP below the fin. Because manifold pressure increases with decreasing altitude, the mixture becomes leaner and so from time to time we richen the mixture to prevent it from getting so lean that the engine starts running rough. If we forget to richen, no problem: The engine will remind us. (Once again, I can skip this step in my turbocharged airplane because manifold pressure remains constant during the descent.)

Because our airplanes aren't equipped with ICP gauges, the red box and red fin can provide only approximate guidance. Without ICP information, we can't know the box or fin boundaries precisely. But as conceptual guidelines, they're close enough. If we keep them in mind and make a conscious effort to stay out of the red zone (and especially out of the purple zone) for more than a few seconds at a time, we will be rewarded with maximum engine longevity and reliability, and minimum maintenance expense.

25
Making Engines Last

There's no big trick in getting your engine to reach published TBO or beyond ... just a bunch of little tricks.

I'm often asked about the best way to operate a piston aircraft engine to achieve maximum engine life, what power settings I recommend, what leaning technique I use, what engine oil is best, what I think about GAMIjectors and lean-of-peak operation, and so forth. Apparently, I'm supposed to be some sort of authority on engine longevity. I have no formal education on this subject—I was trained as a mathematician, not an aeronautical engineer—nor any recognized credentials. I guess my main "claim to fame" in this area is that in the 50-odd years that I've owned piston aircraft, I've never had a problem doing better than published TBO, and therefore (so the theory goes) I must be doing something right.

This gives me a good deal more credit than I deserve. For example, I purchased my current aircraft (a 1979 Cessna Turbo 310) in 1987. It had about 1,300 hours total time on the airframe and engines when I bought it. Published TBO for its TSIO-520-BB engines is 1,400 hours. So, by Continental's reckoning, those engines were pretty much "run out" when I acquired the airplane.

I flew those engines for another 600 hours (TBO+500) with no problems whatsoever, then pulled the engines for major overhaul. Silly me. The overhaul shop reported that all 12 cylinders were still within new limits, as was pretty much everything else. (In other words, the engines could have gone another 1,000 hours—to TBO+500—without breaking a sweat.) Admittedly those were excellent results, for which the previous owner of the airplane—who put the first 1,300 hours on those engines—should have gotten most of the credit. (However, I'll take credit for the decision not to euthanize those engines at 1,400 hours.)

I gave those engines a minimalist (i.e., el-cheapo) major overhaul in 1990, and I then flew behind them for the next 24 years and 3,200 hours before the next overhaul. I'll happily take credit for this run.

While making it to 230% of TBO is certainly an excellent outcome, there's nothing special about making it to TBO or beyond. Plenty of aircraft owners do it. All it takes is a smidgen of common sense, a bit of TLC, and a little attention to detail.

Five golden rules

It all boils down to five basic "golden rules" for making your Continental or Lycoming last a long time:

1 **Don't let it rust.** In the owner-flown fleet, corrosion claims more engines than anything else. If you let your engine rust internally, it doesn't matter if you do everything else right—it won't make TBO.

2 **Avoid cold- and dry-starts.** One start at an OAT of 20°F without a preheat will take a greater toll on your engine than hundreds of hours of normal operation. The same is true of starting an engine after an extended period of disuse. Don't do it.

3 **Keep it clean.** Buildup of dirt or acids in your engine oil can eat away at the innards of your engine. Buildup of exhaust deposits on valve stems can result in burned valves and yanked jugs. Don't let it happen.

4 **Watch the temps.** The aluminum alloy in your cylinder heads starts to weaken dramatically when temperatures reach the 400°F mark. The same is true of the stainless steel alloy in your exhaust system when it reaches 1600°F. You don't want to go there.

5. **Minimize stress.** Some powerplant installations are highly stressed and need to be treated gently (with conservative power settings) to achieve optimum longevity. Other installations are conservatively rated and won't benefit from such "babying." It's important to know which applies to the airplane you fly.

Let's take a closer look at these five rules and how best to follow them.

Rule 1: Don't let it rust

If you fly your plane regularly—at least once every two weeks—internal engine corrosion is probably a non-issue for you. Same thing applies if you base your airplane in a dry desert or mountain climate. You lucky devils are hereby authorized to ignore the following paragraphs and proceed directly to Rule 2.

For the rest of us—and that includes the great majority of owner-flown airplanes (including mine)—corrosion is the number-one reason that engines fail to make TBO. Owners (like me) who live within 100 miles of any large body of water (the oceans, the Gulf, the Great Lakes) are especially at risk. Simply put, our engines often rust out before they wear out.

When it comes to rust, the engine parts we're most worried about are cam lobes, lifter faces, and cylinder walls. The only rust protection these components have is the film of engine oil that was left on them at the end of the last flight. During periods of disuse, gravity sets in and the oil film gradually strips off these critical parts and ultimately leaves them exposed to the atmosphere. And that's bad.

When it comes to rust, the engine parts we're most worried about are cam lobes, lifter faces, and cylinder walls. The lifter on the right has seen better days.

There are several things you can do to protect your engine from rust. The most effective by far is to fly the aircraft regularly—preferably every week or two. Each time you fly, you restore the protective oil film that protects your cam lobes, lifter faces and cylinder walls, and you boil off the moisture inside your engine. That's good.

Many owners believe that hand-turning the prop a few blades or running the engine on the ground will help restore the oil film and preserve the engine. Wrong! Hand-turning the prop accelerates the stripping off of the protective oil film. Ground runs do replenish the oil film to some extent, but they also accelerate the buildup of corrosive moisture and acids in the oil, and on balance they do more harm than good. There's no substitute for flying the airplane once every week or two.

If you can't fly your airplane that often (or even if you can), there are several other things you can do to help protect your engine against internal corrosion. Of these, the most effective is to keep the airplane in a hangar rather than tying it down outdoors. Hangaring drastically reduces the day/night temperature cycling that causes moisture to condense on metal parts and allows corrosion to occur. Consider: A vehicle parked outdoors overnight in a humid climate will be covered with dew the next morning, while one parked in a garage or hangar will remain dry.

Simply put, our engines often rust out before they wear out.

(I've spoken with many owners who believe that the principal reason for hangaring an airplane is to protect the paint and windows from exposure to the sun. Fact is, that's probably the least important benefit of a hangar. The most important reason to hangar an aircraft is to prevent corrosion of metal parts and rotting of rubber and plastic parts.)

Corrosion of cylinder walls can be largely eliminated by installing cylinders that have corrosion-resistant barrel treatments such as chrome or nickel plating. This is particularly advisable for aircraft exposed to extremely corrosive atmospheres (e.g., salt-water floatplanes) and those that are used only seasonally. Of course, installing corrosion-resistant cylinders does absolutely nothing to protect cam lobes and lifter faces.

What about the choice of engine oil? Frankly, this is a controversial subject. Some believe that heavyweight single-grade oil (like Aeroshell W100) provides the longest-lasting protection against corrosion for aircraft that fly irregularly. The oil manufacturers assert that the opposite is true. Shell promotes its semi-synthetic multigrade 15W-50 as providing superior corrosion protection, and Exxon markets its new Elite multigrade as the best corrosion protection on the market. Yet I've spoken with folks at leading overhaul shops who've told me that they find the worst corrosion problems in engines operated on multigrade oil. I continue to use single-grade Aeroshell W100 in my engines year-round—the only exception is if I'm planning a long winter trip to the cold country, in which case I'll temporarily change to multigrade.

Various anti-corrosion additives are pre-blended into some aircraft oils, such as Aeroshell 15W-50, Aeroshell W100 Plus, and Exxon 20W-50 Elite. They do help mitigate corrosion caused by disuse. My personal favorite is an oil additive called ASL CamGuard, which I will discuss more in Chapter 28. I add CamGuard to Aeroshell W100 in a 5% concentration—one pint of CamGuard to 10 quarts of oil—to provide additional anti-corrosion protection.

Finally, any time you know in advance that your airplane will not be flown for 30 days or more (perhaps you're taking a trip to Europe and leaving the plane at home), you should take special steps to preserve the engine from corrosion. Details on how to do this can be found in Continental Manual M-0 and Lycoming Service Letter 180B. If the aircraft will be down for up to 90 days, the procedure involves pulling the top spark plugs, spraying special corrosion-inhibiting oil into each cylinder, and sealing all engine openings with plugs or duct tape. If the aircraft will be down for more than 90 days, a more elaborate preservation procedure is recommended: draining the oil, servicing the engine with special preservative oil (MIL-C-6529 Type II), replacing the top spark plugs with dessicant plugs, and putting dessicant plugs or bags in the various engine openings prior to sealing them.

(But please ignore what Continental says about turning over the prop by hand every five days during periods of disuse. Lycoming says it's a bad idea and does more harm than good. Continental simply got that one wrong, and Lycoming got it right. It's a bad idea. Don't do it.)

By making a conscious effort to protect your engine from rust—whether by simply flying the airplane regularly or by taking some of the other steps discussed here—you'll be doing the most important single thing to make sure your engine reaches TBO or beyond.

Rule 2: Avoid cold- and dry-starts

If the number one enemy of engine longevity is rust, the number two enemy is wear. Here's the most important thing you need to know about wear: The lion's share of engine wear during any flight occurs during the first 60 seconds after engine start. If you make a concerted effort to minimize engine wear during that critical first 60 seconds, the rest of the flight will take care of itself so far as wear is concerned.

While a certain amount of wear is inevitable during every engine start, there are two common situations that exacerbate the problem enormously: cold-starts and dry-starts.

A cold-start simply means starting the engine when it's cold. How cold is cold? There's no well-defined answer—the colder it is, the more abuse your engine suffers immediately after start. As a general rule-of-thumb, starting below 32ºF is a bad idea, and starting below 20ºF is a capital offense. A single start when the engine is below 20ºF can easily produce more wear than 500 hours of normal cruise operation. That's no exaggeration.

Cold-start damage occurs because the various parts of the engine heat up and expand at different rates. The pistons heat up rapidly, while the cylinder barrels heat up much more slowly. To make matters worse, the pistons are made of aluminum alloy which expands twice as fast when heated as the steel from which the cylinder barrels are made. The result: a rapidly expanding piston inside of a lethargically expanding cylinder. It's easy to understand how the necessary clearance between the piston and cylinder can be compromised until the cylinder has a chance to catch up expansion-wise. The result can be serious scoring of the piston skirts and cylinder walls. This is not good.

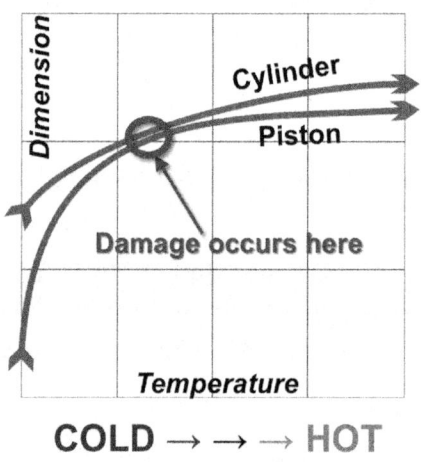

As the piston and cylinder heat up after a cold start, the piston-to-barrel fit may become critically tight, and metal-to-metal scuffing can occur.

The way to avoid cold-start wear is, as every student pilot knows, to pre-heat the engine. The subject of pre-heating has its own dedicated chapter in this book. To make a long story short, it's not enough just to warm up the oil. What's most important is to warm up the cylinder heads. The best method of pre-heating is to put the airplane in a heated hangar overnight. The next best method is to install an electric pre-heating system such as the ones from Rieff or Tanis that have a heating element for each cylinder. Forced-air pre-heats are okay so long as they aren't rushed. Then there's the all-natural method of pre-heating (and a personal favorite of mine): sleep late and wait for Mother Nature to warm things up. But whatever pre-heating method you prefer, just be sure to do it. Cold-starts are engine-killers.

Dry-starts are bad news, too. A dry-start is when you start the engine after a prolonged period of disuse, when most of the residual oil film on the cam lobes, lifter faces and cylinder walls has stripped off. The engine starts running without adequate lubrication, and greatly accelerated wear may occur until enough splash oil reaches the cam, lifters and cylinders to reestablish the necessary oil film. This is not good, either.

You already know the best way to avoid dry-start wear: Fly the airplane regularly, preferably once every week or two. If you know that the airplane will sit unflown for more than 30 days, take the previously discussed steps (as outlined in Continental Manual M-0 or Lycoming Service Letter 180B) to preserve your engine with special preservative oil that sticks like glue and won't strip off.

> One unpreheated cold start can produce more wear than 500 hours of normal cruise operation.

If the airplane has not flown for a month or more and the recommended engine preservation procedure was not performed, there are still steps you can take to minimize dry-start wear. Such steps are appropriate and recommended any time the airplane has been down for extended maintenance, for example. The procedure I use is to remove the top spark plugs, atomize some light oil into each cylinder (LPS-2, Corrosion-X, ACF-50, AeroKroil, etc.), then hook up a battery cart and crank the engine (with the mags off!) until the indicated oil pressure stabilizes. Replace the spark plugs, re-cowl the engine, and start normally. You've just turned a dry-start into a wet-start!

Rule 3: Keep it clean

Once the engine is started and warms up a bit, wear should be minimal for the remainder of the flight. But that assumes that the engine has a good supply of clean oil, free from abrasive material.

If the engine oil becomes contaminated with dirt or other abrasive particulates, it can't do its job of protecting the engine from wear. If it becomes contaminated with water, the resulting acid build-up can accelerate internal corrosion. If combustion byproducts accumulate on exhaust valves and pistons, cylinders can lose compression and require replacement before their time.

Keeping the oil clean isn't difficult. Engine oil should be changed early and often. Change oil after no more than 50 hours or four calendar months, whichever comes first. Some lubrication experts recommend 35 hours or three months. Change it every 25 hours if you operate in dirty or harsh conditions, or if your engine doesn't have a full-flow oil filter.

If your engine doesn't have an oil filter, consider installing one—it's relatively simple and inexpensive, and will repay its cost many times over (because it doubles your oil-change interval). Always change the oil filter whenever you change the oil, and always cut the old filter open and inspect it for metal.

Replace or clean the induction air filter at least annually, and inspect the induction system for air leaks that could permit the engine to breathe unfiltered air. Watch the silicon numbers in your oil analysis reports—if silicon is high, it means dirt is getting into your oil somehow, and you need to find out how and fix it. (You do regular oil analysis, don't you?)

Change oil after no more than 50 hours or four calendar months.

Keeping the engine free of combustion deposits is also not difficult. It's simply a matter of avoiding unnecessarily rich operation. There are only two times you should operate your engine at full-rich mixture: engine start and takeoff. Any other time—idle, taxi, runup, cruise-climb, cruise, descent, landing—the engine should be leaned aggressively. Leaning during ground operations (taxi, runup) is especially important in preventing deposit buildup.

Any time you remove your spark plugs, inspect them for deposit buildup. If there's crud on your plugs (particularly the top plugs), you can bet there's also crud on your valves and pistons. That's a sure sign that you're running too rich.

Rule 4: Watch the temps

Temperature management is crucial to cylinder longevity. Although the CHT redline is set at 500°F for Lycoming engines and 460°F for Continentals, your cylinders won't last very long if they run anywhere near that hot. (During certification endurance tests, the FAA requires the engine to run at red-line CHT for only 150 hours.)

If you want your jugs to have a decent shot at making TBO, try to keep your CHTs no higher than 400°F for Continentals and 420°F for Lycomings. That's probably easy to do if you fly a generously cooled airplane like a Skylane or Dakota, but it might be challenging in a tightly cowled installation like a Mooney M20K or a Cessna P210. To keep CHTs below these values, you may need to use higher airspeeds for climb, open cowl flaps, and use mixtures that are either rich enough or lean enough (depending on whether you're operating rich-of-peak or lean-of-peak). If you still can't keep CHTs well-controlled, a careful inspection of your engine cooling system is in order—perhaps you have a missing or broken cooling baffle or deteriorated flexible baffle seals.

> Keep your CHTs no higher than 400°F for Continentals or 420°F for Lycomings.

If you fly a turbocharged aircraft, you'll also need to watch turbine inlet temperature (TIT) and make sure you keep it well below red-line (generally 1650°F or 1750°F, depending on what turbocharger is installed). The principal reason for this limit is to protect the turbocharger's turbine wheel from "blade creep"—deformation under the stress of centrifugal force when temperatures get too high. If possible, I suggest keeping TIT 50°F below red-line for optimal turbocharger longevity.

Exhaust temperatures above 1600°F will also shorten the life of other exhaust components, most of which are made of type 321 stainless steel with a maximum operating temperature of 900°C (1650°F). If your airplane doesn't have a TIT gauge, you might want to install one. Better yet, install a probe-per-cylinder engine analyzer with TIT capability.

Rule 5: Minimize stress

Finally, there's the matter of what power settings and leaning procedure to use for maximum engine longevity. Will your engine last longer if you cruise at 65% power instead of 75% power? Is high manifold pressure and low RPM better, or vice-versa? Does lean-of-peak operation hurt or help?

The interaction of manifold pressure, RPM and mixture on the combustion process is a complex subject, and the choice of power settings involves complex tradeoffs. If we confine ourselves to the question of engine longevity, it doesn't make a great deal of difference what power setting you use so long as you stay within the envelope authorized by your POH or engine operator's manual, and so long as you follow the other rules we've already discussed: keep CHT moderate, keep TIT below red-line, and lean aggressively enough to prevent deposit buildup.

Does cruising at 65% power offer greater longevity? It depends on what you fly. Some aircraft have highly-stressed engines that will benefit from conservative cruise power settings, while others have de-rated engines that can run all day at 75% power without breaking a sweat.

Here's a rule-of-thumb: Divide your engine's maximum rated takeoff horsepower by the engine's displacement in cubic inches. The resulting "HP per cube" ratio is a good measure of how highly stressed your engine is. If the ratio is 0.6 or greater, you've got a highly-stressed powerplant that will probably benefit from conservative cruise power settings. If the ratio is 0.5 or less, you've got a conservatively rated engine that can be operated at 75% power all day long and still make TBO.

Consider the Mooney 252, for example. Its Continental TSIO-360 engine is rated at 220 horsepower for takeoff. If you divide 220 hp by 360 cu. in., you come up with 0.611 HP per cube—definitely a highly-stressed powerplant. Or consider a RAM-converted Cessna 340 or 414 Series VII with Continental TSIO-520 engines rated at 335 hp each. Divide 335 hp by 520 cubic inches to get 0.644—again, a highly stressed situation. These aircraft will benefit from the use of conservative cruise power settings.

On the other hand, consider the Lycoming-powered Piper Dakota (235 hp/540 cu. in. = 0.435), the Continental-powered Cessna 182 (230 hp/470 cu. in. = 0.489), or the Lycoming-powered Cessna 172 (160 hp/320 cu. in. = 0.500). These are conservatively rated powerplant installations that can be cruised at 75% without the slightest guilt.

Many aircraft have HP per cube ratios that fall somewhere in the middle between these extremes. The twin Cessna that I fly has a pair of Continental TSIO-520 engines rated 285 hp for takeoff—285 divided by 520 yields 0.548. In my airplane, I feel no hesitation about using 75% cruise power when I want to go fast, but I mostly cruise at 65% power just because I'm anal about wringing every last hour out of my engines.

No matter what airplane you fly, it's always wise to be avoid sudden power changes. Apply throttle smoothly on takeoff, don't jam it in. Reduce power smoothly during decent and landing, don't yank the throttle back to idle suddenly. The same applies to the prop and mixture controls: make RPM and mixture changes slowly and smoothly.

Live long and prosper

I've tried to list these five rules for achieving engine longevity in what I consider to be descending order of importance. Your choice of power settings and leaning procedure is far less important than avoiding rust, cold-starts, dry-starts, dirt and deposit buildup, and excessive temperatures.

Getting your engine to reach published TBO or beyond isn't rocket science. By following a handful of simple rules, anyone can do it. But for maximum engine longevity, please don't forget arguably the most important rule of all: If you get to TBO and your engine still shows every sign of being healthy, don't put it to death! With continued TLC and a bit of luck, it might have another 500 or 1,000 or 1,500 hours of life left. Listen to your engine and let it tell you when "it's time."

26
Temperature, Temperature, Temperature

Those three words contain the secret to making your aircraft engine operate trouble-free for a long time.

I've had wonderful luck with piston aircraft engines throughout my 50-odd years as an aircraft owner. My engines have never failed to make TBO, and recently I've learned to take them to well over twice TBO with minimal maintenance along the way.

Early in my tenure as an aircraft owner, I was convinced that the secret of my success was the fact that I "babied" my engines, typically limiting my cruise power settings to no more than 60% or 65% power. I felt that sacrificing a little airspeed in exchange for long engine life and reduced maintenance cost was a good tradeoff.

More recently, I've come to learn that such "babying" is one way to achieve long engine life, but it's not the only way. That's because it's not power that damages our engines—it's temperature. It turns out you can run these engines as hard as you like so long as you are obsessive about keeping temperatures under control. Or as one of my mentors, the late Bob Moseley, once told me, "There are three things that affect how long your engine will last: (1) temperature, (2) temperature, and (3) temperature!"

It's all about the heat

Our piston aircraft engines are heat engines. They have moving parts—notably exhaust valves and valve guides—that are continually exposed to extremely high temperatures in the 1,200°F to 1,600°F range (and sometimes even hotter). Since engine oil cannot survive temperatures above about 400°F, these moving parts must function with essentially no lubrication. They depend on extremely hard metals operating at extremely close tolerances at extremely high temperatures. It's nothing short of miraculous, and a testament to outstanding engineering, that they last as long as they do.

The key to making these critical parts last is temperature control, and the most important temperature is cylinder head temperature (CHT). Moseley monitored and overhauled these engines for nearly four decades, and he claimed that a Continental engine that is operated at CHTs above 420°F on a regular basis will show up to *five times as much wear metal* in oil analysis as an identical engine that is consistently limited to CHTs of 380°F or less. "It's amazing how much a small increase in CHT can accelerate engine wear," he told me.

As critical as CHT is, many owners don't have a clue whether their CHTs are 420°F+ or 380°F-. That's because the original factory engine instrumentation in most legacy airplanes is pathetically inadequate. The factory CHT gauge looks at only one cylinder, and it's not necessarily the hottest one. Further, the factory CHT gauge often isn't even calibrated, and its green arc extends up to a ridiculously hot 460°F (for Continentals) or 500°F (for Lycomings). If all you have is factory gauges, you could easily be cooking your jugs to death while blissfully thinking that all is okay because the CHT gauge is well within the green arc.

To know what's really going on in front of the firewall, you must have a modern multi-probe engine analyzer with a digital readout. Such instrumentation isn't cheap—figure $3,000 for a single or $6,000 for a twin, installed—but if it saves you from having to replace a couple of jugs enroute to TBO, it has more than paid for itself. Installing a digital engine analyzer is probably the best money you can spend on your airplane.

Fuel system setup

For takeoff and initial climb, we normally are at wide-open throttle, full-rich mixture, maximum RPM (if we have a constant-speed prop), and wide-open cowl flaps (if we have those). There's not much we can do from the cockpit to affect CHT during these phases of flight.

What does affect CHT is how our full-power fuel flows are adjusted. Unfortunately, it is shockingly common to see damagingly high CHTs due to improperly adjusted fuel flows, particularly in fuel-injected engines. It is not unusual for the fuel flows to be set wrong from the day an engine is installed, and never to be checked or adjusted all the way to TBO. The owner winds up going through cylinders every 500 hours and never knowing why (or blaming the manufacturer).

In part, the problem lies with mechanics who don't fully understand how critical it is to test and adjust the fuel system setup on a regular basis. Continental recommends that the fuel system setup on fuel-injected Continental engines be checked and adjusted several times a year to account for seasonal changes, but most Continental-powered airplanes go year after year without this being done, and many shops don't even have the necessary test equipment to do it.

Then there's the problem of aftermarket engine modifications. For example, certain PMA replacement cylinders have substantially better "volumetric efficiency" than factory cylinders—in other words, they breathe better. Since they breather more air during every combustion cycle, they need more fuel to maintain the same fuel/air mixture. The full-power fuel flow marked on your fuel-flow gauge may simply not be high enough if you have such cylinders installed.

> The key to making engines last is temperature control, and the most important temperature is cylinder head temperature (CHT).

Even worse are turbocharged engines with aftermarket intercoolers installed. The intercooler reduces the temperature of the air that the cylinder breathes, making it denser. Denser air demands more fuel to maintain the desired fuel/air mixture, so full-power fuel flow must be increased significantly above original factory specifications. Too often this is not done, and the result is fried cylinders.

Many A&Ps are reluctant to adjust takeoff fuel flow above red-line. However, if you have certain PMA cylinders, an aftermarket intercooler, or some other "mod" that allows your engine to produce more power than it did when it left the factory, that's exactly what must be done to keep your CHTs cool and avoid premature cylinder failure.

Enough fuel flow?

How can you tell if your full-power fuel flow is adequate? If you're limited to factory gauges, you probably can't, at least with any precision. About the best you can do is to watch your fuel flow gauge (if you have one).

A good rule of thumb is to multiply your engine's maximum rated horsepower by 0.1 to obtain the minimum required fuel flow in gallons-per-hour, or by 0.6 for pounds-per-hour. For example, if your engine is rated at 285 horsepower, your takeoff fuel flow should be at least 28.5 GPH or 171 PPH; if it's rated 310 horsepower, the minimum should be 31.0 GPH or 186 PPH. If your takeoff fuel flow is less than this, have your mechanic crank it up. And don't forget that if you have certain PMA cylinders or an aftermarket intercooler, your engine might be producing a few percent more horsepower than what the book says, so it might need a few percent more fuel flow.

Now if you have a digital multiprobe engine analyzer, it's easy to tell if your fuel flow is adjusted high enough. Just make sure none of your CHTs exceed 400°F during takeoff and climb—if they're cooler, that's even better.

What about cruise?

Cruise flight represents the lion's share of our flying time. Just as in takeoff and climb, it's essential to keep all our CHTs below 400°F for Continentals or 420°F for Lycomings during cruise to achieve good cylinder longevity, and cooler is better. Hopefully we can do this without pouring excess 100LL on the problem.

There are basically three different strategies for keeping CHTs low during cruise:

- Baby the engine
- Operate very rich
- Operate lean-of-peak

All three strategies work, and conscientious use of any of them will give you a good shot at making TBO with minimum cylinder problems. But each has its pros and cons. Let's take a closer look.

Baby the engine. Many legacy airplane POHs talk about operating at three alternative mixture settings: "best power mixture" (130°F rich-of-peak), "recommended lean mixture" (50°F rich of peak), and "best economy mixture" (peak EGT). In my experience, most pilots tend to operate somewhere between best-power mixture and recommended lean mixture.

It turns out that "recommended lean mixture" (50°F ROP) is just about the worst possible mixture setting for keeping CHT low. CHT reaches a maximum at just about

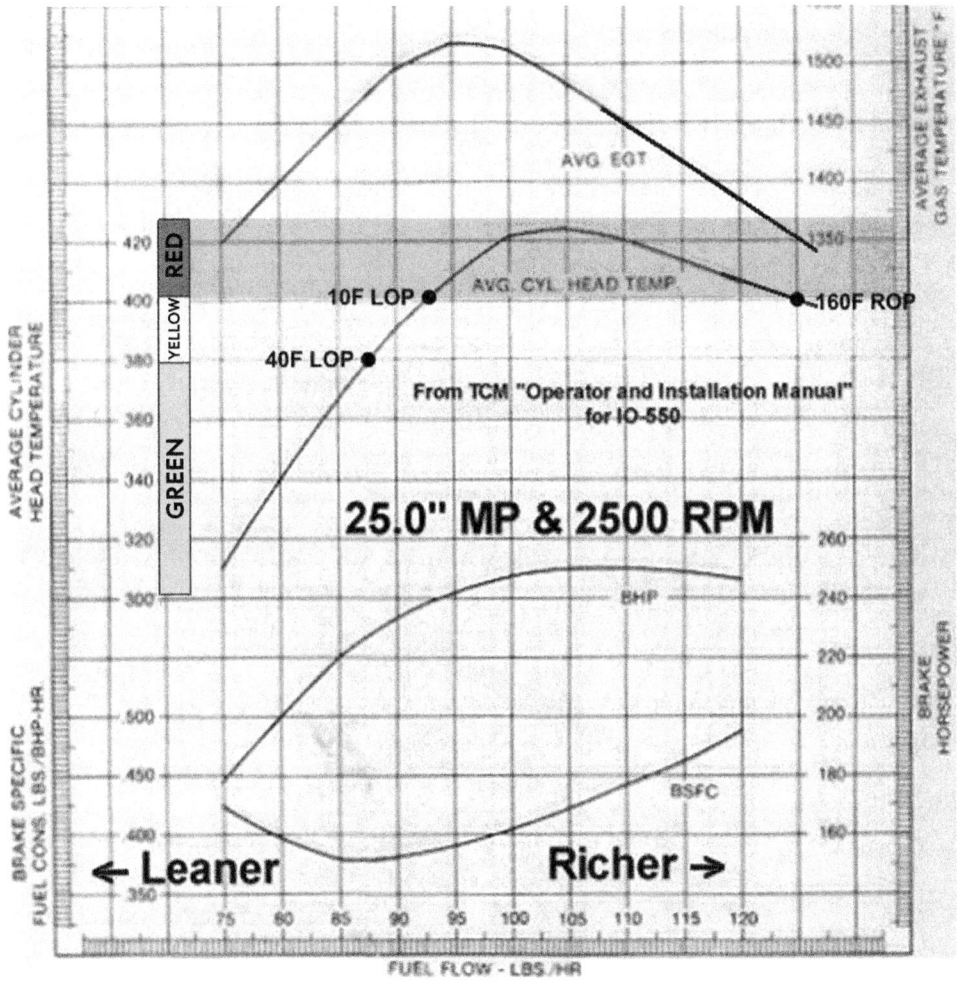

The mixture that most POHs refer to as "recommended lean mixture" is 50°F rich of peak EGT. As this graph shows, using that mixture results in the highest possible CHT. To reduce CHTs to the level required for good cylinder longevity, you need to do one of three things: (1) reduce power, (2) operate very rich, or (3) operate lean-of-peak.

50°F ROP. So, if you want to operate at "recommended lean mixture" and simultaneously keep CHT low, there's only one way to get there: reduce power dramatically (e.g., to 60% power or less). In other words, baby the engine.

Both "best power mixture" (130°F ROP) and "best economy mixture" (peak EGT) result in somewhat lower CHTs than does "recommended lean mixture." At either of these mixture settings, you can usually operate at 65% power or so and keep CHTs in the acceptable range.

In any of these cases, you're trading power and airspeed for reduced temperatures and increased longevity. For most of us, that's a reasonable tradeoff to make.

Operate very rich. But what if you are unwilling to sacrifice power and airspeed? Is it possible to go fast and keep CHTs low?

Sure it is. We already talked about one way to do this in our discussion of takeoff and initial climb: pour lots of 100LL on the problem. In other words, operate very rich.

How rich? To reduce CHTs by 25°F, you need to richen the mixture to about 160°F ROP. For each additional 10°F of CHT reduction, you need to enrichen an additional 50°F ROP. Using such very rich mixtures, you can go fast and still stay cool. But before you decide to go this route, consider the downsides.

The most obvious downside is that this strategy is very fuel-inefficient. Compared to "best economy mixture," the very-rich strategy consumes about 25% more fuel, and reduces range by a similar amount. Advocates of very rich mixtures will tell you that "fuel is cheaper than engines," but don't be so sure. Assuming avgas costs $5/gallon, using 25% more fuel in a 300 horsepower engine can cost more than $33,000 over the engine's TBO, and that's enough to replace a whole bunch of cylinders.

Assuming $5/gallon avgas, using 25% more fuel in a 300 horsepower engine can cost $33,000 over the engine's TBO.

A second and less obvious downside is that very rich mixtures result in "dirty" combustion with lots of unburned byproducts in the exhaust gas. Operating this way for long periods of time tends to cause deposit buildup on piston crowns, ring grooves, spark plugs and exhaust valve stems. Do it long enough and you could wind up with stuck rings, stuck valves, worn valve guides, and fouled plugs.

Operate lean-of-peak. The third way to reduce CHTs is to lean even more aggressively than most POHs recommend and operate on the lean side of peak EGT. You can reduce CHTs by 25°F by leaning to about 10°F LOP. For each additional 10°F of CHT reduction, you need to lean an additional 15°F LOP. Using these very lean mixtures, you can go fast, stay cool, and obtain outstanding fuel economy, all at the same time.

What's the downside of the LOP approach? The only major downside is that if your engine has uneven mixture distribution among its cylinders, it may not run smoothly at LOP mixture settings.

Uneven mixture distribution can usually be corrected in fuel-injected engines by "tuning" the fuel nozzles to eliminate the mixture imbalances. GAMIjectors are tuned nozzles that are now STC'd for the majority of fuel-injected Continentals and Lycomings. Continental now offers its own version of tuned nozzles.

The TSIO-520-BB engines in my airplane are equipped with GAMIjectors. As a result, my mixture distribution is near-perfect and I can usually operate extremely lean (nearly 100°F LOP) without perceptible roughness.

If your engine is carbureted, you have no nozzles to tweak. If your mixture distribution is uneven, your engine probably won't operate LOP without unacceptable roughness, and there's probably not much you can do about it.

There's huge variation in mixture distribution among different makes and models of carbureted aircraft engines. Most carbureted Lycomings have relatively even mixture distributions and are often good candidates for LOP operation. On the other hand, the ubiquitous Continental O-470 engine found in most Cessna 182s is notorious for having uneven mixture distribution, and it can be challenging to lean them beyond peak EGT without getting a serious case of the shakes (although it usually can be done).

A baffling problem?

If your CHTs are running warmer than you'd like, another reason could be that you've got leaky cooling baffles under your cowling. Fixing those leaks is usually simple—and the less air leaks, the more is available to cool the cylinders.

Our modern piston aircraft are powered by tightly cowled horizontally opposed engines. Inside the cowling, a system of rigid aluminum baffles and flexible baffle seals divide the engine compartment into two chambers: a high-pressure area above

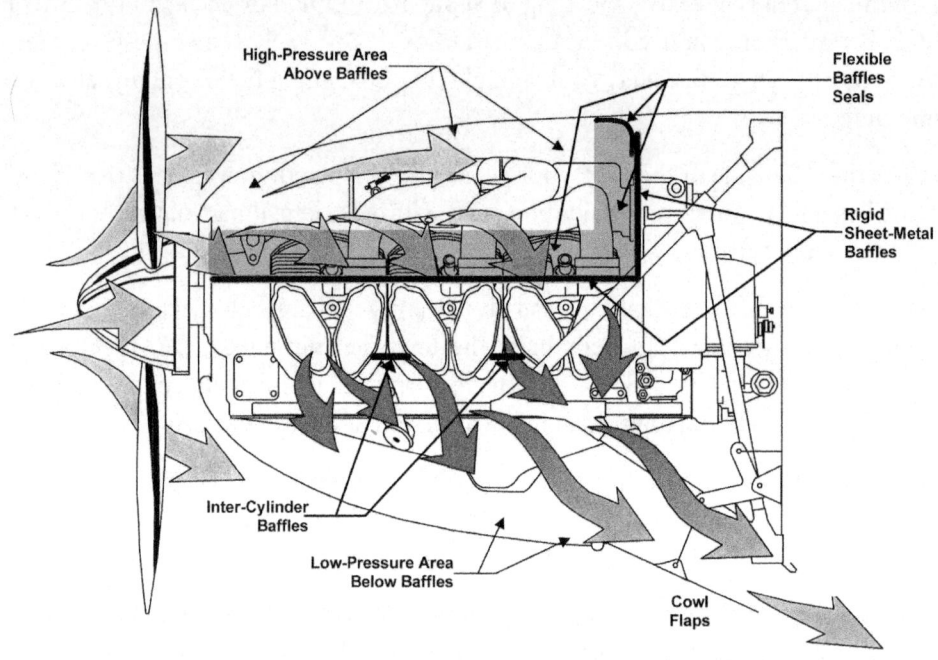

The heart of a modern "pressure-cooled" powerplant installation is a set of rigid sheet-metal baffles and flexible baffle seals that, together with the engine cowling, divide the engine compartment into two chambers: a high-pressure area above the engine and a low-pressure area below and behind the engine. Engine cooling depends upon the vertical airflow from the upper chamber to the lower one. Cowl flaps modulate the cooling by regulating the vacuum in the low-pressure chamber.

the cylinders, and a low-pressure area below the cylinders and behind the engine. Cylinders are cooled by the *vertical* flow of air from the high-pressure above the engine to the low-pressure below it. Cooling airflow is top-to-bottom, not front-to-back.

It's important to understand that the pressure differential between the upper and lower areas is remarkably small: A typical high-performance piston aircraft generally relies on a very small pressure differential, about 0.25 PSI! Aircraft designers try to keep this differential to an absolute minimum to minimize cooling drag.

Because this pressure differential is so very small, even small leaks in the system of baffles and seals can have a serious adverse impact on engine cooling. Any missing, broken, or improperly positioned baffles or seals will degrade engine cooling by providing an alternative path for air to pass from the upper chamber to the lower chamber without flowing vertically across the cylinder cooling fins.

Probably the most trouble-prone part of the cooling system is the baffle seals. These flexible strips (usually high-temp silicone rubber) are used to seal up the gaps between the sheet metal baffles and the cowling. These gaps are necessary because the baffles move around inside the cowling as the engine rocks on its shock mounts. To do their job, the seals must curve up and forward into the high-pressure chamber so that the air pressure differential presses the seals tightly against the cowling. If the seals are permitted to curve away from the high-pressure area—not hard to do when closing up the cowling if you're not paying close attention—they can blow away from the cowling in-flight and permit large amounts of air to escape without doing any cooling.

It is also common to find seals that have developed wrinkles or creases when the cowling is installed, preventing them from sealing airtight against the cowling and allowing air to escape. It's important to look carefully for such problems each time the cowling is removed and replaced, and especially important when new seals have been installed.

Another trouble-prone part of the cooling system is the inter-cylinder baffles. These

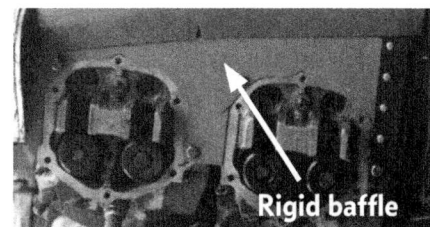

Flexible seals are used to prevent air from escaping through the gaps between the engine-mounted sheet-metal baffles and the cowling. To do their job, they must be oriented so as to curve toward the high-pressure chamber above the engine, so that air pressure pushes them tightly against the cowling.

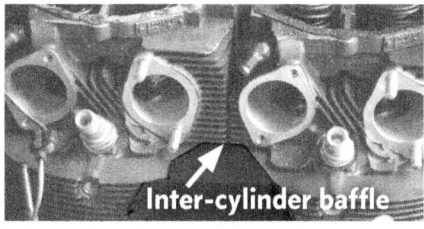

Inter-cylinder baffles are oddly-shaped pieces of sheet metal that mount beneath and between the cylinders, and force the down-flowing cooling air to wrap around and cool the bottom of the cylinders. (This photo was taken looking up from the bottom of the engine, with the exhaust and induction systems removed to make the baffle easier to see.)

are small, oddly-shaped pieces of sheet metal mounted below and between the cylinders. Their purpose is to force the down-flowing cooling air to wrap around and cool the bottom of the cylinders, rather than just cooling the top and sides. These baffles are difficult to see unless you know exactly where to look for them, but they are absolutely critical for proper cooling. It's not at all uncommon for them either to be left out during engine installation or to fall out during engine operation. Either way, the result is major cooling problems.

Chill out!

Whatever strategy you prefer, the important thing is to keep a close watch on your CHTs and ensure that they remain cool. The best way to do this is to install a multi-probe digital engine monitor and program its CHT alarm to go off at 400°F (Continental) or 420°F (Lycoming).

If the alarm goes off during takeoff or initial climb, you're going to have to get your mechanic to turn up the full-power fuel flow. If it goes off during cruise, either richen (if ROP) or lean (if LOP) to bring the CHT down to acceptable levels. If you cannot keep your CHTs in an acceptable range, you just might take a look at your baffling to see if there are any glaring issues with the seal they provide.

If you don't have a multiprobe digital engine monitor, install one. The cost of such instrumentation (including installation) is usually less than the cost of replacing one cylinder. Failure to install such instrumentation is a classic case of "penny wise, pound foolish."

27
The Whys and Hows of Preheating

A single cold start without proper preheating can produce more wear in less than a minute than 500 hours of normal cruise operation.

The first question pilots typically ask is how cold it has to be before preheating is necessary. There's no hard and fast answer to that question. The amount of damage done by a cold start depends on a variety of things, including the type of engine, its age and condition, and what kind of oil is being used. (A brand new or freshly rebuilt engine is more vulnerable to cold start damage than a tired old engine at TBO.)

Generally, I consider any start in which the engine is cold-soaked to a temperature below freezing (32°F or 0°C) to be a "cold start," and any start below about 20°F (-7°C) to be particularly damaging. The colder the temperature, the worse the crime.

Oil pressure isn't enough!

Many pilots believe that the main reason cold starts are bad for engines is that the engine oil is thick and viscous and doesn't flow well. Since it takes longer for oil pressure to come up when the oil is cold, the engine sustains excess wear in the early seconds after start because of inadequate lubrication.

While that's true of single-weight oils like Aeroshell W100, it's not true of the modern multiviscosity oils that are universally used today for cold-weather operations. Multivis oils like Aeroshell 15W-50, Exxon Elite 20W-50 and Phillips X/C 20W-50 flow extremely well even at 0°F (–18°C) or less. Pilots who use multivis oils observe that their oil pressure comes up quickly after starting even in cold weather, and most figure that therefore everything's okay. Wrong!

It's the clearance, Clarence...

The biggest culprit in cold-start damage is the fact that our aircraft engines are made of dissimilar metals with very different expansion coefficients. The crankcase, pistons and cylinder heads of your engine are made from aluminum alloy, while the crankshaft, connecting rods, piston pins and cylinder barrels are made from steel. Aluminum expands about twice as much as steel when heated, and aluminum contracts about twice as much as steel when cooled. Therein lies the problem.

Consider your steel crankshaft, which is suspended by thin bearing shells supported by a cast aluminum crankcase. As the engine gets colder, all its parts shrink in size, but the aluminum case shrinks twice as much as the steel crankshaft running through it. As ambient temperature goes down, so does the clearance between the bearing shells and the crankshaft—and that clearance is where the oil goes to lubricate the bearings and prevent metal-to-metal contact. If there's not enough clearance, then there's no room for the oil, regardless of oil pressure.

The Continental IO-520 overhaul manual lists the minimum crankshaft bearing clearance as 0.0018 inch (that's 1.8 thousandths) at room temperature. What happens to this clearance in cold temperatures? Tests performed in 1984 by Tanis Aircraft Services in Glenwood, Minnesota (where it gets mighty cold) showed that an IO-520 loses 0.002 inch (2.0 thousandths) of crankshaft bearing clearance at –20°F. In other words, a tight new engine built to Continental's minimum specified bearing fit at room temperature would have negative bearing clearance at –20°F; the crankshaft would be seized tight!

You've probably noticed how difficult it is to pull the prop through by hand before starting in cold weather. Now you know why. It's not that the oil is thick (and if you use multivis oil, it's not). It's that the clearance between the crankshaft and bearings is tighter than normal. If it's cold enough, you might not be able to pull the prop through at all.

Start an engine in this condition and you're likely to experience accelerated bearing wear and possible scuffing of the crankshaft journals in the first minute or two of engine operation. In the extreme, it's even possible for the bearing shells to shift in their saddles (a so-called "spun bearing"), misaligning the oil feed holes and starving the bearing from lubricating oil.

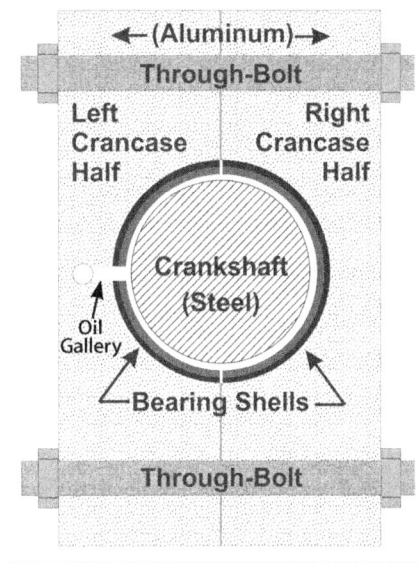

The clearance between crankshaft main journals and their bearings may become critically low (even zero) at frigid OATs.

Ironically, this problem is at its worst with a fresh-from-the-factory engine built to the tightest new-engine tolerances. A tired, loose, high-time engine with worn bearings might well have plenty of clearance even at subzero temperatures. But, even if your engine is approaching TBO, you can't afford to be complacent about cold starts. That's because inadequate bearing clearance is only one of several evils associated with cold starting.

Piston clearance is a big deal, too

Consider what happens to your pistons and cylinders when you cold-start an engine. Here, the situation is the opposite of the one we just talked about: instead of a steel crank inside an aluminum case, we have an aluminum piston inside of a steel cylinder barrel. So the clearance situation is reversed: piston-to-cylinder fit is loose when the engine in cold, and tightens up as the engine comes up to full operating temperature. (This is why compression tests are normally done when the engine is hot.)

When an engine is started cold, the piston heats up very rapidly after start, while the cylinder barrel may take quite a long time to warm up. Why? The piston is small and has relatively low thermal mass, so it heats up quickly. The cylinder is big, massive, and covered with cooling fins bathed in frigid ambient air, so it warms up quite slowly (as you can easily see on your CHT gauge).

The result is that the piston expands to its full operating dimension quite quickly after start, while the cylinder takes a lot more time to expand to its full operating diameter. The fit of the piston in the cylinder bore may quickly become tighter than normal shortly after starting when the piston has come up to temperature but the cylinder still has a way to go. If it's cold enough, the piston-to-cylinder clearance can wind up going to zero, resulting in metal-to-metal scuffing between the piston and cylinder barrel. Because most cylinders are designed with a taper ("choke"), the most severe cold-start-induced scuffing usually occurs at the top of stroke.

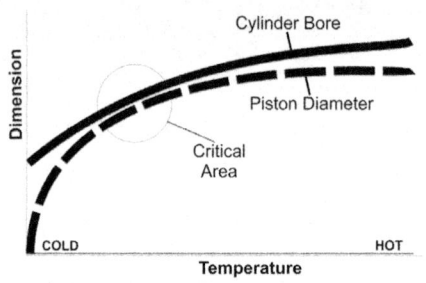

As a cold-soaked cylinder heats up after engine start, the piston expands faster than the cylinder bore, and there's a risk the clearance will go to zero.

As you can see from this discussion, warming up the engine oil is not enough to avoid cold-start damage. All the warm oil in the world won't help if the crank-to-bearing or piston-to-cylinder clearances go to zero. To avoid this, it's essential for a preheat to warm up the crankcase and the cylinder barrels—especially the top of the cylinder barrels near where they join to the heads.

The world's best preheat

The best way to accomplish this is to put the airplane in a heated hangar overnight. Why? Because this preheats every part of the airplane to an even temperature. After 8 to 12 hours in a 40°F hangar, the oil is at 40°F, the crankcase is at 40°F, the cylinder heads are at 40°F, the gyro instruments are at 40°F (gyros have their own cold-starting problems), the windshield is at 40°F (so it won't fog up the minute you breathe), and even the pilot's seat is at 40°F (which solves yet another problem).

I'm based on the California coast where the weather rarely gets below freezing, but when I travel to the cold country, I always try my best to use the overnight-in-a-heated-hangar method of preheating. Most FBOs seem to charge anywhere between $25 and $100 to store my twin in their heated hangar overnight. Even at

$100, I figure it's quite a bargain compared to the alternative (accelerated wear of my two expensive engines).

If I'll be staying at a cold-weather airport for awhile, I'm generally too much of a skinflint to pay for the airplane to be hangared for the whole duration. Instead, I'll arrange with the FBO to pull the airplane into the heated hangar the night before my scheduled departure. If it's really, really cold out on the morning of departure, I've been known to preflight the airplane in the hangar, climb into the cockpit, secure the door, and then have the line crew open the hangar door and tow the airplane out onto the ramp with me in it. As soon as they unhook the tug, I start the engines before they've had a chance to get cold-soaked.

Multipoint electric heaters

Short of overnight in a heated hangar, the best preheating method is a multipoint electric heating system that has individual heating elements attached to the oil pan, the crankcase, and each cylinder. By plugging such a system into AC power a few hours before departure—overnight is even better—you can at least be assured of warm cylinders, a warm case, and warm oil when you start up.

The best-known multipoint electric preheating systems come from Tanis Aircraft Services in Glenwood, Minnesota. The Tanis systems consist of electric heating elements connected by a wiring harness. 50- or 100-watt cylinder heaters screw into the threaded CHT-probe bosses in each cylinder head. The heating elements are available with built-in CHT thermocouple probes that work with most digital engine monitor systems. A flat silicone rubber heating pad is glued to the crankcase with high-temp RTV, and another is glued to the bottom of the oil pan. The wiring harness terminates at an ordinary AC plug that is usually mounted near the oil filler door in the cowling. You simply run an extension cord out to the airplane, plug in the preheating system, and let it cook for 4 to 6 hours.

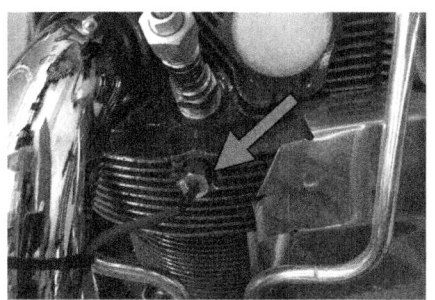

Tanis TAS100 electric preheating system.

Reiff Preheat Systems of Atkinson, Wisconsin offers a similar product called the "HotBand" system. In lieu of cylinder

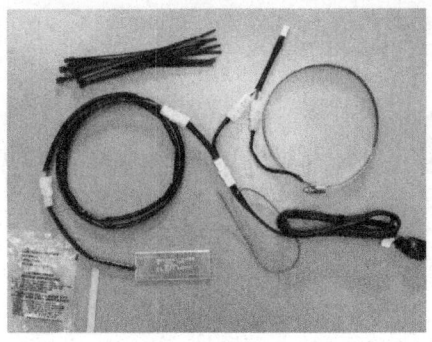

Reiff HotBand electric preheating system.

head heaters, the Reiff system uses 50-watt heating elements mounted on large stainless-steel clamps that mount on the non-finned portion of each cylinder barrel. As a result, there's no interference with existing CHT instrumentation. The Reiff system also includes an oil pan heater, but not a crankcase heater (because the crankcase receives sufficient heat by conduction from the cylinder barrel heaters). The Reiff system is less expensive and easier to install than the Tanis, but takes somewhat longer to fully heat-soak the cylinder heads.

I like the Reiff system for use in typical cold climates, but prefer the Tanis for aircraft based where it's truly frigid.

Engine and prop covers

If the temperature is not too cold and the aircraft is being preheated in a hangar or other protected area, then a multipoint electric heating system is all you need to do the job. But if it's really frigid, or if you have to preheat outside on an exposed ramp (particularly if it's windy), then you also need some means of insulating the engine compartment and keeping most of the heat from escaping.

At the very minimum, you'll need an insulated engine cover. Although you may be able to make do with a quilted blanket, custom-fitted insulated covers are available from Kennon Products, Inc. (Sheridan, Wyo.), Bruce's Custom Covers (Morgan Hill, Calif.), and a few other firms. In intense cold or windy conditions, the propeller becomes a major source of heat loss during preheating. Kennon and Bruce's offer insulated propeller and spinner covers to solve this problem.

Another compelling advantage of insulated engine and prop covers is that using them may eliminate the need for a preheat altogether if you're going to be making a quick-turn. By installing the covers promptly after shutting down, engine heat can be retained for three or four hours even when the airplane is parked outside on a cold, windy tiedown.

Insulated engine and propeller covers.

Leave it on all the time?

There has been considerable controversy about whether it's a good idea to leave an electric preheating system plugged in continuously when the airplane isn't flying. Both Continental and Shell have published warnings against leaving engine-mounted electric preheaters on for more than 24 hours prior to flight.

Their concern is that heating the oil pan will cause moisture to evaporate from the oil sump and then condense on cool engine components such as the camshaft, crankshaft or cylinder walls, resulting in accelerated corrosion of those parts. In Continental engines, the starter drive adapter is particularly vulnerable. However, if the entire engine is heated uniformly by means of a multipoint heating system, and especially if the engine and propeller are covered with insulated engine and prop covers, such condensation is very unlikely to occur.

In fact, using an insulated cover and a multipoint preheating system that is plugged in continuously is one of the most effective methods of eliminating internal engine corrosion, particularly if the aircraft is kept in an unheated hangar rather than outdoors. If the entire engine is maintained above the dewpoint, condensation simply cannot occur.

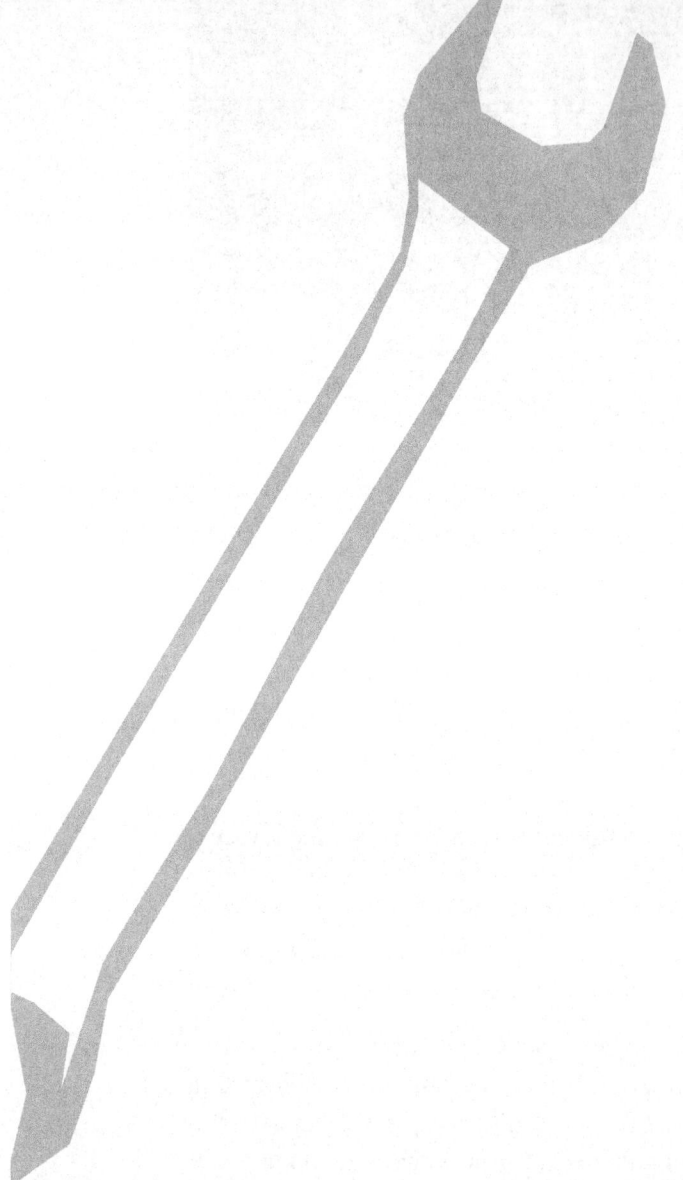

28
The CamGuard Chronicles

I tested ASL CamGuard in my airplane and I liked what I saw.

For decades, I've been a staunch skeptic when it comes to aftermarket oil additives. I've often referred to them sarcastically as "miracles-in-a-can." I've investigated various additives-du-jour—Marvel Mystery oil, Microlon, Slick 50, AvBlend, etc.—but I've never seen any convincing evidence that they offer any benefit. This always struck me as unremarkable, because I figured that if an oil additive was demonstrably beneficial, Aeroshell, Exxon and Phillips would already be blending it into its aviation oils.

Then I started hearing about yet another miracle-in-a-can called ASL CamGuard. I naturally assumed that it was just one more alchemical admixture of snake oil, food coloring and perfume designed to separate aircraft owners from their hard-earned cash.

Then a funny thing happened: I started corresponding with the developer of ASL CamGuard. His name is Ed Kollin, and the more we corresponded, the more I thought to myself, "Gee, this guy really sounds like he knows what he's talking about!" I did a little Googling and discovered that Ed was a research chemist who previously worked for Exxon Research & Engineering as director of its engine laboratory. I couldn't help thinking that this didn't sound like the resume of your typical snake-oil salesman, and started wondering whether perhaps ASL CamGuard might be an exception to the rule.

So one year at EAA AirVenture, I made a point of seeking out Ed Kollin at the ASL CamGuard booth, and I spent nearly a half-hour grilling him about his product. It was a fascinating conversation.

The way Ed tells the story, he was working at the engine lab at Exxon Research when Exxon decided to come out with an advanced-technology aviation oil. Ed was tasked to develop an initial formulation. After much research, he developed an additive package that he believed would perform head and shoulders above the competition. Then, much to his disappointment, Exxon decided (according to Ed) to take a more conventional route. Ed subsequently left Exxon and over the next year developed a no compromises additive package as an aftermarket product called ASL CamGuard.

After talking to Ed, I was sufficiently impressed that I flew home from Oshkosh with a case of ASL CamGuard in the baggage compartment of my Cessna 310, and decided to give it a try. My intention was to use it for 100 hours and see what impact (if any) it had on my oil analysis results.

Shortly after my return from Oshkosh, I did an oil change on both Continental TSIO-520 engines and added a pint of CamGuard to each for the first time. After 83 hours, the oil analysis results were promising. But I considered them somewhat inconclusive, so I decided to continue using ASL CamGuard for a while longer. After another 100 hours, I felt I had enough data to draw some conclusions.

Why CamGuard?

My interest in CamGuard stems from my long-held belief that by far the biggest threat to piston engine longevity in owner-flown airplanes is not wear but corrosion. I've amassed a great deal of evidence showing that it's damn near impossible to wear out the bottom end of most piston aircraft engines because they're incredibly robust. Case in point: The two Continental TSIO-520s on my own Cessna 310 made it to 3,200 hours SMOH, and upon being torn down they proved to be in great shape with no significant corrosion.

It can be a real challenge to keep these engines corrosion-free in owner-flown airplanes like mine. Let me show you what I mean.

About 15 years ago, I developed a serious health problem that put me out of action for many months and actually landed me in intensive care for a period of time. During that scary episode in my life, my airplane sat unflown in its hangar for about four months. My oil analysis clearly shows the toll this took on the engines:

		6/4/01	11/29/01	2/28/02	10/13/02	3/9/03	8/10/03
	Sample Date						
	Days on sample		175	89	223	146	151
	Hours on sample		47	47	17	36	40
	Hours/day		0.27	0.53	0.08	0.25	0.26
Left Engine	Iron PPM		46	37	28	31	34
	Iron PPM/hr		0.98	0.79	1.65	0.86	0.85
Right Engine	Iron PPM		62	50	43	40	49
	Iron PPM/hr		1.32	1.06	2.53	1.11	1.23

There's a clear correlation between inactivity and corrosion (as revealed by elevated levels of iron in oil analysis).

While I normally flew the airplane 150 to 250 hours a year, you can see that between March and October on that report, the airplane only flew 17 hours in a period of 7½ months. Not good. You can also see that because of that inactivity, the normalized quantity of iron in the oil (measured in parts per million per hour) doubled from the normal values during the preceding and succeeding periods of normal activity. A graph of this data (shown on the following page) makes this even more obvious.

Look how clearly iron production—rust, actually—correlates with lack of activity. Irregular usage is simply a fact of life for most owner-flown airplanes. Mine is a perfect example. While I fly more hours a year than the average owner, my usage tends to be in fits and starts. One month I'll put 30 hours on the airplane. The next month, it's a hangar queen. This is clearly not good for the engines, but as Walter Cronkite used to say, "that's the way it is." There's not much I can do about it.

Or is there? Ed Kollin claimed that he developed ASL CamGuard specifically to address the problem of corrosion during periods of engine disuse. Does it actually work? After talking to Ed at AirVenture, I decided to find out for myself.

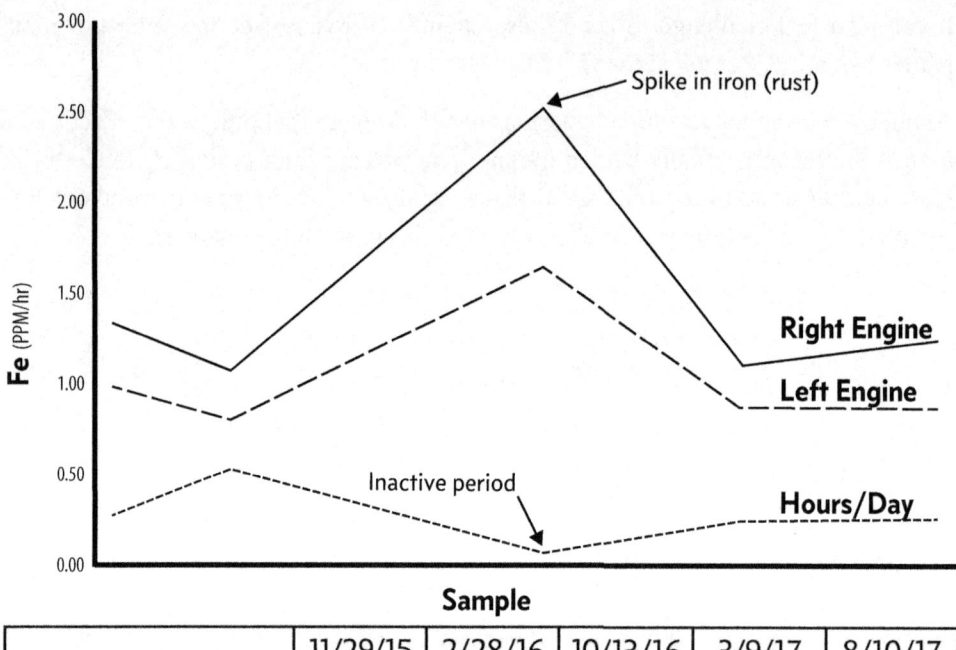

	11/29/15	2/28/16	10/13/16	3/9/17	8/10/17
——— RE Fe PPM/hr	1.32	1.06	2.53	1.11	1.23
- - - - LE Fe PPM/hr	0.98	0.79	1.65	0.86	0.85
········ Hours/Day	0.27	0.53	0.08	0.25	0.26

The correlation between inactivity and elevated iron is obvious here.

So how'd it do?

I started using CamGuard in my engines at the beginning of August 2007, adding one pint to 10 quarts of the Aeroshell W100 that I've used in these engines for decades. After flying over 180 hours with the stuff, and performing four oil changes and oil analysis samples during a period of just over a year, I compared the normalized wear metals for three oil change intervals before starting CamGuard with three oil change intervals after starting CamGuard (leaving out the last non-CamGuard and the first CamGuard intervals to eliminate transition effects). The graphs on the following page present these results.

As you can see, there was a modest decrease of all major wear metals both engines, with the sole exception of nickel which increased slightly on the left engine. It seems

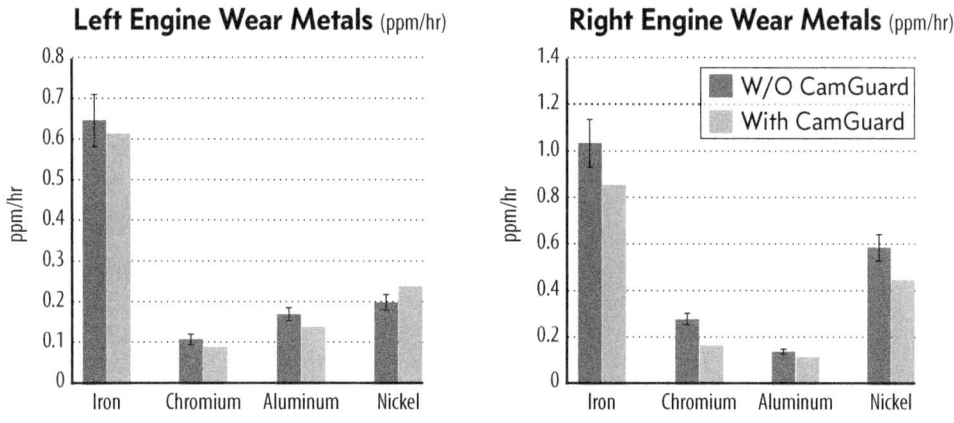

The addition of CamGuard to Aeroshell W100 resulted in a modest decrease of wear metals in both engines.

to me that such a decrease in wear metals, even though relatively modest, easily justifies the cost of using CamGuard.

Wait! It gets better!

What I find even more significant, however, is the evidence that CamGuard is clearly doing a spectacular job of reducing corrosion damage during periods of disuse. How do I know that? Watch closely, because this gets interesting.

Over the more than 30 years that I've owned my airplane (and done regular oil analysis on its TSIO-520 engines), I've noticed a fascinating seasonal pattern to the levels of iron in the oil. Specifically, the first oil change I do after each annual inspection shows a significant spike in iron compared with other samples throughout the year.

There's a very good reason for this. Because my airplane is a fairly complex beast (turbocharged known-ice twin), and because I do all the work on it myself, and because I'm probably the world's slowest mechanic, the airplane is always down for at least a month and sometimes a month and a half before I get it put back together again. This prolonged annual ordeal is usually the longest period of inactivity that the airplane has all year. The chart on the following page clearly reflects the iron levels in the oil.

The left engine normally makes between .6 and .8 ppm/hr of iron, except for the first oil change after each annual inspection where it makes more than 1.1 ppm/hr. Simi-

	Sample Date	3/14/05	7/7/05	9/17/05	11/2/05	2/19/06	6/11/06	7/31/06	10/6/06	3/6/07	7/29/07	10/14/07	3/12/08	6/26/08	8/25/08
	Hours on sample	50	36	32	59	34	36	48	47	56	58	41	42	42	55
Left Engine	Iron PPM	31	40	27	31	26	40	35	33	32	65	37	26	29	30
	Iron PPM/hr	0.62	1.11	0.84	0.53	0.76	1.11	0.73	0.70	0.57	1.12	0.90	0.62	0.69	0.55
Right Engine	Iron PPM	40	44	37	43	32	54	50	50	57	104	59	35	39	45
	Iron PPM/hr	0.80	1.22	1.16	0.73	0.94	1.50	1.04	1.06	1.02	1.79	1.44	0.83	0.93	0.82
		WITHOUT CAMGUARD										**WITH CAMGUARD**			

The reduction of iron during the post-annual-inspection oil samples shows the convincing benefits of CamGuard as an anti-corrosion agent.

larly, the right engine usually makes between .8 and 1.1 ppm/hr, except for the first oil change after each annual where it makes 1.2 to 1.5 ppm/hr.

[Note that the unusually high number of 1.79 ppm/hr right after the 2007 annual was due to the fact that I changed a cylinder, and breaking it in generated even higher iron than normal.]

Ah, but look at the post-2008-annual numbers—the first such results after I started using CamGuard. The iron numbers for those samples not only did not spike up (as they always had before), but were actually below average for each respective engine! To me, this was a very significant change from the long-established pattern of annual iron spikes after annual inspections, and convinced me that CamGuard was doing a very effective job of corrosion protection. So I kept using it, and through the years my observations have only served to strengthen my conviction that CamGuard offers a worthwhile improvement to engine longevity.

PART VI
Condition Monitoring

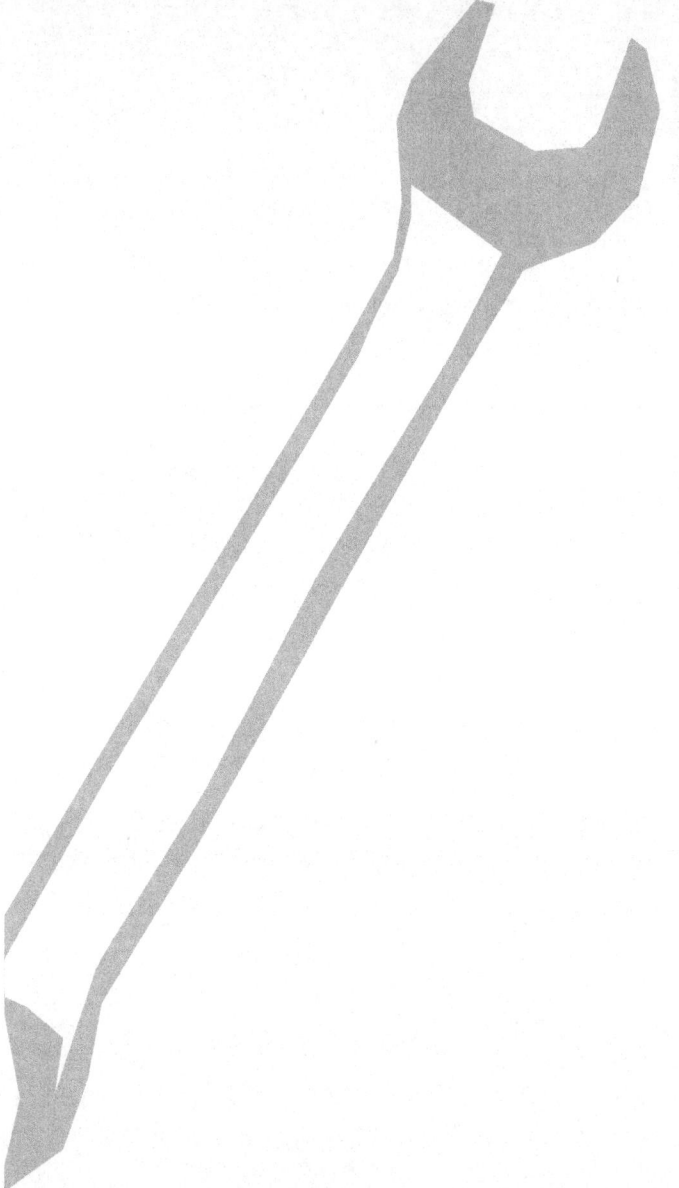

264 ENGINES

29
How Healthy is Your Engine?

How should we assess whether a piston aircraft engine is airworthy?

My friend and mentor Bob Moseley was far too humble to call himself a guru, but he knew as much about piston aircraft engines as anyone I'd ever met. That's not surprising, because the man had been rebuilding Continental and Lycoming engines for over four decades, so there's not much about these engines that he hadn't seen, done, and learned.

From 1993 and 1998, "Mose" (as his friends called him) worked for Continental Motors (then TCM) as a field technical representative covering Missouri, Kansas, Iowa, Nebraska, North and South Dakota, Minnesota, and the portion of Canada north of those states. "Then I made someone at the factory mad," he said, "so they gave me Arkansas." (Not really, but it always got a laugh.)

In 2011, Mose was in Beijing, China running the new Superior Air Parts engine factory there when he unexpectedly passed away in his sleep at age 64. I miss his wit and his wisdom a lot.

James Robert "Bob" Moseley (1947-2011)

Mose was generous to a fault when it came to sharing his accumulated powerplant expertise with others. For years, he'd been an active participant on various type club forums. I often referred owners to him when they needed an expert second opinion on some engine-related subject. He also helped educate mechanics by teaching at FAA-approved IA renewal seminars.

Which engine is airworthy?

During his IA renewal seminars, Mose would often challenge a roomful of hundreds of A&P/IA mechanics with a hypothetical scenario that went something like this:

> Imagine you are doing annual inspections on two similar airplanes—let's say they're Bonanzas—with like-type engines—let's say Continental IO-550s.
>
> Bonanza A's engine is a brand new engine that the owner installed just 40 hours ago. As you'd expect, all cylinders exhibit near-perfect compressions of 77/80 or better. The owner is complaining that since the new engine was installed, the oil consumption has been high—a quart every three hours or so. He also claims that the aircraft seems sluggish in cruise, the takeoff roll somewhat longer, and the climb rate slower than what he considers normal for a Bonanza. He takes you up on a brief test flight, and sure enough the plane does strike you as being a bit of a dog. The owner swears that it performed better before the new engine was installed.
>
> Bonanza B's engine has 1500 hours since field overhaul by some shop you never heard of. This engine is also using a quart of oil every three hours. You check the compressions and find that they're marginal, all in the low 50s and high 40s, just barely squeaking by the compression test criteria as set forth in Continental Manual M-0. The owner is not happy with the high oil consumption and the all oil on the belly, but claims that the airplane seems to be performing okay.
>
> Okay, you're the IA, so you have to make the call. Which engine is airworthy?

The correct answer, of course, is that Bonanza B's engine is airworthy, and Bonanza A's is unairworthy.

Bonanza B's engine meets Continental's minimum compression specs (even if just barely), and oil consumption of a quart every three hours also is high but still well within specified limits. In contrast, Bonanza A's engine may have superb compression readings, but it's clearly unairworthy because it's not making full rated power—and that's far and away the most important airworthiness criterion for any engine.

What's wrong with Bonanza A's engine? There are all sorts of things that can cause an engine with near-perfect compression readings not to make full-rated power: improperly timed ignition, misadjusted fuel injection system, collapsed lifters, incorrect pushrod length, bad cam, or incorrect pistons (just to name a few possibilities).

Many aircraft owners and mechanics have placed too much emphasis on compression test readings as a measure of engine airworthiness. The truth is that an engine can have relatively low compression readings while continuing to run smoothly and reliably and make full rated power all the way to TBO and beyond.

In this case, suggested Mose, the combination of high compression readings and high oil consumption suggests that the problem is most likely glazed cylinder walls due to improper break-in procedure. It's possible that flying the engine at high power for a few hours might cure the problem; if that doesn't do the trick, the cylinders might have to come off for honing and then break-in repeated using the proper procedure.

Who has the best engine?

Now, just to make things more interesting, Mose presented the roomful of IAs with another hypothetical scenario.

> Four good-looking fellows, coincidentally all named Bob, are hanging out at the local Starbucks near the airport one morning, enjoying their usual cappuccinos and biscotti. Remarkably enough, all four Bobs own identical Bonanzas, all with Continental IO-550 engines. Even more remarkable, all four engines have identical calendar times and operating hours.

While sipping their overpriced coffees, the four Bobs start comparing notes. Bob One brags that his engine only uses one quart of oil between 50-hour oil changes, and his compressions are all 75/80 or better. Bob Two says his engine uses a quart every 18 hours, and his compressions are in the low 60s. Bob Three says his engine uses a quart every 8 hours and his compressions are in the high 50s. Bob Four says his compressions are in the low 50s and he adds a quart every 4 hours. Who has the best engine? And why?

This scenario always provoked a vigorous discussion among the IAs. One faction typically thought that Bob One's engine was best and Bob Four's was worst. Another usually opined that Bobs Two and Three had the best engines, and that the ultra-low oil consumption of Bob One's engine was indicative of insufficient upper cylinder lubrication and a likely precursor to premature cylinder wear.

Mose took the position that with nothing more than the given information about compression readings and oil consumption, he considered all four engines equally airworthy. While many people think that ultra-low oil consumption may correlate with accelerated cylinder wear, Continental's research didn't bear this out, and Mose knew of some engines that went to TBO with very low oil consumption all the way to the end.

Oil consumption is an even less important factor than compression when assessing if an engine is airworthy. Lycoming and Continental publish maximum acceptable oil consumption values, usually something like 0.15 pounds/hour/horsepower; for a 300 hp engine, that works out to a whopping 1.5 quarts/hour!

While the low compressions and high oil consumption of Bob Four's engine might suggest impending cylinder problems, Mose said that in his experience engines that exhibited a drop in compression and increase in oil consumption after several hundred hours might still make TBO without cylinder replacement. "There's a Twin Bonanza that I take care of, one of whose engines lost compression within the first 300 hours after overhaul," Mose once related. "The engine is now at 900 hours and the best cylinder measures around 48/80. But the powerplant is running smooth, making full rated power, no leaks, and showing all indications of being a happy engine. It has never had a cylinder off, and I see no reason it shouldn't make TBO."

Lesson of a lawnmower

To put these issues of compression and oil consumption in perspective, Mose liked to tell the story of an engine that was not from Continental or Lycoming but from Briggs & Stratton:

> Years ago, I had a Snapper lawnmower with an 8 horsepower Briggs on it. I purchased it used, so I don't know anything about its prior history. But it ran well, and I used and abused it for about four years, mowing three acres of very hilly, rough ground every summer.
>
> "The fifth year I owned this mower, the engine started using oil. By the end of the summer, it was using about 1/2 quart in two hours of mowing. If I wasn't careful, I could run out of oil before I ran out of gas, because the sump only held about a quart when full. The engine still ran great, mowed like new, although it did smoke a little each time I started it.
>
> The sixth year, things got progressively worse, just as you might expect. By the end of the summer, it was obvious that this engine was getting tired. It still ran okay, would pull the hills, and would mow at the same speed if the grass wasn't too tall. But it got to the point that it was using a quart of oil every hour and was becoming quite difficult to start. The compression during start was so low (essentially nil) that sometimes I had to spray ether into the carb to get the engine to start. It also started leaking combustion gases around the head bolts and would blow bubbles if I sprayed soapy water on the head while it was running. In fact, the mower became somewhat useful as a fogger for controlling mosquitoes.

If this one-cycle engine can perform well while using a quart of oil an hour, surely an aircraft engine with 50 times the displacement can too.

BUT IT STILL MADE POWER and would only foul its spark plug a couple of times during the season when things got bad.

Now keep in mind that this engine was rated at just 8 horsepower and had just one cylinder with displacement roughly the size of a coffee cup, was using ONE QUART OF OIL PER HOUR, and had ZILCH COMPRESSION. Compare that to an IO-550 with six cylinders, each with a 5.25-inch bore. Do you suppose that oil consumption of one quart per hour or compression of 40/80 would have any measurable effect on an IO-550's power output or reliability—in other words, its airworthiness? Not likely.

Let's use common sense

I really like Bob Moseley's commonsense approach to aircraft engines. Whether we're owners or mechanics (or both), we would do well to avoid getting preoccupied with arbitrary measurements like compression readings and oil consumption that have relatively little correlation with true airworthiness.

Instead, we should focus on the stuff that's really important: Is the engine "making metal"? Are there any cracks in the cylinder heads or crankcase? Any exhaust leaks, fuel leaks, or serious oil leaks? Most important, does the engine seem to be running rough or falling short of making full rated power?

If the answer to all those questions is "no," then we can be reasonably sure that our engine is healthy and we can fly behind it with well-deserved confidence.

30
Assessing Cylinder Condition

Compression testers often lie; borescopes never do.

It happens every year: We put our aircraft in the shop for its annual inspection. The IA pulls out the compression test gauges and measures each cylinder while we hold our breath and pray silently until the verdict is rendered. If the readings are good, we can smile and relax; if not, we brace ourselves for the inevitable sticker shock.

I learned an important lesson about compression tests shortly after I became an A&P, when I did something dumb: I pulled a perfectly good cylinder off my engine!

Had I known then what I know now, I wouldn't have touched that cylinder. But at the time, I thought I was doing the right thing. (It does seem like most of the wisdom I've gained over the years came from first exhausting all other alternatives.)

I had downed my airplane for its annual inspection, and the first items on my checklist were to drain the hot oil and perform a hot compression test. All cylinders measured in the 70s except one; that one measured 60/80 with air obviously leaking past the exhaust valve.

The differential compression test first became popular during World War II as a means of determining cylinder condition. Today, we have much more reliable methods.

At the time the applicable guidance was Continental Service Bulletin M84-15. That SB instructed mechanics that it was okay for a jug on a Continental engine to leak lots of air past the rings (or what Continental called the "dynamic seal") and still be airworthy, but that NO leakage past the valves ("static seal") was permissible.

My cylinder clearly was leaking at the exhaust valve. So off it came.

Pulling the jug was a PITA. It took me more than two hours to remove the cooling baffles, exhaust and induction manifolds. It took another hour to remove the rocker cover, rocker shafts, rocker arms, pushrods and pushrod housings. Finally, I used cylinder base wrenches, a big breaker bar, and considerable brute force to coerce the eight cylinder base nuts loose. Four hours into the project, I held the offending jug in my arms and carried it over to my workbench to survey the damage.

I inspected the cylinder carefully, with special attention to the exhaust valve. Try as I might, I couldn't find anything wrong with it. The cylinder looked normal. The exhaust valve looked fine, with no evidence of heat distress or metal erosion. The valve seat looked fine, too.

I was frustrated. After all this time and effort, I wanted to see a smoking gun. I couldn't find one.

I drove over to the local cylinder shop. They examined my cylinder and couldn't find anything wrong either. On general principles they dressed the seat, replaced the exhaust valve, gave the barrel a light hone to restore the crosshatch pattern, and gave me an invoice for $500 and change.

I installed a new set of rings on the piston, then re-hung the cylinder and reinstalled all the stuff I'd previously had to remove. By the time everything was back together I had about 10 hours of sweat equity into the project, plus about $800 in parts and outside work. (If I hadn't been doing the work myself, I'd be out the better part of two grand.)

Terrible timing

That sordid affair turned out to be a classic case of bad timing. Nine months later, Continental radically changed its guidance about when a cylinder should come off. Service Bulletin SB03-3 titled "Differential Pressure Test and Borescope Inspection Procedures for Cylinders" explicitly superseded M84-15, and differed from it in two crucial respects.

First, SB03-3 completely did away with the earlier distinction between leakage past the "dynamic seal" and the "static seal." Under the new guidance, compression is permitted to be as low as the mid-40s, and there's no distinction whether the leakage is past the rings or the valves.

Second, SB03-3 requires that mechanics perform a borescope inspection of the cylinder in conjunction with each compression test. It says that if the cylinder looks good under the borescope, the mechanic should not remove the cylinder regardless of how low its compression is. Instead, SB03-3 directs the mechanic to try several tricks (such as "staking the valves" and rotating the prop backwards) to obtain a better compression reading. If that still doesn't raise the reading to an acceptable value (typically mid-40s or higher), then SB03-3 says that the aircraft should be flown "for at least 45 minutes" and then the compression test be repeated, hot.

The fundamental message of SB03-3 is that the compression test is not a reliable way of assessing cylinder condition, and that the borescope inspection is much more reliable. If a jug flunks the compression test but looks good under the borescope, then the compression measurement must be considered suspect, and heroic measures should be taken to raise it before a decision is made to pull the jug.

Borescopy ascendancy

I was an early adopter of borescopy. Having gone through the painful experience of pulling cylinders due to low compression readings, only to find nothing physically wrong with them, I was anxious to adopt this more enlightened way of evaluating cylinder condition. I borrowed a Lenox Autoscope from a shop on my field and began inspecting the 12 cylinders on my Cessna 310. It was literally an eye-opening experience, almost as if I could climb inside each combustion chamber… or at least stuff one eyeball inside.

By inserting the scope through the top spark plug hole and twisting and turning it, I could get a decent view of the intake and exhaust valves, the cylinder walls, and the piston crown. I found it spellbinding. Direct inspection of the combustion chamber provided a much better picture and deeper understanding of the true condition of the cylinder, compared with the crude indirect assessment provided by the differential compression test. A compression test could tell you that air was leaking past the exhaust valve, but with the borescope you could tell whether it was because of a benign glob of lead on the seat that would quickly self-resolve next time the engine ran or a malignant warped or eroded valve likely to fail catastrophically in the next ten hours. How cool was that?

Over the years, the compression test has proved quite untrustworthy and prone to false positives, resulting in tens of thousands of cylinders being removed unnecessarily (including a few of mine). That's why the SB03-3 guidance calls for any disqualifying compression test that is not corroborated by borescope evidence be retested after flying for at least 45 minutes. That's excellent advice. I've been personally involved in many cases where a cylinder that flunked the first compression test easily passed the second one. In one notable case involving a Cirrus SR22 whose maintenance my company managed, a cylinder that tested at 38/80 (and that the shop doing the annual wanted to yank) wound up measuring 72/80 on the re-test after a one-hour flight.

Now, SB03-3 did not go so far as to recommend that borescope inspections should replace the venerable compression test. Continental couldn't do that, because the requirement to perform a compression test is written right into the FARs (Part 43 Appendix D to be exact). But SB03-3 did all it could to convey that Continental is no big fan of the compression test for determining cylinder airworthiness. (I once had a senior Continental executive confess to me that if they could have dropped the compression test altogether, they would have.)

I consider SB03-3 to be the best thing ever written about how to decide whether or not a cylinder needs to come off. It has saved owners of Continental engines millions of dollars in reduced maintenance costs. In my view, it's high time that Lycoming followed suit and revised its archaic guidance on the subject. (In 2017, Continental incorporated the contents of SB03-3 into its new Standard Practice Maintenance Manual M-0, so it no longer exists as a separate service bulletin.)

Nowadays, my company manages the maintenance of nearly 1,000 piston airplanes, more than 80% of them Continental-powered. We always ask that the shops performing annual inspections on "our airplanes" follow the SB03-3 protocol to the letter.

We want every cylinder to be borescoped at every annual—something very few shops do unless we request it explicitly. If a cylinder looks good under the borescope, we move heaven and earth to make sure it isn't removed, no matter what the initial compression test says.

Scoping the jug

Why does the compression test criteria for cylinder airworthiness refuse to die? One reason, I think, is because the compression test produces a numerical score that gives the illusion of precision. (It's an illusion, because compression readings are notoriously non-reproducible, and can vary all over the place as you can see in the compression history graph below.) In contrast, the borescope inspection requires a subjective evaluation of what the IA sees through the 'scope, and that requires some training, experience and judgment.

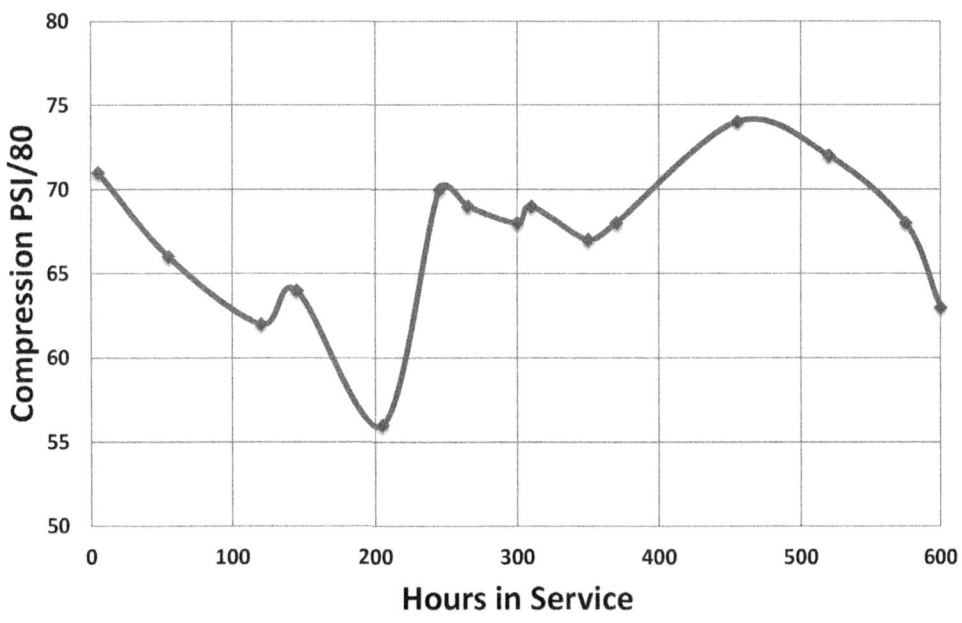

The compression test is less than reliable. Look at this cylinder's compression history obtained during a 600-hour endurance run at the factory. The readings vary all over the place.

A healthy exhaust valve is symmetrical in appearance under the borescope. This one is not. It has a profound hot spot in the 9:30 position and was probably just 5 to 10 hours from failing.

Few A&Ps are adequately trained in how to interpret what they see through the borescope. It's just now beginning to be taught in A&P schools. I haven't been able to find any textbooks or training materials on the subject. Anything an A&P knows about borescope inspection has been learned through on-the-job training.

The value of a borescope inspection depends on whether the IA knows what to look for. Some do; others are clueless. That's not to suggest that borescope inspection is difficult. You can learn most of what you need to know in about 30 minutes, simply by looking at a bunch of borescope images of good cylinders and bad cylinders until you learn to recognize what a bad one looks like. It's not rocket science.

The most important thing to learn is how to detect a failing exhaust valve, because exhaust valve failure is by far the most common safety-of-flight failure mode for cylinders. It's easy to tell the difference between a healthy exhaust valve and a sick one: A healthy exhaust valve has a symmetrical appearance under the borescope—rather like a bullseye—indicating that it's operating at a uniform temperature around its

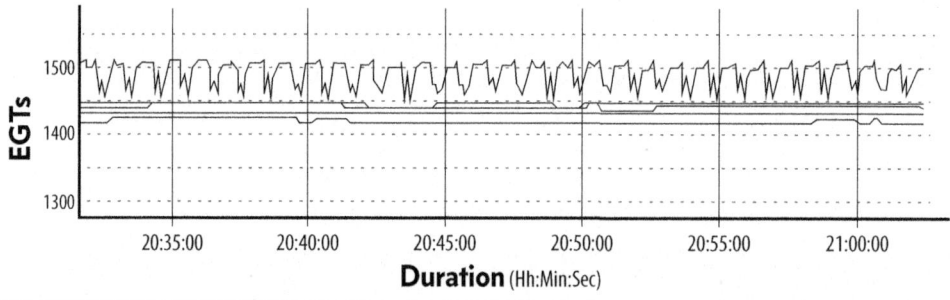

A failing exhaust valve can usually be detected by analyzing digital engine monitor data. The classic symptom is a slow, rhythmic EGT oscillation with a period on the order of one cycle per minute.

entire circumference. A sick exhaust valve has an asymmetrical, lopsided appearance that shows the valve has a hot spot. The hot spot will get progressively worse until the valve fails catastrophically in-flight and shuts down the cylinder.

A failing exhaust valve can also often be detected by a telltale oscillation of EGT, which can be readily identified in data recorded from a digital engine monitor (if one is installed). Almost all new-production piston aircraft are factory-equipped with such monitors, and I estimate that nearly half of the legacy fleet is now so-equipped. All sorts of cylinder problems—including failing exhaust valves—can be readily detected and diagnosed through engine monitor data analysis, as we'll discuss in greater detail in the next Chapter. If characteristic EGT oscillations suggest a burned exhaust valve, the suspect valve should be examined with a borescope as soon as practicable to verify the diagnosis.

Lack of adequate mechanic training has hampered the acceptance of borescopy as the gold standard for assessing cylinder condition. But I'm optimistic that in time it will supplant the old compression test.

Today's scopes: wow!

In the computer industry, "Moore's Law" (named after Intel co-founder Gordon Moore) states that the number of transistors packed on an integrated circuit would double ever two years. Something similar has taken place in borescope technology in the years since SB03-3 was published. Today's borescopes use tiny, cheap solid-state CCD cameras to replace the costly optics that were previously required. The result is the current crop of scopes is both vastly better and an order of magnitude cheaper than the $2,300 Lenox Autoscope that Continental recommended in 2003 when it first published SB03-3.

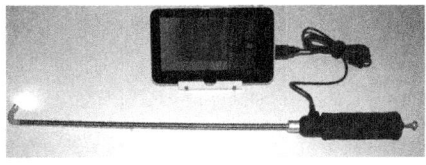

ViVidia Ablescope VA-400 ($200).

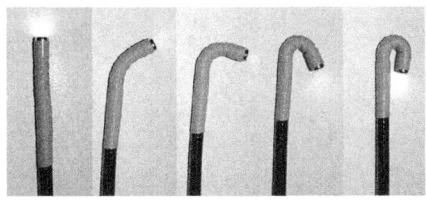

VA-400 viewing angle can be adjusted from 0° to 180° by pressing the plunger on the handgrip.

As I write this in 2018, my current favorite is the ViVidia Ablescope VA-400 scope from Oasis Scientific, which you can purchase on Amazon.com for less than $200. It comes with a USB cable that can be plugged into any notebook PC or Android tablet, and with software for both Windows and Android that can capture both still photos and videos. In addition to its impressive image quality and excellent lighting, the ViVidia scope has the unique ability to adjust its viewing angle from 0° (looking straight down at the piston) to 180° (looking backwards at the valves) or anything in between. (By contrast, the Lenox Autoscope has a fixed 90° viewing angle and no capability for capturing images.)

Lycoming O-360 exhaust valve. Note the symmetrical "bullseye" appearance.

This is a very sick exhaust valve right on the verge of failing catastrophically. Note the asymmetrical appearance: That's bad!

Closeup of the healthy O-360 exhaust valve, with a good view of the seat and valve sealing surface.

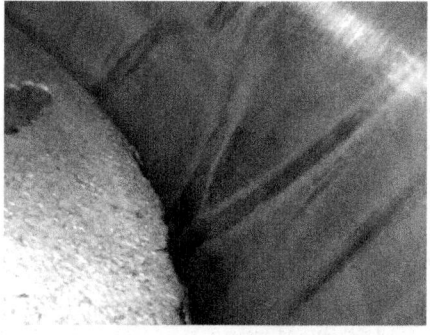

O-360 cylinder barrel and piston. This jug is very worn: Note the vertical scoring and lack of crosshatch.

Pictured at left are some VA-400 images that a colleague captured inside some Lycoming O-360 cylinders. The quality is spectacular for a $200 scope.

With scopes of this quality available for $200, there's no excuse for any A&P not to own one and to use it as his primary means of assessing cylinder condition. In fact, anyone who does owner-performed maintenance should consider buying one.

What about Lycomings?

Continental literally "rewrote the book" on evaluating cylinder condition in 2003 when they first issued SB03-3. If we could just persuade our IAs to follow its guidance—and rely more on the borescope than the compression tester—we'd eliminate the epidemic of inappropriate top overhauls and unnecessary cylinder removal that has plagued piston GA for as long as I can remember.

But what if you fly behind a Lycoming?

Lycoming's equivalent to SB03-3 is Service Instruction 1191A. It's just two pages long and makes no mention of borescopes or engine monitors or any of the things we've learned about cylinder evaluation over past 50 years. It perpetuates the WWII-vintage notion that compression readings of 70/80 or better are satisfactory, readings below 65/80 are worrisome, and readings below 60/80 are unacceptable.

If you own a Lycoming-powered aircraft, this service bulletin is bad news and will cost you a lot of money.

The good news—if there is any—is that guidance in Lycoming S.I. 1191A is couched in "squishy" language. It doesn't say that a cylinder that measures less than 60/80 "must be removed." It says removal and overhaul of such a cylinder "should be considered." This gives your IA wiggle room, if he's brave enough to take it.

But don't be surprised if he isn't.

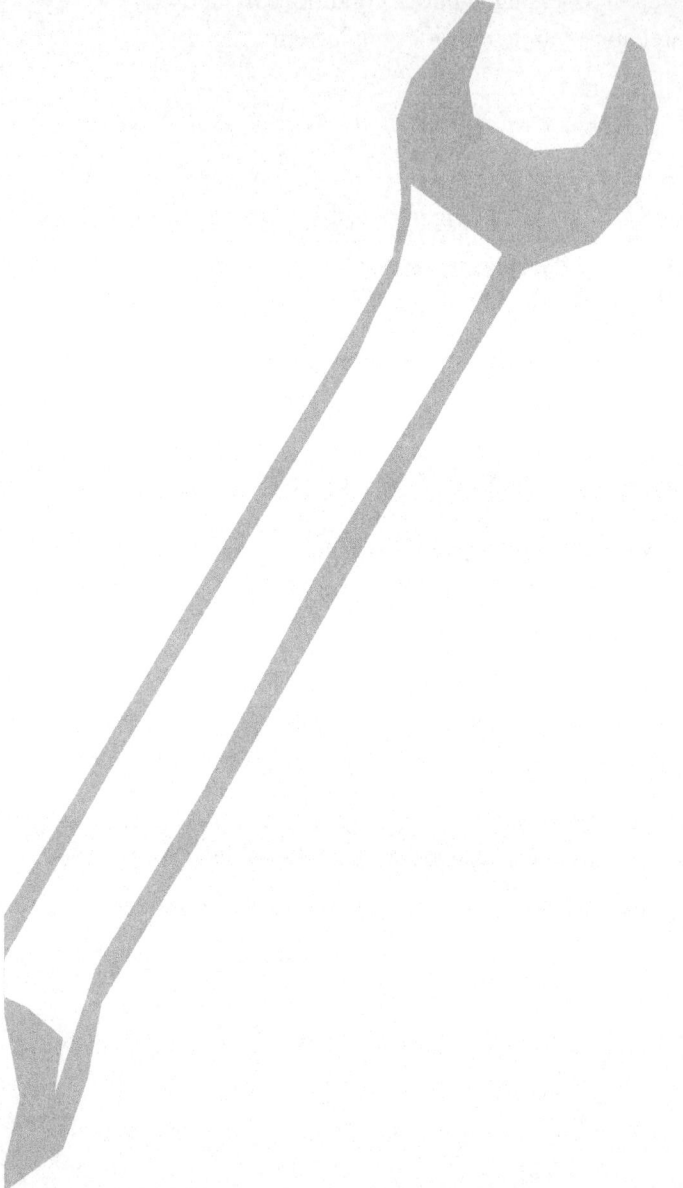

31
Oil Analysis

The oil in your engine provides important clues for assessing the health of your powerplant.

The oil system of every piston aircraft engine provides two levels of filtration. There's a relatively coarse suction screen at the oil pickup tube whose job is to catch large chunks of metal before they can get to the oil pump (and possibly damage it). Then there's either a fine pressure screen or a full-flow oil filter after the oil pump whose job is to catch tiny pieces of metal before they can get to the engine's bearings (and possibly contaminate them).

When implementing a comprehensive engine condition-monitoring program, it's crucial to understand that there are three distinct sizes of metal particles that we're looking for:

- **Large particles or flakes** that cannot pass through the coarse suction screen.

- **Tiny particles** that are too small to be caught by the coarse suction screen and get trapped in the fine pressure screen or oil filter.

- **Microscopic particles** that are too small to be trapped by the pressure screen or oil filter.

Therefore, our condition monitoring program should comprise three distinct elements.

Since **microscopic particles** are too small to be trapped by the oil filter (and too small to see with the naked eye even if some were trapped), it's wise to place the engine on a spectrographic oil analysis program (SOAP) in order to detect abnormal wear events that throw off such microsopic metal particles. An oil sample should be captured at every oil change and sent to the lab for analysis.

To detect **tiny particles**, we must remove and cut open the oil filter (or inspect the pressure screen) at every oil change. Tiny particles can be hard to see, so it's essential to cut the filter media off its spool, spread it out flat, and carefully inspect each pleat under a bright light (and preferably with a magnifying glass).

To detect **large particles or flakes**, we cannot rely on filter inspection or oil analysis, because large stuff never makes it to the filter or into the sample jar. For Lycoming engines, we need to remove and inspect the suction screen at every oil change. This

Lycoming Engines employ a suction screen at the oil pickup in the sump that should be inspected at every oil change. The largest metal particles are filtered by this screen. Unfortunately, these screens are not easily accessible on Continental engines.

step is often neglected because the screen can be a bit hard to get to, and shockingly some A&Ps don't even know about it!

Unfortunately, Continental engines do not permit the suction screen to be removed and inspected. (To gain access to the suction screen, you must drop the oil pan, something that usually can't be done while the engine is mounted in the aircraft.) So for Continental engines, about the best we can do is to (1) drain the oil through a piece of window screen or cheesecloth and then inspect it for any large particles or flakes of metal, and (2) run a magnetic pickup tool around in the oil drain bucket and see if it picks up any pieces of ferrous metal. (This isn't a bad idea for Lycomings, too.) Alas, very few A&Ps or aircraft owners perform these steps, either.

The result is that the worst engine problems—the ones that throw off large chunks or flakes of metal—often go undetected until it's too late. There's no excuse for this if we're doing our condition-monitoring job correctly.

Oil filter inspection

Oil filter inspection is probably the most important single tool we have for monitoring the health of a piston aircraft engine's bottom end. If your engine isn't equipped with a full-flow filter, it's worth adding one simply for the diagnostic value. (It'll also pay for itself quickly by doubling your oil change interval from 25 to 50 hours.) Relatively inexpensive retrofit kits are available for most engines that came from the factory equipped with a pressure screen instead of a full-flow oil filter.

The filter should be changed at every oil change, and should always be cut open and inspected. Tossing an old filter in the trash without cutting it open for inspection is a capital offense. In fact, whenever I visit a maintenance shop, I always make a point of sneaking a peek in the shop's trash can to see if there are any uncut filters there; it's an easy way to know if they are serious about engine condition monitoring.

Every shop—and every owner that does his own oil changes—needs a good oil filter can cutter with a sharp cutting wheel. Cheap cutters tend to leave shards of the can in the filter medium, which can really confuse the filter inspection. A variety of cutters are available from the typical online purveyors of aircraft supplies.

After cutting open the filter and cutting the pleated paper filter media from its spool with a sharp knife or hacksaw, take it outdoors and examine it in direct sunlight. If

At every oil change, the oil filter should be cut open and inspected for metal.

It pays to use a good quality filter cutter to make sure that shards from the filter can don't wind up contaminating the filter media.

that's not practical, inspect the media under the brightest light you can find. Small metallic particles embedded in the filter media are reflective and will generally glisten when viewed in direct sunlight, but may well be invisible under ordinary indoor lighting.

Sometimes it's difficult to determine whether flakes in a filter are metal or carbon. Here's an easy way to tell them apart: place some between your fingertips and rub your fingers while squeezing hard. Carbon flakes will break apart, while metal flakes won't.

A newly overhauled engine or one that has just had one or more cylinders replaced will often have a small amount of fine metal particles in the oil filter, but once the break-in has been completed and the break-in oil replaced, any appreciable amount of metal in the filter should be cause for concern.

How much metal is considered "appreciable"? There's no hard-and-fast rule. Lycoming Service Instruction 1492D states that less than 10 small pieces of metal in the filter is normal, and doesn't recommend grounding an engine for anything less than ¼ teaspoon of small metal particles or large metal chunks the size of a broken pencil point. My own view is that any substantial increase in metal above what has been the norm for your particular engine certainly deserves a closer look.

The first step in that "closer look" is to rinse the filter media in a clean jar or can using clean solvent to wash the particulate matter out of the pleats. Then slowly pour the now-dirty solvent through a large, clean coffee filter. This will allow you to examine the particles much more clearly.

Next, pass a strong magnet underneath the filter paper to determine whether the metallic particles are ferrous (steel) or non-ferrous (aluminum, chrome, tin, bronze, etc.) A small amount of non-ferrous metal is nothing to worry about; ferrous particles are of greater concern.

What to look for in the filter

Carbon particles. A certain amount of carbon in the filter is normal, and turbocharged engines generally exhibit more carbon than do normally aspirated ones. An unusually large amount of carbon in the filter suggests that oil is getting excessively hot and coking. This can be caused by several things. One is excessive blow-by past the rings, and is usually accompanied by elevated oil consumption and marginal compression readings in one or more cylinders. Another cause is one or more badly worn exhaust valve guides, usually accompanied by carbon build-up under the cylinder rocker covers, heat-damaged valve springs, and/or valves that move more than a very small amount in a "wobble test." Yet another cause of carbon occurs in turbocharged engines that are shut down without a reasonable turbocharger cool-down period.

Steel. Steel is readily identifiable because it is magnetic. Any significant quantity of steel particles or flakes in the filter is cause for concern. If the source of the metal is not readily apparent, you may want to consider sending your filter contents to an expert for microscopic examination, which often can pinpoint the source. (We typically use Aviation Laboratories in Houston or Second OilPinion in Tulsa.) Lycoming Service Instruction 1492D gives good guidance for how to react to various quantities of metal in the oil filter, and it's good guidance even if your engine is a Continental.

Aluminum. These are silver-colored non-magnetic particles that dissolve when exposed to a dilute solution of lye, the active ingredient in common household drain cleaners like Drano and Red Devil brands. Small amounts of aluminum are normal in some engines, but significant quantities warrant further investigation. In my experience, the most common cause of aluminum flakes is piston pin plug scuffing, which isn't usually an airworthiness issue and often self-resolves. (Look for aluminum smearing on a cylinder barrel using a borescope.) Other more serious possible sources of small aluminum particles include fretting crankcase halves (check torque on spine bolts and through-bolts), or a loose valve guide (check rocker boxes for metal). Larger aluminum chunks suggest damaged pistons, possibly caused by preignition (check with borescope).

Chrome. Chrome is shinier than aluminum, much harder, and is often found as flakes (rather than particles) that feel sharp to the touch. Any amount of chrome in the filter is not normal except possibly during break-in. The most common source is from chrome-plated piston rings abraded by a rough or pitted cylinder, or from chrome-plated cylinder barrels that are developing a problem (check with borescope). Another source is abnormal wear of chrome-plated exhaust valve stems.

Brass/copper/bronze. Identified by distinctive yellow color. In Continental engines, the presence of long bronze slivers often indicates failure of the starter adapter spring. Smaller particles may come from worn bushings, or older aluminum/bronze valve guides.

Dry particle analysis

If you find metal in the filter that prompts concern something might be coming apart inside your engine, it's often a good idea to seal the filter contents in a plastic bag and overnight it to a lab for microscopic dry particle analysis.

An expert can often tell from the size, shape and appearance of the particles or flakes, as seen under the microscope, whether they came from a spalled lifter, a damaged cam, an oil-starved main or rod journal, a defective gear, or a scored cylinder barrel. Both Continental and Lycoming operate metallurgy labs that will perform microscopic dry particle analysis of filter contents from their engines, as does Aviation Laboratories in Houston Texas. Another excellent resource is Howard Fenton's Second OilPinion service (7820 S. 70th East Ave. Tulsa, Oklahoma, phone 918-492-5844).

Spectrographic oil analysis

While oil filter inspection is usually the best way to determine if something is coming apart inside your engine, spectrographic oil analysis can be thought of as an early warning system capable of giving you advance notice of certain kinds of incipient problems, often long before they reach the safety-critical stage.

The key to understanding oil analysis is that it focuses on only the tiniest of wear metal particles that are so small that they can't be trapped by your oil filter or even seen readily under a microscope. We're talking here about particles that are 5 microns (about .0002 inches) or smaller in diameter. (A "micron" is one-millionth of a meter.)

Unlike the larger particles that are visible during oil filter inspection (typically .001 inches or larger in diameter), it is perfectly normal for an engine to continually shed micron-scale particles during normal operation as the unavoidable byproduct of wear. The quantity of such micron-scale particles produced during normal operation varies widely from one engine to another.

The purpose of spectrographic oil analysis is to establish a historical trend of wear particle production for a particular engine, and then to provide warning of any significant departures from that historical norm. Let's look at a specific example, an IO-520 engine in a Cessna 206 that was placed on an oil analysis program right after major overhaul.

Sample Date	Total Hours	Parts Per Million						
		Al	Cr	Fe	Ni	Cu	Sn	Si
2/1	10	23	28	115	6	13	5	8
5/5	55	12	22	62	4	9	3	8
8/27	103	9	19	48	4	8	2	10
11/15	150	8	20	45	4	7	2	9
2/9	197	9	18	43	3	7	2	12
4/30	245	13	25	92	4	9	2	27

Elevated silicon indicated dirt getting into the engine of this Cessna 206, and explained the other elevated wear metals.

At 10 hours, the original break-in oil was changed, and was relatively high in wear metals—particularly aluminum (Al), chromium (Cr) and iron (Fe)—just as you might expect. Some metal particles were also seen in the oil filter, but we know that this is also normal during the initial break-in process.

The engine oil was then changed approximately every 50 hours, the filter inspected, and an oil sample sent to the lab for analysis. No significant metal was found in the filter after the initial 10-hour break-in interval, and there were no other signs of engine problems.

The oil analysis results also looked perfectly normal during the first year. Wear metal levels came down quickly and stabilized within the first 100 hours, establishing a good baseline for comparison of future oil analysis results.

However, the sample taken on 4/30 at 245 hours total time revealed a big increase in iron (Fe) to 92 parts per million (compared to mid-40s during the preceding 150 hours). Aluminum (Al) and chromium (Cr) also increased, although not quite as dramatically.

What's going on here? In this case, the smoking gun can be found in the oil analysis results themselves: specifically, the three-fold increase in silicon (Si) from historical levels. When silicon is found in engine oil, it's usually caused by abrasive, silica-laden dirt getting into the engine. (High silicon can also be caused by silicone sealants or fragments of silicone gaskets getting into the oil, but typically that occurs only after major engine maintenance.)

Tipped off by the high silicon readings in the oil analysis, the owner of the 206 alerted his mechanic, who checked and found the inside of the engine's induction ducts coated with a gritty-feeling substance. The induction air filter appeared to be in good shape (although the mechanic replaced it anyway on general principles).

> When elevated silicon is found in oil analysis, it usually means that dirt is getting into the engine.

After a bit more investigation, the mechanic discovered a broken spring that prevented the alternate air door from sealing properly. This had been allowing the engine to breathe unfiltered air, which accounted for the grit in the induction system and the high silicon level in the oil. The abrasive grit in turn accounted for the higher-than-usual levels of wear metals.

The bad alternate air door spring was replaced, the airplane was returned to service, and the next oil sample 50 hours later showed wear metal and silicon levels returning to the historical norms.

Interpreting the reports

This example underscores several very important concepts about spectrographic oil analysis. First, the numbers seen in the report on the 4/30 sample at 245 hours were not inherently remarkable, and might very well have been perfectly normal for some other engine. The only thing that made them noteworthy in this case is that they were significantly higher than in the previous three or four oil samples taken from this engine. That's why you really can't tell anything from one or two oil analysis reports; it's a trend-monitoring tool, and you need to give it time to establish a trend before you can derive useful information from it.

Second, the accelerated wear rates caused by dirt ingestion through the faulty alternate air door were still quite modest. Had the engine not been on oil analysis and the air leak been missed at annual inspection (which could easily happen), the engine would undoubtedly have continued to run fine for hundreds of more hours before any major symptoms of the accelerated wear showed up (most likely as bad compression on one or more cylinders). The oil analysis provided an early warning of a problem that would typically have been found much later through other means.

This is rather typical of the kind of benefit you can expect from doing regular oil analysis. Finding the problem early undoubtedly saved the 206 owner a lot of money in the long run by possibly sparing him the expense of a mid-TBO top overhaul. It probably didn't do anything dramatic like save his life, because the dirt getting into his engine and causing accelerated wear was not likely to make his airplane fall out of the sky.

Occasionally, an oil analysis report will come back showing a huge increase in wear metals, enough to create concern of some sort of impending catastrophic failure. Usually, however, such reports will be accompanied by visible metal in the oil filter that would also provide warning that something serious is going on.

Usually, that is, but not always. There are plenty of cases where a major internal engine problem was missed during filter inspection but caught due to oil analysis. Even more common are cases where a major problem never showed up in oil analysis (because it was making only big chunks of metal) but was caught during filter inspection.

If you care about the health of your aircraft engine, you really need to do both… religiously.

What the numbers mean

The data shown on oil analysis reports vary somewhat from one lab to another, so you generally should pick one lab and stick with it. I use and recommend Blackstone Laboratories in Fort Wayne, Indiana (www.blackstone-labs.com). The lab was founded by an A&P mechanic and the staff really know aircraft engines.

Most of the information contained in the oil analysis reports by Blackstone and other labs shows the results of spectrographic elemental analysis of wear metals and other elements of interest. These include:

- **Aluminum:** High levels of aluminum generally come from abnormal wear on piston skirts, piston pin plugs, and fretting crankcase halves.
- **Chromium:** Normally from abnormal wear of chrome-plated piston rings. If chrome-plated cylinders are installed, may also indicate cylinder barrel wear.

Here's a Blackstone Labs oil analysis report on the left engine of my Cessna 310 after an unusually long (58-hour) oil-change interval. At the time of this report, the engine was 150 hours past TBO; it's still going strong more than 1,600 hours later.

- **Copper:** Typically from bronze bearing shells and bushings. Occasionally from the core of a deteriorating oil cooler.

- **Iron:** Iron is the principal wear element in most piston aircraft engines, since most of the major wear components are made of steel. The iron comes mainly from steel cylinder walls, but can also come from cam lobes, lifter faces, crankshaft and camshaft journals, and gears.

- **Nickel:** Used in high-temperature alloys, such as exhaust valve guides. Also found in nickel-based cylinder coatings.

- **Silicon:** Usually indicative of abrasive silica from dirt getting into the engine despite the induction air filter. Can also come from silicone sealants and gaskets or from glass beads (usually only after overhaul or major engine maintenance).

- **Tin:** Typically from the babbitt alloy layer used on the main and rod bearings. Bronze parts also contain tin alloyed with copper.

For all these elements (and several others of marginal interest), Blackstone includes not only sample results (measured in parts-per-million), but also what they call "universal averages" and "unit/location averages." The "universal averages" attempts to show normal values as derived from Blackstone's entire database of engines of a particular type (such as my Continental TSIO-520-Bs), while the "unit/location averages" are historical moving averages for my particular engines. While the universal averages are interesting, there is so much variation from one engine to the next (even if they're the same make and model) that I tend to pay much more attention to the unit/location averages.

In addition to this elemental analysis, Blackstone and various other labs also include data on:

- **Insolubles:** Shows the quantity of insoluble solids found in the oil, tested by centrifuging the oil sample. High levels of insolubles are usually an indication of poor oil filtration.

- **Viscosity:** Shows how much the rated oil viscosity has deteriorated during the oil-change interval. Very low viscosity can indicate fuel contamination of the oil, while higher-than-normal viscosity can indicate that the oil became overheated.

- **Water:** Shows the amount of moisture in the oil. High levels of water generally mean the aircraft isn't being flown enough, and suggests that the oil should be changed more frequently.

Getting consistent numbers

Oil analysis is a trend monitoring technology where you're watching for unexplained departures from a historical norm. Therefore, it's essential to make sure that the numbers you get from one oil sample to the next are comparable.

I've already mentioned one important element in this regard: Pick one lab and stick with it. Due to differences in equipment and procedures, the results you get from one lab will not necessarily be consistent with those from another lab. If you must change labs for some reason, wait until the new lab has analyzed three or more of your periodic oil samples before you start drawing any major conclusions from the data.

Another important point is to try to take the samples at a consistent time interval, like every 30 hours or every 50 hours. Oil that has been in service longer will naturally have higher wear metals. Realistically, of course, it's sometimes necessary to go a bit longer or shorter than usual between oil changes—and when that happens, simply keep this in mind when interpreting the analysis results.

Although less important, it's also a good idea to pick one kind of engine oil and stick with it. Changing from one kind of oil to another can sometimes cause odd fluctuations in the oil analysis results. This can be particularly true when changing from mineral-based oil (like Aeroshell W100) to semi-synthetic oil (like Aeroshell 15W-50), but can also be true in other situations.

For example, I once got back a report from Blackstone that showed a big jump in phosphorus levels from what I'd been used to seeing. That concerned me for a moment, until I realized what had caused it: My local oil jobber had run out of Aeroshell W100 oil, so I wound up using a few quarts of Aeroshell W100 Plus as make-up oil. The main difference between W100 and W100 Plus is that the latter has an antiwear additive called butylated triphenyl phosphate (bTPP), and that's what was causing the higher phosphorus numbers on my oil report.

Finally—and perhaps most important—you should be very consistent about how you take your oil samples that you send to the lab. For best results, samples should be taken immediately after coming in from a flight, when the oil is still hot and particulates have not had a chance to settle out.

There are two main methods used for obtaining oil samples: catching a sample as the oil is being drained from the engine, or aspirating a sample through the oil filler using a long tube and vacuum bulb. Both methods work fine, but pick one and stick with it. I personally prefer catching a sample while I'm draining the oil.

With both methods, it's important to avoid getting sludge in your oil sample. If you use the catch-while-draining method (as I do), try to catch your sample about midway through the oil draining process, avoiding the first or last oil that comes out of the pan. If you use the aspiration method, make sure the end of the pickup tube does not touch the bottom of the oil pan while you're drawing your sample.

Oil filter inspection and spectrographic oil analysis are both important tools for monitoring the health of your powerplant. Don't leave home without them.

32
Big Data

It's amazing what can be learned by analyzing data from a million and a half GA flights.

Pilots have long suspected that when it comes to headwinds and tailwinds, the deck is stacked against them. I think the late Bob Blodget, Senior Editor of FLYING Magazine, captured how most pilots feel about this when he wrote that "we all come to the conclusion that there are always more headwinds than tailwinds; and that the headwinds are always stronger than reported or forecast, and the tailwinds weaker."

Blodget wrote those words in 1968 and died in 1973. In his time, we could only speculate about such things. But in today's era of computerized avionics and big data, we can prove them.

My colleague Chris Wrather is a longtime friend, pilot, Bonanza owner, and A&P mechanic with a Ph.D. in operations research. Among other things, he oversees the SavvyAnalysis division of my company. Over the past four years, we've built a database of digital engine monitor data comprising 1.5 million flights flown by about 9,500 piston-powered general aviation airplanes.

Digital engine monitors typically capture data from myriad sensors between 10 and 60 times a minute. This means that for each of those million-plus flights, we've captured tens of thousands of data samples involving dozens of instrumented parameters. We're talking big data.

Chris has been doing interesting research using this big data. When I mentioned the old more-headwinds-than-tailwinds conjecture to him the other day, he decided to do some number crunching to see if he could prove or disprove it empirically.

The deck is stacked

A couple of days later, he shared with me the results of his study of about 3,500 flights of late-model Cessna 182S Skylanes powered by Lycoming IO-540 engines. Chris said he chose this particular "cohort" for this study for two reasons: First, the Skylane cruises at a modest 140 knots, so the effects of headwinds and tailwinds should be greater than for faster aircraft. Second, every 182S is equipped with Garmin G-1000 avionics that captures both true airspeed (TAS) and groundspeed (GS), precisely the two parameters he wanted to study.

Chris wrote software to analyze the data from these 3,500 Skylane flights, locate the longest stable cruise segment of each flight, calculate the average TAS and GS during those stable cruise segments, and then plot those values and calculate the median TAS and GS.

As the chart demonstrates, the median GS was 7.5 knots slower than the median TAS. This means that if you fly a 140-knot Skylane, the deck may be stacked against you by more than 5%. Ouch! (Compare that with the "house edge" in Las Vegas blackjack or video poker, both of which are less than 0.5%.)

Why it's stacked

In my 50+ years as a pilot and flight instructor, I'd always attributed this phenomenon to the obvious fact that we spend more time in headwinds than in tailwinds because the flights into headwinds are slower and take longer, while the ones with tailwinds are faster and finish more quickly. While that's true enough, it doesn't explain the results of Chris' Skylane study. His analysis looked only at the average TAS and GS

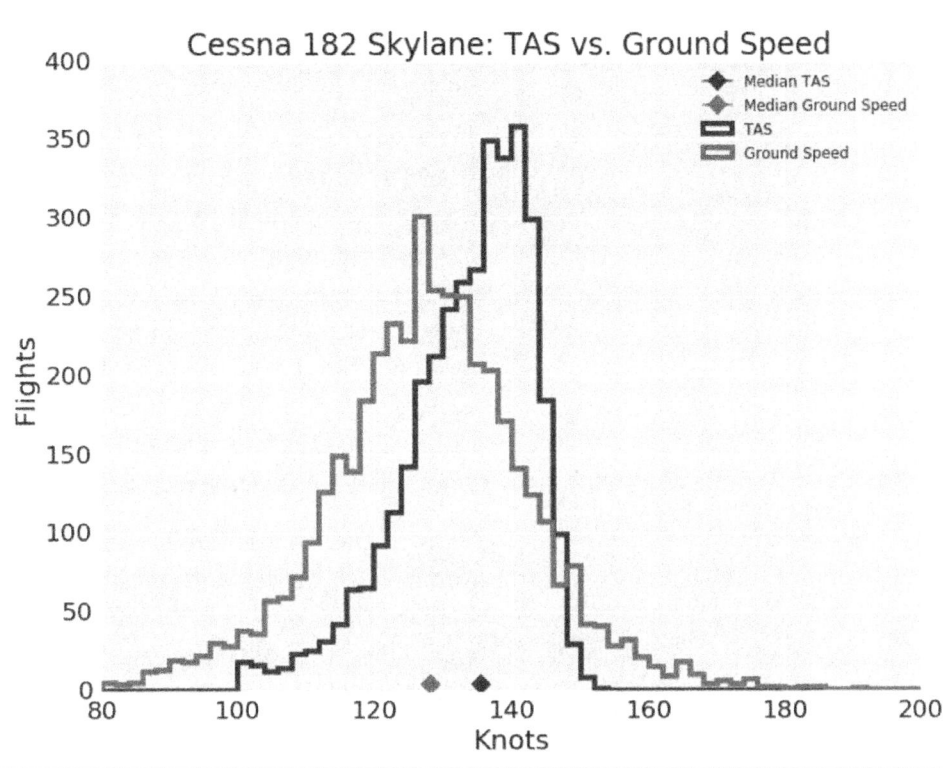

In a study of 3,500 Cessna 182S flights, median GS was 7.5 knots slower than median TAS.

during each of the 3,500 flights and ignored the length of those flights. (If he'd used time-weighted averages, the results would undoubtedly have been far worse.)

So, what exactly is going on here? Here's how Chris explained it to me…

Imagine you're cruising on a constant heading on a CAVU day. Visualize dividing the horizon into two 180° arcs, one extending 90° to either side of the airplane's nose and the other extending 90° to either side of the airplane's tail.

We might define wind blowing from a direction within the forward arc as a "headwind," wind blowing from a direction in the aft arc as a "tailwind," and wind blowing from exactly 90° to our track as a direct crosswind (neither a headwind or a tailwind). We might go further and postulate that if wind direction relative to our track is randomly distributed around the entire 360° horizon, we should expect the likelihood of having a headwind or tailwind to be a toss-up.

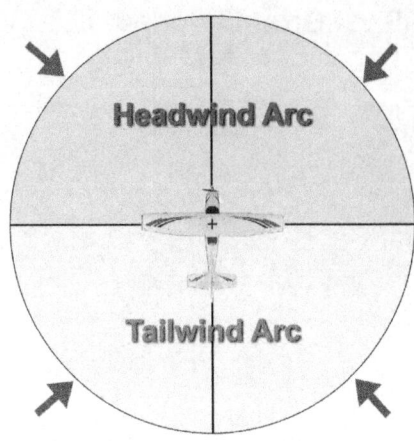

Imagine two arcs drawn around your aircraft defining headwinds and tailwinds.

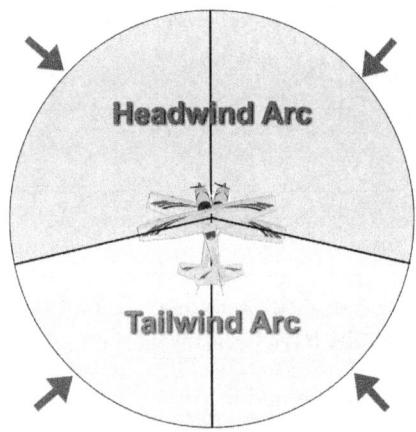

By crabbing into a crosswind we cause the headwind arc to be larger than the tailwind arc.

But this mental picture is simply wrong unless the wind is negligible. Here's why…

If the wind aloft is significant and blowing from 90° to our track, we must crab into the wind to maintain that track. Crabbing into the wind reduces our GS, resulting in a GS that's less than TAS, and transforming what we thought was a direct crosswind into a headwind. The wind direction would have to shift noticeably aft of the 90° point before it changes from a headwind to a tailwind and permits GS to exceed TAS.

Thus, if the wind is significant compared to our TAS, the forward "headwind arc" is more than 180° and the aft "tailwind arc" is less than 180°, causing headwinds to be more likely than tailwinds. The stronger the wind is, the more pronounced this effect is. This explains the 7.5-knot difference revealed by Chris' Skylane study, and confirms that what we pilots have long suspected about the deck being stacked against us is not an old wives' tale after all.

CHT studies

Our "big data" repository has been used to do lots of other interesting studies, too. Some have focused on what I consider to be one of the most important factors affecting engine reliability and longevity: cylinder head temperature (CHT).

For example, I've long noticed that Lycoming engines seem to run hotter CHTs than Continentals, and I wanted to see if we could verify and quantify that. Chris crunched the numbers from several hundred thousand flights and came up with some interesting results.

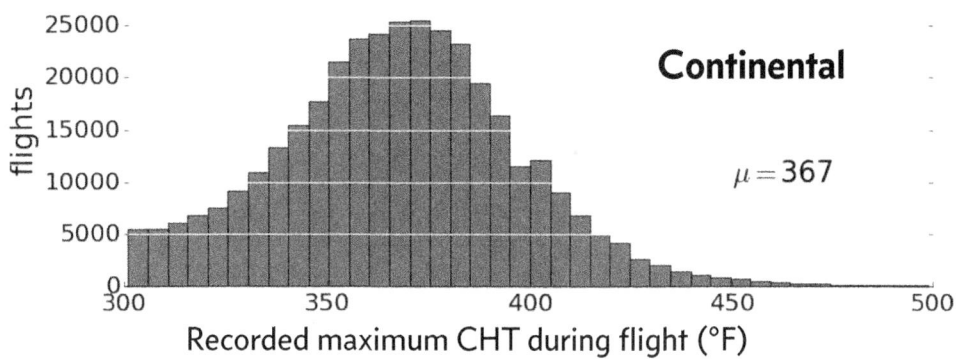

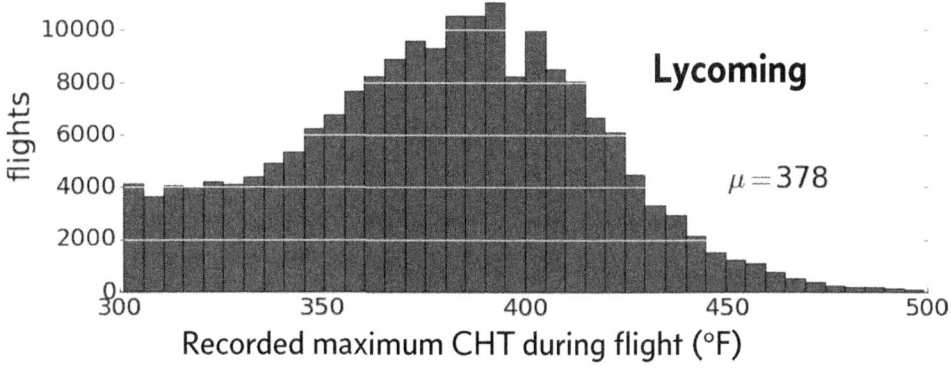

Distribution of maximum CHT during flight (Lycomings vs. Continentals).

These charts demonstrated clearly that Lycoming CHTs do indeed run a good 20°F hotter than Continental CHTs. There are some good engineering reasons for this, notably the fact that Lycomings use sodium-filled exhaust valves that transfer heat from the valves to the cylinder heads more efficiently than do Continental valves. Fortunately, Lycoming cylinders employ a more robust head-to-barrel junction design that can handle higher temperatures, which is why Lycoming sets its CHT red-line at 500°F compared to Continental's more conservative 460°F.

One interesting aspect of this study which had me puzzled was the odd-looking double peak on the Lycoming CHT histogram. Drilling down further into the data revealed that the Lycoming IO-360-series was responsible for the double peak (we have a lot of RV flights in our database). It turns out that Lycoming originally spec'd the IO-360 ignition timing at a rather aggressive 25° BTDC (resulting in hot CHTs), and then later changed its mind and retarded the timing to a more conservative 20° BTDC (resulting in much cooler CHTs). It appears that there are now a mix of IO-360s in the field—some timed to 25° and some to 20°—accounting for the double peak.

Another thing I've noticed is that piston airplanes designed in the past 20 years or so (e.g., Cirruses, Columbias, Diamonds) have much more efficient cooling systems and much more even CHT distribution than legacy aircraft (e.g., Bonanzas, 210s, Saratogas). To find out just how much more efficient, we did a comparative study of normally-aspirated Cirrus SR22s and Cessna 210s, looking at the spread between the hottest and coolest CHT. The results speak for themselves.

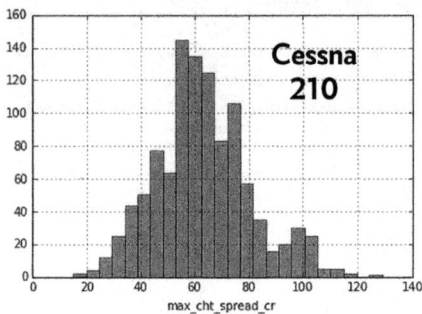

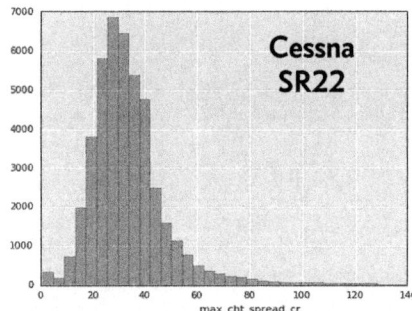

CHT spread, Cessna 210 vs. Cirrus SR22.

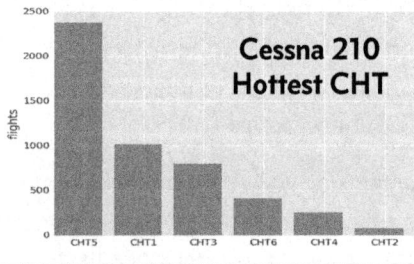

Cylinder with the hottest CHT, Cessna 210.

The modern cooling system design of the SR22 provides much tighter CHT spreads than the 60-year-old design of the cooling system in the Cessna 210. Drilling down further, it's clear that the 210 cooling system's problem is systemic, not random. This chart makes it clear that the left (even-numbered) bank of cylinders have much better cooling than the right (odd-numbered) bank, and the right-front cylinder (#5) is almost always the hottest-running problem child. That suggests a fundamental design issue with the 210's cooling system. Conventional wisdom holds that the problem is related to #5 being right behind the engine's right-front-mounted oil cooler, making it difficult to obtain adequate cooling air. More recent engine designs from Continental—like the one used by the Cirrus SR22—relocate the oil cooler to the back of the engine behind cylinder #2.

It's fascinating and fun to draw general conclusions across aircraft types from big data. More important, however, is how big data analysis is providing aircraft owners with specific actionable intelligence about their individual airplanes, helping them uncover problems with both hardware and operating technique.

Report cards

We generate regular "report cards" to SavvyAnalysis subscribers who have uploaded enough engine monitor data to support statistically significant findings. These are emailed to the owner on a regular basis and analyze various critical flight parameters related to aircraft performance, efficiency, and engine longevity. The report cards also compare those parameters for the particular aircraft with all the other aircraft we follow of the same type. Below is an example of what such a report card looks like. This one is for a normally-aspirated Cirrus SR22. It covers 27 flights that were uploaded to the SavvyAnalysis platform in the course of one year and compares them with the corresponding parameters for 46,886 flights in the database for the "cohort" of 741 normally-aspirated Cirrus SR22s that we follow.

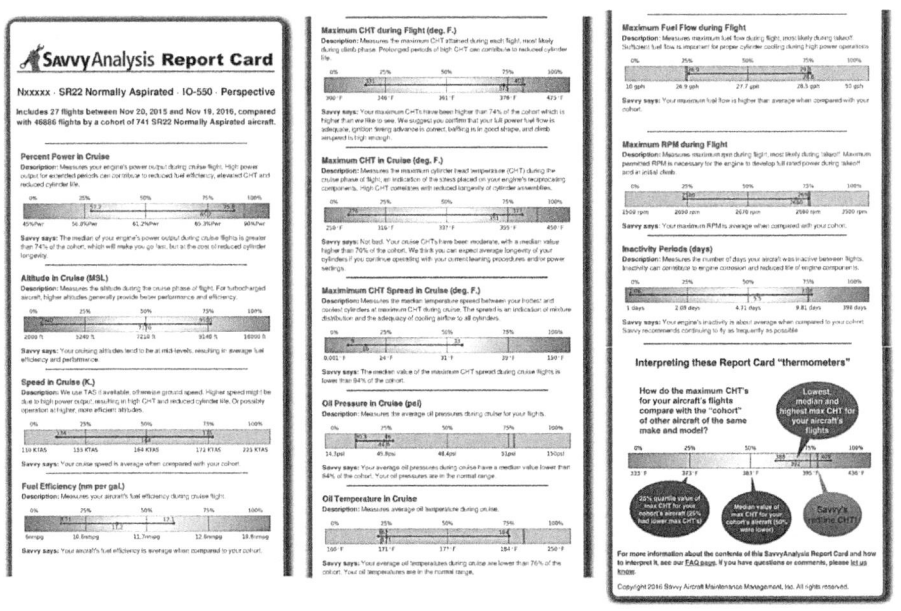

Report card for a Cirrus SR22.

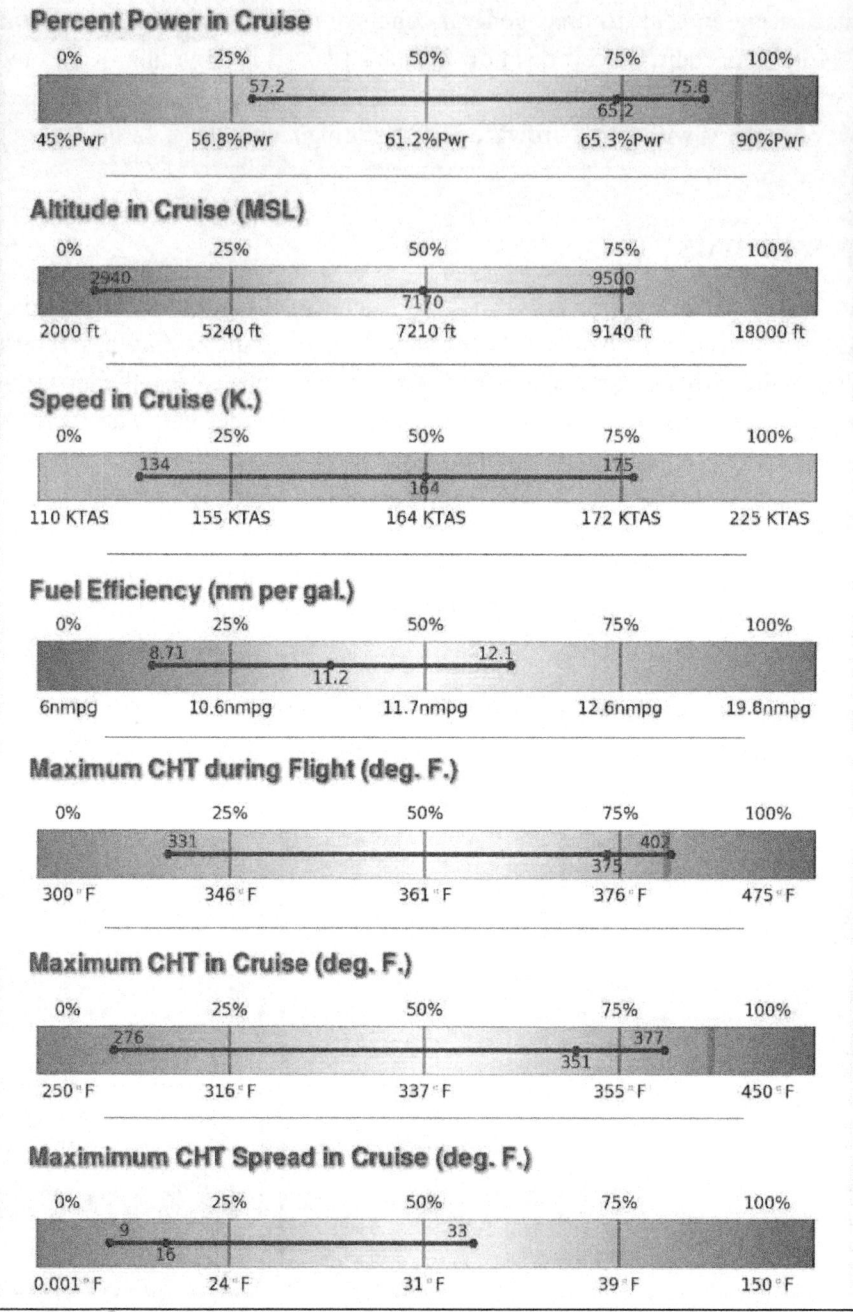

SR22 report card thermometers.

For each of the dozen analyzed parameters, the report card presents a "thermometer" that graphically shows the minimum, maximum, and median value of the parameter for the particular airplane, and how those compare with the distribution of median values observed in the entire cohort. Each thermometer is followed by textual commentary that advises whether we think the values for the subject aircraft are normal or abnormal, and (if abnormal) offering suggestions for corrective action.

Let's examine some of the thermometers from this Cirrus's report card. This aircraft cruised at a median 65.2% power, higher than 70% of the cohort, so I'm guessing the pilot wanted to fly fast. Interestingly, however, the aircraft's median cruise speed of 164 KTAS was right in the middle (50th percentile) of the cohort, and the median cruise altitude of roughly 7,000' MSL was also right in the middle. The median fuel efficiency of 11.2 nm/gallon wasn't exactly stellar, falling in the 38th percentile. The aircraft's CHTs both in climb and cruise are higher than normal (75th and 70th percentiles, respectively). These elevated CHTs don't seem to be due to any flaw of the aircraft's cooling system, because the CHT spread among the six cylinders is exceptionally low (20th percentile).

So this owner is cruising at higher power than most Cirrus SR22 pilots, getting mediocre fuel economy, running high CHTs, and yet not getting any extra airspeed to show for it. Maybe he should have his aircraft rigging checked and work on his engine operating technique (as in "LOP"). This is what I mean by "actionable intelligence."

Alerts

SavvyAnalysis emails report cards to owners on a regular basis, the frequency depending on how active the aircraft is and how much data the owner uploads. It also allows owners to generate report cards on-demand for any time period. But sometimes the software sees something in the data that is urgent and just can't wait. That's when the system emails an "alert" to the owner. Here's an actual (de-identified) one that was sent to the owner of a Mooney M20J alerting him that his CHT and oil temperature were way too high (96th and 97th percentile of the cohort), and suggesting he take corrective action right away.

SavvyAnalysis ALERT

Nxxxxx · M20 (F/J) · IO-360 · EDM-830

Based on 35 flights between May 16, 2016 and Nov 12, 2016, compared with 5066 flights by a cohort of 142 M20 (F/J) aircraft.

We are alerting you to conditions that we believe warrant attention.

Maximum CHT during Flight (deg. F.)

Description: Measures the maximum CHT attained during each flight, most likely during climb phase. Prolonged periods of high CHT can contribute to reduced cylinder life.

Savvy says: ALERT: Your maximum CHTs have been higher than 96% of the cohort. We think this is too high. Savvy suggests you confirm that your full power fuel flow is adequate, ignition timing advance is correct, baffling is in good shape, and climb airspeed is high enough.

Maximum CHT in Cruise (deg. F.)

Description: Measures the maximum cylinder head temperature (CHT) during the cruise phase of flight, an indication of the stress placed on your engine's reciprocating components. High CHT correlates with reduced longevity of cylinder assemblies.

Savvy says: ALERT: Your cruise CHTs have been higher than 96% of the cohort which does not bode well for the longevity of your cylinders. We suggest that you adjust your leaning procedures and/or power settings to reduce your CHTs in cruise. ALERT: We are concerned because your median cruise CHT was over 400°F, and we think that's way too high.

Oil Temperature in Cruise

Description: Measures average oil temperature during cruise.

Savvy says: The median of your average oil temperatures during cruise are higher than 97% of the cohort. ALERT: Your oil temperatures are on the high side. We recommend keeping an eye on it.

The data uploaded by this Mooney M20J generated an automated alert for high CHT and high oil temperature.

Trend analysis

As the SavvyAnalysis database has grown we've expanded our analysis to include a capability to identify trends in the data for a specific aircraft. Is oil pressure decreasing over time? Is CHT increasing? How about cruise speed and fuel economy? Are there statistically significant trends in any of these parameters?

Savvy's Trend Analysis charts build on the information contained in the Report Card. Rather than summarizing an aircraft's performance measures as minimum, median and maximum values compared to a cohort population, the Trend Analysis charts plot data points for each flight of a specific aircraft along a timeline so that trends can be visualized.

The timeline is shown on the horizontal axis at the bottom of the chart. The value of the parameter for each flight is plotted as a diamond along the timeline. The vertical axis on the left side of the chart represents the values of the parameter—in this case maximum CHT during cruise—while the vertical axis on the right side of the chart shows the 25th, 50th, and 75th percentile values of the parameter for the cohort. A heavy horizontal line represents the median value of the parameter for the aircraft.

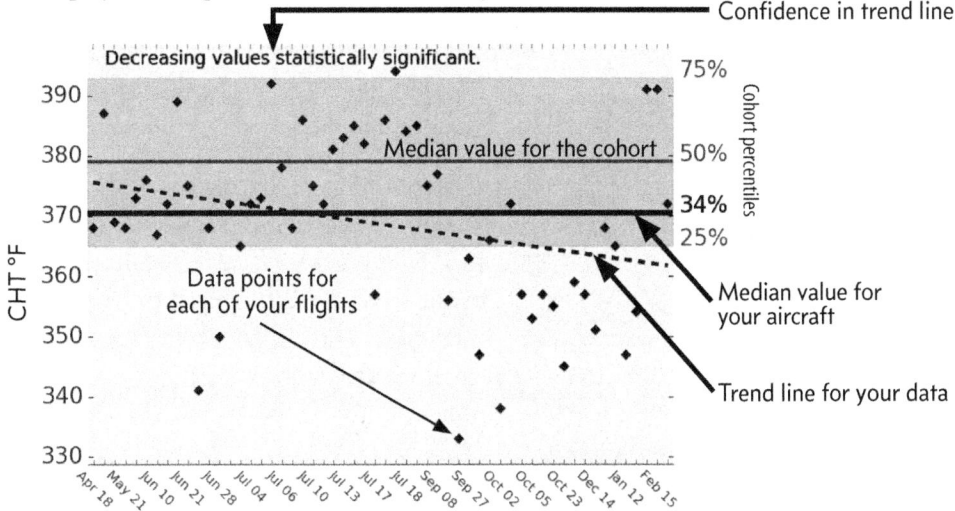

Example trend analysis report for maximum CHT in cruise

Savvy's software performs a mathematical test on an aircraft's individual data points, testing whether an increasing or decreasing trend in the data over time is statistically significant—that is, whether it's likely due to real underlying changes in the data or just the result of random variations. If the test shows a statistically significant increasing or decreasing trend, a dotted line is drawn to show that trend. If the trend is deemed not statistically significant, no dotted line appears on the chart.

Exhaust valve failure warnings

Many years ago, I discovered an interesting phenomenon: When exhaust valves start to burn and leak in piston aircraft engines, they often produce an oscillation in the cylinder's EGT that can be used to detect incipient exhaust valve failure. The characteristic signature is subtle: a slow rhythmic EGT oscillation with a period of roughly one cycle per minute and an amplitude of 30 to 50 degrees Fahrenheit (just 2% to 3% of a typical 1500°F EGT).

Our SavvyAnalysis platform scans all uploaded data for such signatures, enabling us to alert aircraft owners when one is detected. We call this FEVA ("Failing Exhaust Valve Analytics"). Shown on the next page is a case involving a Cirrus SR22.

Those EGT traces are a mess, very noisy and hard for a human analyst to make much sense of them. But if we split the EGT for cylinder #5 from the pack (opposite page, bottom figure), its obvious that there's something different about that cylinder.

Physical evidence that this exhaust valve was burning and well on its way to failure.

Cylinder #5 is exhibiting precisely the slow rhythmic oscillation that's characteristic of a failing exhaust valve, while the other cylinders aren't. We alerted the owner, and he subsequently emailed us a photo of his exhaust valve (shown left).

Ouch! Clearly if that valve had remained in service much longer, it would have failed catastrophically in flight.

Another example of a FEVA alert from a Cessna 340 is shown on page 309. Here

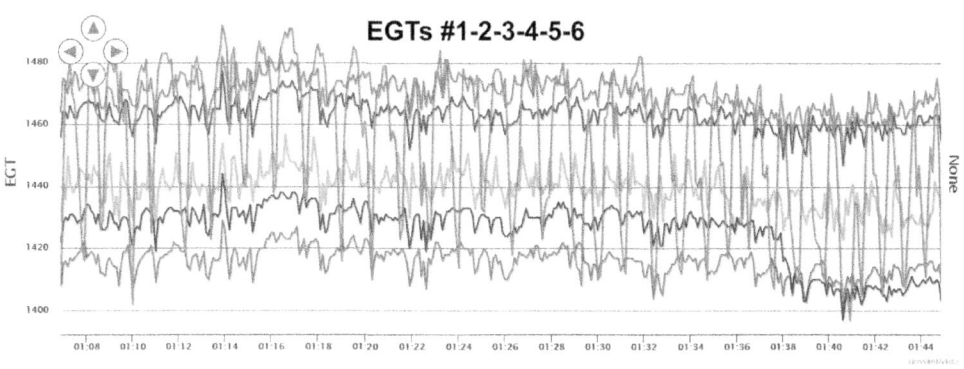

EGT traces for a Continental IO-550-N engine in a Cirrus SR22. What a mess!

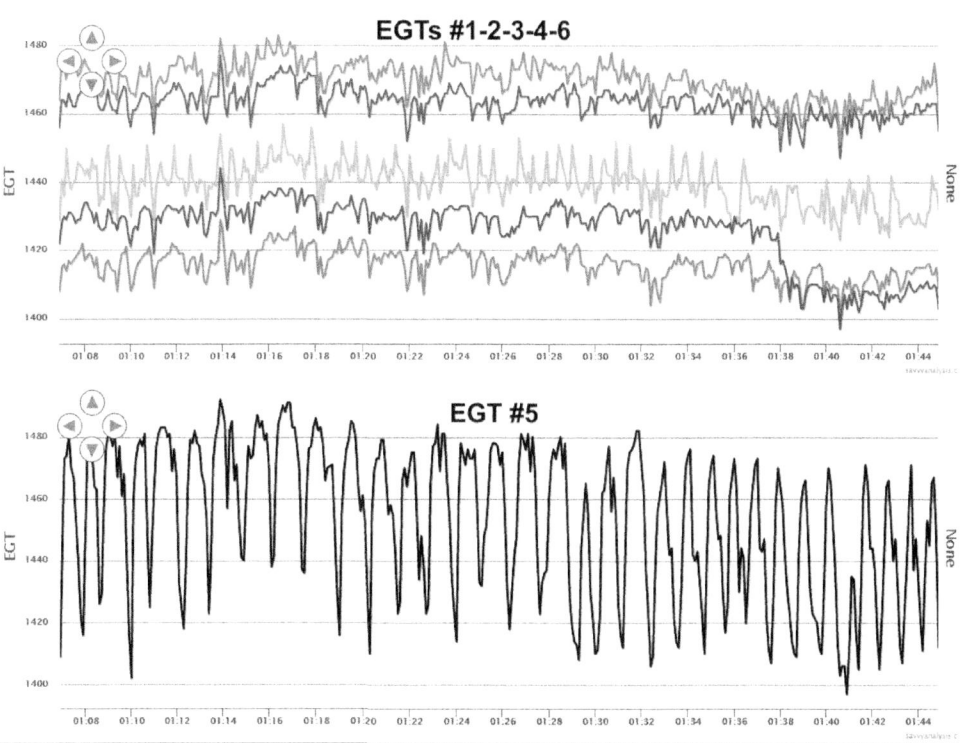

When EGT #5 is isolated from the others, it becomes obvious that cylinder #5 is exhibiting the slow rhythmic EGT oscillation characteristic of a failing exhaust valve.

This exhaust valve burned and eventually failed; too bad the owner didn't upload his engine monitor data more frequently!

the slow rhythmic oscillation of EGT #3 (topmost trace) is quite obvious and is also reflected in a more subtle oscillation of TIT (second trace from the top), while the other five cylinders look normal.

Pictured left is the photo the owner took when cylinder #3 was pulled. Unfortunately, this Cessna 340 owner wasn't in the habit of uploading his data on a regular basis. He didn't upload this data until after his engine started running rough, so didn't receive a FEVA alert in time to head off the valve failure. The moral of this story is obvious.

Big data doesn't help if it's not analyzed

I estimate at the time of this writing that roughly 50% of the piston GA fleet is equipped with recording digital engine monitors capable of capturing real-time engine and flight data for analysis. I'm convinced the potential benefits of collecting and studying this data are enormous.

Sadly, however, most captured data is never extracted for analysis, and winds up in the proverbial bit bucket, forever lost. It amazes me how many aircraft owners have fancy data monitoring equipment but have never retrieved the recorded data and don't even know how. One of my challenges is to convince aircraft owners to dump their captured data for analysis on a regular basis, much as they have oil filters inspected and oil samples sent to the lab. My hope is that the next generation of avionics will be able to "phone home" autonomously, disgorging captured data over the Internet without manual intervention.

Analysis of big data isn't limited to studying groups of aircraft. By comparing data obtained from one aircraft against data obtained from a "cohort" of similar aircraft, it's possible to provide vital and actionable information to aircraft owners. Software can automatically detect trends, exceedances, and outliers, and identify mechanical and operational problems that might otherwise go undetected.

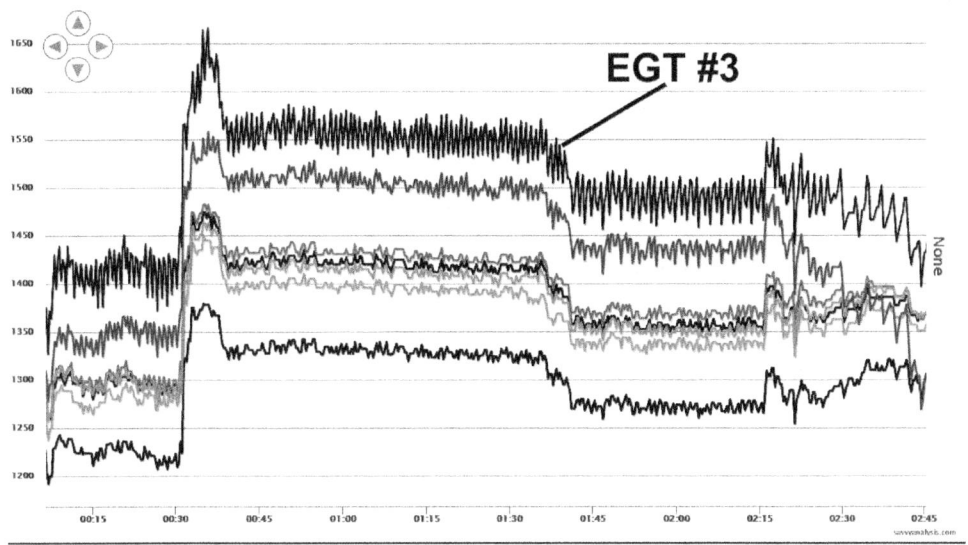

FEVA signature (topmost trace) of a failing #3 exhaust valve on a Cessna 340.

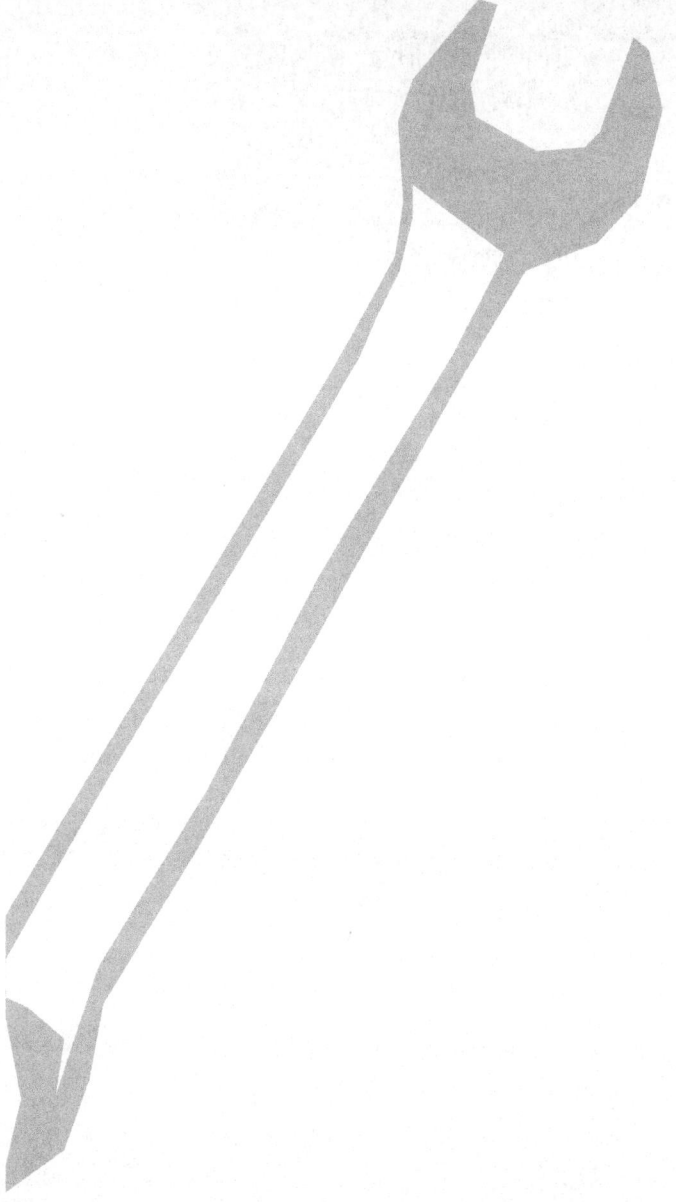

PART VII
Troubleshooting

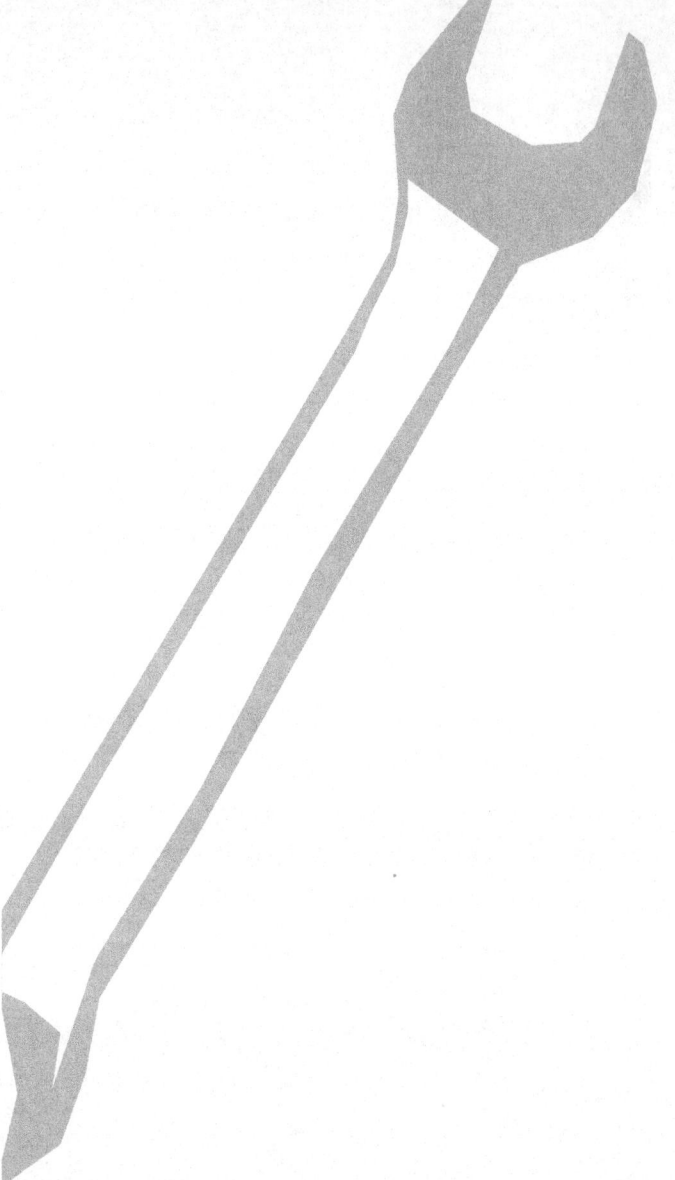

33
The Art of Troubleshooting

Fixing is usually the easiest part of aircraft maintenance. Figuring out what's wrong is often a lot harder.

A funny thing happened on my way to Milwaukee… I was flying my trusty 1979 Cessna Turbo 310 to speak at the annual national convention of the Flying Physicians Association in Milwaukee. My talk to the flying docs was to be about the art of troubleshooting. Little did I know that my troubleshooting skills were about to be put to the test.

After an uneventful overnight in Denver, I taxied out for the final leg to Milwaukee. I was cleared for takeoff, released the brakes, smoothly advanced the throttles for my takeoff roll on runway 17.

Uh oh! Something felt wrong. It was taking way too much right rudder to hold the centerline. I scanned the engine gauges and noted a big split in the manifold pressure needles. The right engine was showing 32 inches as it should, but the left engine was only 24 inches. Not good. I retarded both throttles to idle, advised the tower I was rejecting the takeoff, and pulled off the runway at the next intersection.

The right engine manifold pressure was showing 32 inches as it should, but the left engine was only 24 inches.

I taxied back to the ramp and tried a full-power runup with the brakes set. The same thing happened: 32 inches on the right engine, 24 inches on the left. A moment of reflection revealed that 24 inches is roughly what one would expect from a normally-aspirated engine on takeoff from a field elevation of 5,500 feet.

Hmmm… It appeared that the left engine was operating normally-aspirated. Its turbocharging system had stopped working, seemingly overnight. Why? I had no clue.

What the @#$% could be wrong?

As I taxied back to the ramp in search of a shop where I could borrow some tools, I started making a mental list of all the things I could think of that could produce these symptoms. (Making such a list should always be the first step in troubleshooting.) Visualizing how the turbosystem works (see Chapter 22) and considering its various failure modes, I came up with six possibilities:

1. Failed turbocharger
2. Stuck-open wastegate
3. Failed controller
4. Big induction system leak
5. Big exhaust system leak
6. Big oil leak in controller/wastegate system

I taxied past the open door of a hangar inside which I saw two guys in work suits wrenching on airplanes, each with a big red Craftsman tool cabinet. I shut down the engines, climbed out of the airplane, introduced myself, and explained my predicament. They seemed happy to help.

My first step was to remove the top cowling from the left engine and look around for something obviously wrong. I saw no signs of oil or exhaust stains where they didn't belong, so I crossed items #5 and #6 off my suspect list. I went over the induction system with a flashlight and mirror and couldn't see anything untoward, so I crossed off item #4, too.

I popped open the cover of the induction air filter canister, removed the filter element, and inspected the turbocharger's compressor. It looked pristine, with no hint of foreign object damage. I reached in and spun the rotor with my fingers and wiggled it to check its radial and axial play. The rotor turned freely and felt normal. I crossed item #1 off my list.

My first step was to uncowl the engine and look around for something obviously wrong.

Isolating the fault

Now I'd whittled my list of six suspects down to two likely culprits: the wastegate and the controller. I spent a few minutes thinking about how I could figure out which one was the bad guy, and came up with a plan: By removing the oil return line from the controller to the engine and capping it off, I could effectively disable the controller and force maximum oil pressure to the wastegate actuator. If the wastegate was working, then this should result in a fully closed wastegate and maximum turbo boost. If red-line manifold pressure still wasn't available, that would prove that the wastegate was bad.

We rummaged around the toolboxes until we found some suitable AN fittings to cap off the oil line. I reinstalled the top cowling, climbed into the cockpit, and started the left engine. When I advanced the throttle, manifold pressure climbed up to the red-line at 32 inches. I grinned, then shut the engine down and climbed back out of the cockpit.

I'd just proven that the wastegate was fine. In fact, I'd proved that all the turbosystem components were fine except for the controller (which I'd temporarily disabled). It was now the only remaining item on my list that hadn't been crossed off. By the process of elimination, it must be the culprit.

Fix-it time

This was both good news and bad news. The good news was that I was now confident I knew what was wrong: the controller was malfunctioning. The bad news was that it would take at least 48 hours to get a replacement controller overnighted to me in Denver, and by then it would be too late to make my speaking engagement in Milwaukee. It was looking like I might be forced to abandon my trusty Cessna in Denver and (gasp!) catch an airline flight to Milwaukee, then return to Denver on the airlines and replace the controller then.

Ugh! I really didn't want to do that…

To avoid this unspeakable fate, I started giving serious consideration to removing the controller from the airplane and disassembling it, hoping I could figure out what was wrong with it and maybe even coax it into working. This is not something an A&P would normally do as it's considered a highly specialized procedure. But I figured since the controller was already broken, I had little to lose by taking it apart.

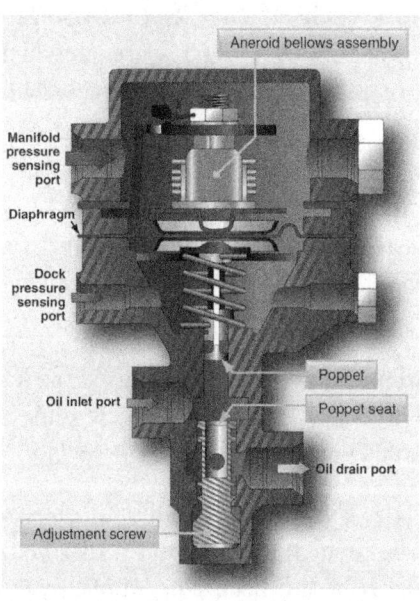

If the turbocontroller poppet valve couldn't close, the controller wouldn't work.

Then I had another thought: What if the only problem with the controller was that it had become contaminated with some sort of debris that got stuck in its poppet valve and prevented the valve from closing? If the valve couldn't close, the controller couldn't work. That would account for the symptoms I was seeing. I wondered whether there might be a way of cleaning any debris from the poppet valve without taking the controller apart.

I disconnected both oil lines from the controller and asked my new mechanic friends whether I could borrow their air compressor for a few minutes. Using a rubber-tipped air nozzle, I hit the controller's oil output port with several shots of 80 PSI air while holding a shop rag over the oil inlet port to catch any expelled oil. My idea was to "backflush" the poppet valve with some air blasts and hopefully dislodge any debris that might be stuck. Sure enough, some flakes of what looked to be carbonized oil wound up in the rag.

I reconnected the oil lines to the controller, reinstalled the top cowling, climbed into the cockpit, crossed my fingers and toes, and started the left engine. After letting it warm up for a few minutes, I slowly advanced the throttle while watching the manifold pressure gauge. The needle advanced smoothly up to…YESSSS!!!…32 inches.

I said my thanks and goodbyes to my mechanic friends, re-filed my IFR flight plan from my iPhone, and taxied out for takeoff. Everything worked as advertised, and I arrived at Milwaukee only two hours behind my original schedule.

Differential diagnosis

What I just described to you is a textbook example of what doctors call "differential diagnosis." To quote Wikipedia:

> Differential diagnosis (abbreviated DDx) is a systematic method of distinguishing a disease or condition from others presenting similar symptoms. This method is essentially a process of elimination that excludes candidate conditions until a single probable diagnosis remains.

Every physician receives extensive training on this DDx technique in medical school. It typically involves five steps:

1. Gather information about symptoms.

2. List candidate conditions consistent with these symptoms.

3. Prioritize the list of candidate conditions.

4. Rule out candidate conditions (through testing or therapy) until a definitive diagnosis has been established through the process of elimination.

5. Verify that the surviving diagnosis is correct.

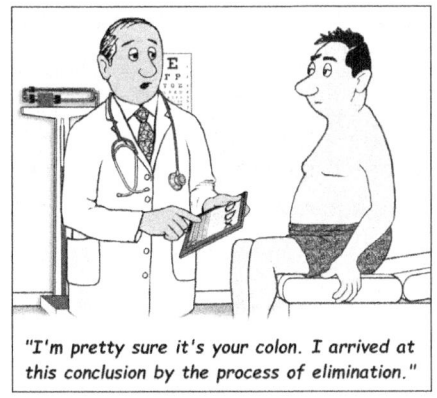

"I'm pretty sure it's your colon. I arrived at this conclusion by the process of elimination."

Differential diagnosis involves making a list of what could be wrong and then narrowing that list through the process of elimination.

The application of the DDx technique is hardly limited to medicine. It is, in fact, the way all troubleshooting of complex systems should be done, whether dealing with human bodies, household plumbing, cars, boats, or... airplanes.

Alternative approaches

Unfortunately, A&Ps tend not to be nearly as well-trained in DDx as MDs are. This is an area of aircraft mechanic training that is greatly in need of improvement. A&Ps could learn a lot about troubleshooting from their doctors. In lieu of performing proper DDx, I see a lot of mechanics using alternate techniques.

One common troubleshooting technique that occurs way too often for my taste is "shotgunning." This is when a mechanic—instead of methodically analyzing and eliminating possible failure modes—simply replaces components on a trial-and-error basis, hoping to get lucky. If the mechanic guesses right the first time, he comes out looking like a hero. If he doesn't, the aircraft owner often winds up passing out from sticker shock. The mechanic's initial guess might be (1) whatever component turned out to be the culprit last time he saw a similar problem, (2) whatever component he happens to have on hand, (3) whatever component is easiest to replace, or (4) whatever component is most expensive. Such shotgunning often makes good sense in airline and commercial fleet operations where downtime must be minimized at any cost, and replacement parts are readily available off-the-shelf. In the case of owner-flown GA aircraft, however, a little extra downtime is often more tolerable than a sky-high repair bill. Whatever the exact algorithm, shotgunning is based on guesswork, not analysis, and that's not a good thing.

A variant of shotgunning is "educated guesswork" where the technician relies on experience to make an educated guess about what might be wrong. "I've seen these spurious overvoltage trips many times before, and usually they're due to a bad regulator—so I replaced your regulator and I'm guessing it'll solve the problem." If the mechanic guessed right, the owner is happy. If not, the airplane returns to the shop for another guess.

Educated guesswork is okay if the parts replaced are cheap and easy to change. But in aviation, "cheap part" is an oxymoron, and even a $15 sparkplug can cost $100 in labor to replace by the time you include uncowling and re-cowling the engine. Personally, before I spend a couple of hundred bucks to replace a part on my airplane, I'd like to be pretty darn sure that the part actually needs replacing. Maybe that's just me.

Yet another approach that is even worse is "overkill" where a mechanic engages is inappropriately expensive or invasive maintenance in order to guarantee that the problem is resolved. I can't count how many times I've seen A&Ps recommend doing a $10,000+ "top overhaul" just because an engine has marginal compression in one cylinder. Or even worse, recommended a $40,000+ major overhaul just because a few flecks of metal were found in the oil filter. Such overkill by A&Ps strikes me as serious malpractice. I've even seen a few cases of owners who sued their mechanics over this sort of thing and prevailed, but most owners just wind up taking it in the wallet.

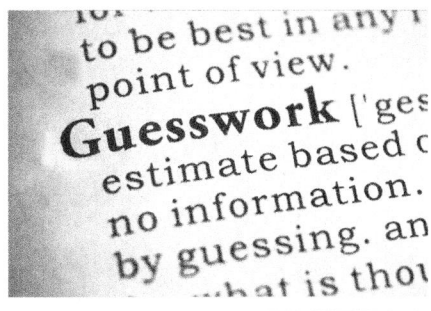

Try never to put a mechanic in a position where he has to resort to guesswork.

The diagnostic approach

Consequently, I've become something of a nut about diagnostic troubleshooting whose mantra is "don't try to fix a problem until you're pretty sure you know its cause." An obvious corollary is "never replace a costly part until you've proven the part is faulty." Under this philosophy, the maintenance process is divided into two distinct phases—diagnosis and therapy—performed strictly in that order.

The tools used in the diagnostic phase are not screwdrivers, pliers and wrenches, but rather service and parts manuals, schematic diagrams, test instruments, and brains... especially brains. Good diagnosis consists overwhelmingly of headwork, not handwork. It's cerebral, not physical.

In a perfect world, doctors and mechanics would be both brilliant diagnosticians and skilled therapists. In real life, this is rarely the case. The best person to diagnose a problem with your aircraft isn't always the same person who fixes the problem. The medical profession recognizes that diagnosis and therapy (or troubleshooting and repair, if you prefer) are dramatically different activities that require dramatically different skill sets. We don't expect our neurosurgeons to interpret MRIs or analyze tissue samples or evaluate blood labs—we rely on radiologists, pathologists and hematologists for those things. Perhaps it's time to consider a new aviation maintenance specialty, "A&P diagnostician."

Perhaps 'diagnostician' should be recognized as an aviation maintenance specialty.

Good problem diagnosis is a blend of science and art, logic and judgment, deduction and intuition. To some extent, troubleshooting skills are universal—if you're skilled at troubleshooting aircraft problems, you can probably do a decent job troubleshooting problems with clocks, lawnmowers, computers or plumbing. The best troubleshooters are the ones who have a deep understanding of the systems they're troubleshooting, but you can usually compensate for lack of experience through diligent study—if you don't know the answer, look it up or ask somebody who does.

I attribute my skill as a troubleshooter largely to my training as a mathematician and my 30-year career as a professional software developer, both fields that deal with complex abstraction and absolutely demand strong rational and objective thinking. It's rare to find career A&P mechanics with these sorts of backgrounds.

Why is my CHT so hot?

While I was speaking to the Flying Physicians in Milwaukee about troubleshooting, half a world away a client in Tzfat, Israel was in a panic. He and his partner had just spent nearly $50,000 to replace the IO-550-N engine in their 2003 Cirrus SR22 with a Continental factory rebuilt engine. On the first few flights after the engine change, the CHTs were quite a bit higher than what the owners were accustomed to seeing—especially cylinders #2 and #6—but they understood that higher-than-normal CHTs are normal during the initial break-in period.

At four hours SMOH, most of the CHTs (including #2) seemed to be coming down, but #6 seemed to be getting worse. Then on a takeoff from a small airport 100 miles away from home base, CHT #6 rose smoothly to an alarming 470°F, while the other five CHTs remained in the 360°F to 380°F range. The pilot immediately reduced power and returned to the airport. By the time the airplane touched down, CHT #6 had come back down to about the same as the other cylinders.

The pilot attempted another takeoff, but CHT #6 started to run away again, so he aborted the takeoff and stopped on the remaining runway. According to the pilot, the #6 EGT remained normal throughout, about the same as the EGTs for the other five cylinders.

Based on the pilot's verbal report, I suspected that the high CHT #6 was some sort of indication problem, because a "real" cylinder problem would have been confirmed by an anomalous EGT. But experience has made me skeptical of such eyewitness accounts, so I told the owner I'd feel a lot more comfortable giving him advice if I could see the actual engine data. The owner arranged to dump the data from his Avidyne MFD and upload it to our SavvyAnalysis platform, where I reviewed it in my Milwaukee hotel room.

My review of the data immediately confirmed my suspicions that the #6 CHT indication was bogus. At one point, its value changed more than 40ºF between two adjacent data samples spaced six seconds apart. A 40ºF change in CHT in six seconds is simply not possible…it defies the laws of physics. I've studied literally thousands of flights, and the most rapid CHT change I've ever seen—during a destructive pre-ignition event that wound up melting the piston—was about 1.5ºF per second. This one changed 7ºF per second, and that's simply not plausible. Clearly, the CHT #6 indication was lying, due either to a faulty probe or a bad connection.

The CHT probes in Cirrus aircraft are notorious for failing and providing erroneous CHT data.

I asked whether it was possible to have the #6 and #4 CHT probes swapped to verify that the anomalous CHT indication remained with the bad probe. Unfortunately, there were no services at this airport, and no mechanic available to swap the probes. After reviewing the engine monitor data, I was sufficiently confident in my remote diagnosis that I recommended that the owner fly the airplane home and have the CHT indication problem debugged there. The flight was uneventful, and the new engine broke in just fine.

Identify, verify, reproduce

The first step in troubleshooting any problem is to identify the problem and verify that what you think is wrong is actually wrong. This may seem so obvious as to be hardly worth mentioning, but it's amazing how often this essential step is omitted.

This Israeli SR22 is a perfect example. The owner knew something appeared wrong, and suspected dire situations such as a cylinder that wasn't breaking in properly. In the end, it turned out the CHT indication was faulty and the cylinder was perfectly normal. Had the owner not performed due diligence to investigate the problem, he might have been convinced to remove the "defective" cylinder to "fix" the problem.

One good way to help avoid situations like this is to make a habit of SHOWING problems to your mechanic rather than just telling him about them. Ask him to sit in the copilot seat so you can show him the symptoms. This guarantees that he knows how to reproduce the problem. If a mechanic can't reproduce a problem, there's no way he can troubleshoot it systematically, so he'll have to resort to guesswork and you don't want that.

Maybe when your mechanic actually sees the symptoms, he'll draw different conclusions from them than you did. If so, take a few minutes to discuss the situation. See

Make a habit of showing problems to your mechanic rather than just telling him about them.

if you can come to agreement on what the problem is before you hand over the keys and sign the work order.

Dealing with intermittents

Intermittent problems are invariably the toughest to troubleshoot. You encounter a problem, but by the time you get your plane to the shop, the problem is gone. You can't reasonably expect your mechanic to fix a problem that he can't reproduce. If you ask him to do that, you're forcing him to resort to guesswork.

Often the best strategy for dealing with such hard-to-reproduce intermittent problems is simply to sit tight, wait awhile, and see what happens. One of two things is likely to happen: the problem will get worse, or the problem will go into spontaneous remission and may not recur for years (or perhaps ever). If the problem gets worse, it becomes a lot easier to troubleshoot. If the problem goes into spontaneous remission, it's probably not worth worrying about unless and until it comes back (although it's undeniably aggravating that the cause remains unknown).

This "sit tight and wait awhile" philosophy has served me well over the years, but it's appropriate only when dealing with problems that are clearly not safety-of-flight issues. If you observe major fluctuations in engine oil pressure for 30 seconds, after which everything returns to normal, this is ***NOT*** a good candidate for the "sit tight and wait awhile" approach. Prudence would demand landing at the first opportunity and inspecting the engine oil filter and propeller governor gasket screen for the presence of metal. Transient oil pressure fluctuations are most often caused by a chunk of something getting caught in the oil pressure relief valve and interfering with its ability to regulate oil pressure and must be considered a Very Bad Thing.

On the other hand, if your electric trim doesn't seem to work during 5% of your landings, but always works perfectly when you test it on the ground, this is probably NOT the sort of thing you should be in a hurry to take to your avionics shop. If you can't make the system fail on the ground, your avionics technician probably won't either, and that's an engraved invitation for him "to throw your money at the problem" by replacing expensive components that he guesses (but can't prove) might be the cause of the intermittent problem. You might be better served by waiting until the electric trim system fails more often before you put your plane in the shop.

Gather fault-isolation data

As the owner/operator, you play a key role in the troubleshooting process. You are the first line of defense for observing the symptoms and collecting data that will hopefully lead to an accurate diagnosis. Lots of aircraft problems only occur in flight, and can't be reproduced in the maintenance hangar where your mechanic can see them. The mechanic's diagnosis is often totally dependent on the details of the symptoms you report to him. You're his eyes and ears.

Try to narrow the possible causes by testing all possible modes or control combinations that might affect the system in question. Change one thing at a time and see what the effect is (if any) on the problem. Keep good notes of everything you do so you can go over it later with your A&P.

Suppose you start hearing a strange noise on your comm radio. Does it occur on all comm frequencies or just certain ones? Does it occur on both comm radios or just one? Does it affect the nav radios or just the comms? Does the noise change as you change engine RPM? Does it go away if you shut off an alternator? How about if you shut off a magneto? Or the strobe? Can you hear it through the cabin speaker, or only through your headset? Do you have another headset you can try? You get the idea...

Develop theories

Once you've tried everything you can think of to isolate the problem, it's time to develop a theory—or perhaps several alternative theories—about what kind of component failure(s) could be responsible for producing the observed symptoms. Here's where the serious headwork begins.

At this stage, it's essential to have a thorough understanding of how the system works and what components are involved. Unless you are sure you know the system cold, this is the time to read the service manual and study the schematic diagrams. If you're still unclear about exactly how the system works, talk to an expert until you're convinced you understand.

Now consider all the possible components in the system that might have failed. Would a failure of that component produce the symptoms you're seeing? What sort of test could be performed to determine whether that component was the culprit?

Another case of too-high CHT

Another client flew his airplane to a top-notch California service center for its annual inspection. The shop repaired a number of discrepancies and sent out both magnetos for a scheduled 500-hour IRAN. Two weeks later, the owner came back to pick up his aircraft. He paid the shop's invoice, fired up, taxied out, performed a normal engine runup, and took off.

About 90 seconds after takeoff, the airplane's digital engine monitor alarmed, and the pilot saw that CHT #2 had reached 395ºF and was still rising. He immediately reduced power and told the tower that he was returning to land. He taxied back to the shop and described the problem to the Director of Maintenance (DOM).

Because both magnetos had been removed and reinstalled, the DOM's first thought was that the magneto timing was adjusted incorrectly and was too far advanced. He had his lead IA re-check the mag timing (while he watched), and it was spot-on (within 0.2º of spec). The IA checked the #2 fuel nozzle and found it clean and unobstructed. He also checked the spark plugs and they were fine.

More than an hour later, the owner attempted a second takeoff and this time the CHTs were even worse. CHT for cylinders #1, #2 and #3 reached 400ºF, 408ºF and 411ºF, respectively, before the pilot reduced power and came back to land. At this point, he dumped his engine monitor data, uploaded it to our SavvyAnalysis platform, and phoned me. I pulled up the data on my computer while he was still on the phone.

I compared the engine data for the last pre-annual flight to the data for the two aborted post-annual flights. What I found was that at full takeoff power from approximately sea level, both post-annual flights had CHTs about 50ºF higher and EGTs about 50ºF lower than the pre-annual benchmark flight. I explained to the owner that this could mean only one thing: advanced magneto timing. I understood that the shop had just checked the mag timing and found it spot-on, but I also knew that the engine data wasn't lying, so I urged the owner to have the shop check the mag timing again. I said that the high-CHT/low-EGT condition I was seeing in the data would require that at least one magneto had timing advanced by 3º to 5º.

Two hours later, I spoke with the DOM and the shop. He explained that he re-checked the mag timing, but this time instead of using the shop's fancy $360 digital inclinometer, he used the old mechanical "flower pot" timing tool that he'd used for more than 30 years before buying the digital unit. The flower pot revealed that the timing of both mags was advanced three degrees beyond the proper timing specification. The shop's $360 digital inclinometer was off by three degrees. (I later learned that one of the technicians had accidentally dropped it on the concrete hangar floor, and it was never quite the same after that.)

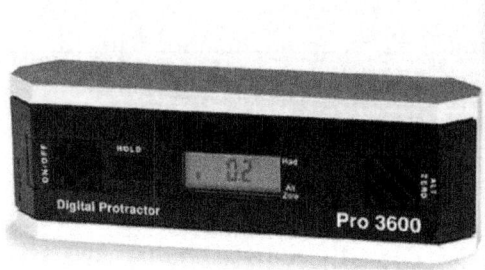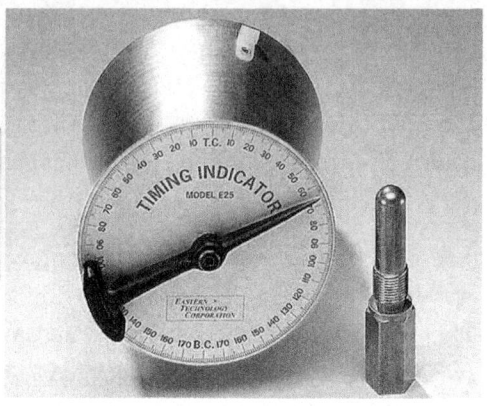

The $360 Pro 3600 digital inclinometer (left) turned out to be off by 3 degrees. The old-fashioned E25 "flower pot" timing indicator (right) got the job done.

You know the rest of the story. The shop readjusted the mag timing using the old flower-pot method, and the owner flew home with his EGTs and CHTs right where they were supposed to be.

Troubleshooting is a methodical approach to understanding what's wrong. When done properly, it involves gathering as much data as possible about the symptoms, identifying candidate failure modes consistent with those symptoms, systematically narrowing down the possibilities through a process of elimination until just one or two plausible diagnoses remain, and finally testing to confirm that the diagnosis is correct. As the owner/operator, you can help your A&P technician by describing your symptoms in detail and documenting the conditions under which they occurred. By carefully thinking through the problem before you permit anyone to attack your aircraft with tools, you can often minimize both downtime and the impact on your wallet. Above all, never put your mechanic in a position where he's forced to resort to guesswork.

34
Flight Test Profiles

The best place to diagnose an engine problem is usually in the air.

"My engine started running rough about halfway home yesterday," my client reported, "so I dropped it off at the service center. Could you please work with my mechanic to troubleshoot this problem?"

Arggghhh!!!

So many aircraft owners have a knee-jerk reaction to put their airplane in the shop whenever problems arise. Apparently, they assume that diagnosing the problem is the job of a mechanic. That's like having a bellyache and making an appointment to see a surgeon. Like surgeons, aircraft mechanics are primarily in the business of fixing things that aren't working properly. But before you go to a surgeon or a mechanic, you need to figure out what's wrong. You need a diagnosis.

Every diagnosis starts with data. If you're feeling unwell, the initial data would probably come from a Q&A session with your primary-care physician asking you for a detailed account of your symptoms. Additional tests such as a physical exam, blood work, electrocardiogram, imaging studies, needle biopsy, etc., might be used to gather additional data to refine and confirm the diagnosis.

Similarly, if your aircraft seems unwell, the initial data would typically come from your detailed account of its symptoms ("squawks") and a dump of digital engine monitor data. Additional tests such as oil and oil filter analysis, compression test, borescope inspection, etc., might be used to gather more data to refine and confirm the diagnosis.

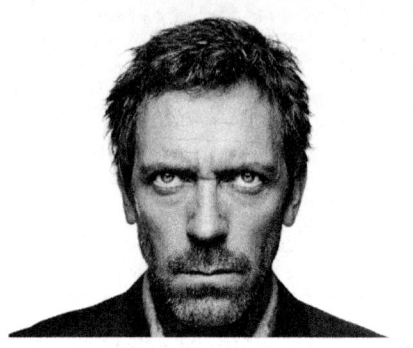

If you can't analyze your engine monitor data, you'll probably need to send it to someone who knows how. Sadly, few mechanics have any training or experience in doing this.

Your doctor can ask you about your symptoms, but your mechanic can't ask your airplane about its symptoms, because airplanes can't talk. If the mechanic's role is that of a surgeon, the aircraft owner's role is that of primary-care physician. It is up to you to "interview your airplane" by carefully observing and documenting its symptoms and downloading its engine monitor data. Chances are that you'll also need to analyze that data—or to send it to a specialist who knows how to analyze it—since very few mechanics have any training or experience in engine monitor data analysis. (This is something that hopefully will change gradually as mechanic training is dragged kicking and screaming into the 21st century.)

The owner as test pilot

When it comes to engine problems, the initial data required for diagnosis has to be gathered while the airplane is airborne. Only the grossest and most obvious problems—like a serious oil leak or a dead magneto—can be diagnosed in the maintenance hangar or during ground runs. The responsibility for gathering that data will most likely fall upon you, not your mechanic. Like it or not, you're a test pilot.

Typically, when one of our clients reports an engine problem, the first thing we ask for is a test flight during which one or more specific flight-test profiles is performed, followed by a dump of the engine monitor data for us to analyze. Eventually, we hope to get the owner trained to conduct the flight-test profiles without us having to ask, eliminating the need for a separate test flight.

The most useful of these flight-test profiles include:

- Ignition stress test
- Mixture distribution test
- Induction leak test

Let's look at each of these tests in detail.

Ignition stress test

Also known as the "in-flight LOP mag check," the ignition stress test is the best way to evaluate the performance of your engine's ignition system. Every student pilot is taught to perform a mag check during the pre-flight engine run-up, but many pilots have never shut off a magneto in flight and are uncomfortable with the idea. That's a shame, because the in-flight ignition stress test is far more exacting and revealing of ignition system problems than the usual run-up. The typical pre-flight mag check can detect only the grossest ignition system defects, while the in-flight stress test will reveal much more subtle ignition issues.

The best ignition system test you can perform is an in-flight mag check with the engine leaned aggressively lean of peak EGT (LOP). If your engine is not capable of operating smoothly when LOP, then do the test at the leanest mixture at which the engine can run without uncomfortable roughness. We want the mixture to be as lean as possible because the leaner the air-fuel mixture, the more difficult it is to ignite. Therefore, if your ignition system performance is marginal, it will show up during a lean in-flight test long before it becomes apparent in any other phase of operation.

If your engine monitor has a user-programmable sampling rate, it's best to set it to the highest rate possible. For example, many JPI EDM-series monitors have a default sampling interval of 6 seconds, but can be programmed to use a 2-second interval. When you're done with the flight-test profiles, you can reset the sampling rate to the default value if you like. (NOTE: This applies to all of the flight-test profiles discussed in this chapter.)

The ignition stress test should be performed in normal economy cruise configuration with the mixture set as lean as possible consistent with reasonably smooth engine operation (preferably LOP). Place your engine monitor in "normalize mode" which sets all the EGT bars to mid-scale and increases the sensitivity of the bar-graph display. Now go through the usual mag check procedure: BOTH-LEFT-BOTH-RIGHT-BOTH for singles or shutting off one mag toggle switch at a time (left to right) for twins.

Do not rush this procedure. Perform it very slowly, making sure that you run the engine on each individual magneto for at least 10 engine monitor sample intervals before moving on to the next phase. If the monitor samples every 6 seconds, run on each individual magneto for a minimum of 60 seconds; if the monitor samples every 2 seconds, run on each mag for at least 20 seconds. Return to two-mag operation for a similar length of time between single-mag runs.

If your ignition system is healthy, when you switch to single-magneto operation you should see all EGT bars rise by 50°F to 100°F. They may not all rise the same amount; in fact, it's perfectly normal to see even-numbered cylinders rise more than odd-numbered ones (or vice-versa). What's important is that all EGT bars rise, and that all remain stable at their elevated values. You will feel a small but perceptible loss of power during single-mag operation and a small but perceptible increase in roughness, but the roughness should not be alarming. (The engine will always run slightly less smooth on one magneto than on two, but roughness sufficient to get your non-pilot passengers to ask "what's wrong with the engine" is too much!)

Once on the ground, dump the engine monitor data and analyze it using our free SavvyAnalysis platform or your favorite analysis software. You will be able to see the test results much more clearly by graphing the engine monitor data than by looking at the instrument during the test flight. The graphic on the next page shows a good example of an ignition stress test—note how #5 EGT goes unstable when the engine is running on the left mag only, indicating that the bottom spark plug in cylinder #5 has a problem. Also notice that all EGTs rise more during left-mag operation than during right-mag operation, suggesting that the two mags are timed differently.

I recommend performing the ignition stress test on a regular basis—I do it on most every flight, generally at the end of the cruise phase just before starting my descent. It should certainly be performed any time any sort of engine anomaly is suspected.

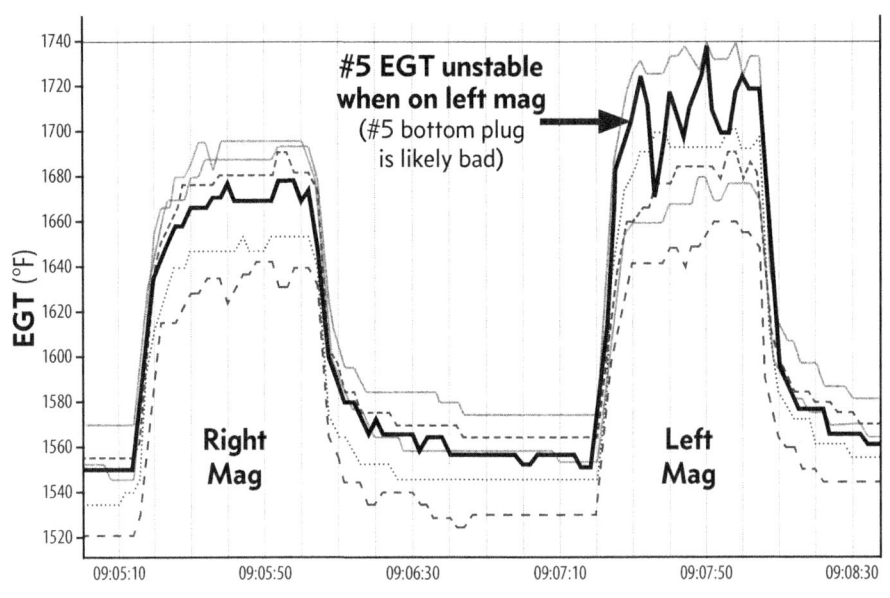

Ignition stress test reveals a problem with the bottom spark plug on the #5 cylinder, and also split ignition timing between the two mags.

Mixture distribution test

Also known as the "GAMI lean test," the mixture distribution test determines how much mixture variation exists among the cylinders of your engine. It can detect dirty or wrong-sized fuel nozzles, intake valve problems, induction leaks, and other engine anomalies that can cause uneven mixtures among cylinders.

The test is best performed at 65% cruise power or less. For normally-aspirated engines, you should perform the test at a high enough altitude that the throttle can be wide-open.

If your engine monitor captures fuel flow, then you can perform the test and analyze the data later. If it doesn't capture fuel flow, then you'll need to write down fuel flow readings during the test procedure. (The following description assumes that the monitor does not capture fuel flow and that the data has to be recorded manually.)

Starting with a full-rich mixture, write down the fuel flow and the EGT of each cylinder. Now lean very, very slowly—a vernier mixture control really helps—until the first cylinder reaches peak EGT. Note the peak EGT value for that cylinder and the exact fuel flow (to the nearest 0.1 GPH or 1 PPH) at which that peak EGT was achieved. Continue leaning very, very slowly until each cylinder reaches peak EGT, and again write down the peak EGT value for each cylinder and the exact fuel flow at which each peak was achieved.

Next, reverse the procedure, richening the mixture very, very slowly and making note for each cylinder the peak EGT value and the exact fuel flow at which peak EGT was obtained. It's usually best to repeat this rich-to-lean-to-rich mixture sweep procedure several times to ensure reliable results.

Once the data has been gathered, you can derive two valuable pieces of information. The first is the difference between full-rich EGT and peak EGT for each cylinder (referred to as the "lean range" for that cylinder), and the second is the difference in fuel flow between the first cylinder and the last cylinder to reach peak EGT (referred to as the "GAMI spread").

For most engines, the lean range of each cylinder—the EGT rise from full-rich to peak—should typically be around 250°F to 300°F. If any cylinder has a substantially lower lean range than the others, it may be operating too lean at takeoff power and might be vulnerable to overheating or detonation. (Suspect a clogged fuel nozzle or an induction leak.)

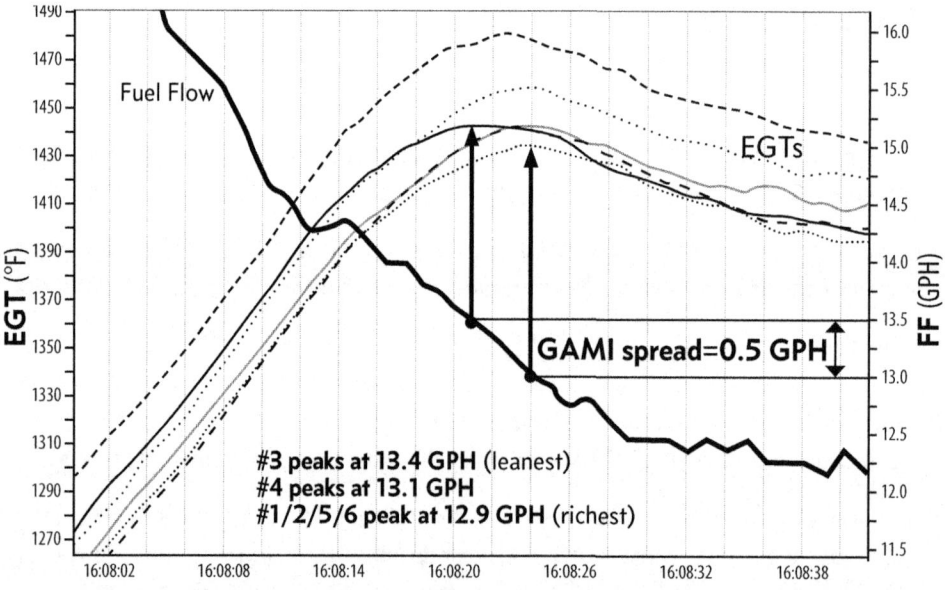

Mixture distribution test reveals an excellent GAMI spread of 0.5 GPH.

The "GAMI spread" is a measure of uneven mixture distribution. The smaller the spread, the better. A perfectly tuned engine would have a zero spread—all cylinders would reach peak EGT at exactly the same fuel flow—but that never happens in real life. Realistically, a fuel-injected engine with properly tuned fuel nozzles should exhibit a GAMI spread of 0.5 GPH or less. Using stock (non-tuned) nozzles, injected Lycoming and crossflow Continental engines typically have a spread around 1.0 GPH, and injected bottom-induction Continentals often have a spread of 1.5 GPH or more. Some carbureted engines (e.g., Continental O-470-series) can have spreads in the 2–3 GPH range. Experience indicates that if your engine has a GAMI spread above 1.0 GPH, it probably won't be able to run smoothly at LOP mixtures.

The graphic shows a mixture distribution test that reveals a GAMI spread of 0.5 GPH, which is quite respectable. Note that this engine monitor captured fuel flow information, making the analysis very easy and eliminating the need to record the fuel flow information manually.

I recommend performing this test every 12 months or 100 hours, whichever comes first, and whenever any sort of engine anomaly is suspected.

Induction leak test

This in-flight test is an effective method for detecting leaks in the engine's induction system. It is best accomplished in level cruise flight at about 5,000 feet MSL. It consists of a pair of test runs, one at high manifold pressure (MAP) and the other at lower manifold pressure.

For the high-MAP run, start with a relatively high power setting—wide-open throttle for normally aspirated engines, or MAP equal to outside ambient pressure for turbocharged engines—and full-rich mixture. Write down the EGT for each cylinder.

For the low-MAP test, reduce MAP by about 10 inches and again write down the EGT for each cylinder.

Disregard the absolute EGT values. Instead, calculate the change in EGT for each cylinder between the high-MAP and low-MAP tests. Ideally, the amount of EGT change should be roughly the same for all cylinders. If one cylinder (or two adjacent cylinders) exhibit(s) significantly less change than the others, suspect an induction system leak affecting that cylinder (or those adjacent cylinders).

Here's the principle behind this test: During the high-MAP test, the induction manifold pressure is very close to outside ambient pressure, so any induction leak will have little or no effect on engine operation. During the low-MAP test, the manifold pressure is significantly lower than outside ambient (by about 10 inches), so any induction leak will flow lots of air and cause the affected cylinder (or cylinders) to run substantially leaner than the others, resulting in a smaller drop in EGT than the others.

I recommend performing this test any time a MAP or mixture distribution anomaly is suspected.

Now put on your test-pilot cap and go fly some of these flight-test profiles!

35
Rough Engine

Understanding the underlying causes of engine roughness is the key to dealing with it.

One of the most common squawks reported by pilots and aircraft owners is a rough-running engine. If the roughness can be duplicated during a ground runup, then the mechanic might have a chance to troubleshoot it systematically.

Most of the time, however, the roughness is apparent only in-flight. Since the mechanic can't reproduce the problem on the ground, he's usually reduced to shotgunning—throwing parts, labor and money at the problem and hoping to get lucky. This is something aircraft owners should try to avoid at all costs (pun intended). The way to avoid it, of course, is never to put your mechanic in a position where he has to guess what's wrong.

I can't count the number of times I've seen expensive components—especially magnetos and fuel pumps—overhauled or replaced in an attempt to cure engine roughness. Those are usually desperation moves on the part of a mechanic who has run out of ideas. It's unfortunate, because there's almost no way a fuel pump problem can cause an engine to run rough, and it's extremely rare for a magneto issue to cause an engine to run rough.

What causes roughness?

Engine roughness is a very specific symptom that is almost always due to a very specific reason: *The cylinders aren't all producing the same amount of power.* So the underlying cause of roughness has to be something that can affect some cylinders differently than other cylinders.

If your engine is running rough, it's almost certainly NOT the fuel pump's fault!

If you think about it for a moment, there's no way a fuel pump problem could affect different cylinders differently. If the fuel pump is not producing enough fuel pressure, this could cause the engine to lose power or run too hot, but it would affect all cylinders equally, and therefore would not cause roughness. The same is true of all fuel system components, with the sole exception of fuel nozzles on a fuel-injected engine. (Since a clogged, dirty or otherwise defective fuel nozzle affects only one cylinder, it's a prime suspect when an injected engine runs rough.)

Similarly, almost all magneto issues affect all cylinders equally, so mag problems are very seldom implicated in engine roughness. If the roughness is related to the ignition system, the culprit is almost always going to be something that would affect only one cylinder. Most often it's a bad spark plug; sometimes it's a bad ignition harness lead or contact spring.

Usually the quickest way to diagnose such issues is to dump the aircraft's digital engine monitor data and analyze it (or have it analyzed). You do have a digital engine monitor installed, don't you?

Mixture maldistribution

One common complaint is an engine that runs smoothly at rich mixtures but becomes rough at lean mixtures, particularly lean-of-peak (LOP) mixtures. The cause is invariably that different cylinders are running at different air-fuel ratios, a phenomenon called "mixture maldistribution."

In a perfect world, all cylinders would run at the exact same air-fuel ratio, but in the real world that almost never happens. There's always one cylinder that runs leanest, another that runs richest, and the rest are somewhere in between the two outliers. We measure this maldistribution by calculating the GAMI spread and we want it to be as small as possible.

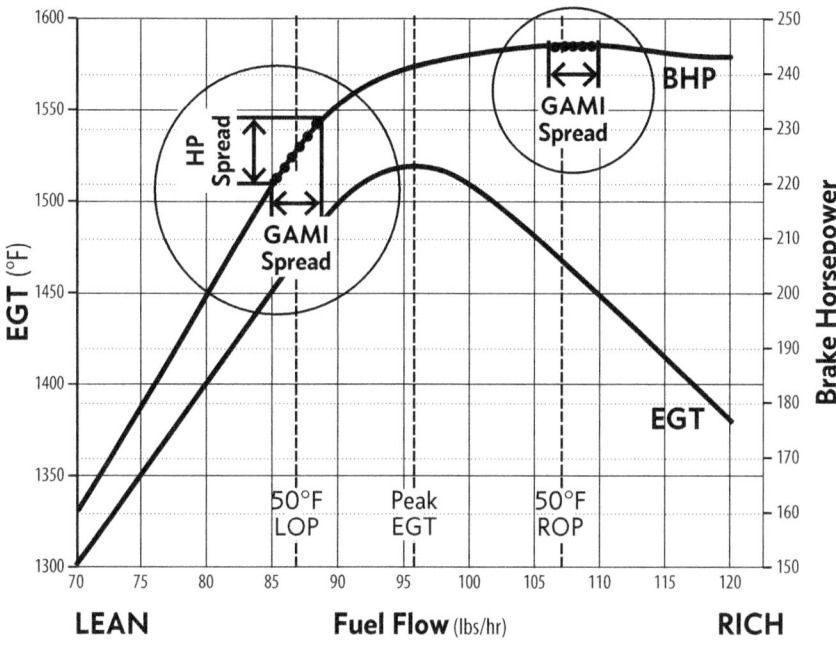

Mixture maldistribution has very little effect when ROP, but can cause big HP differences among the cylinders when LOP. That's one reason engines run rougher at lean mixtures than at rich mixtures.

Look what happens if the mixture maldistribution is significant—as it typically is with most carbureted engines and with injected engines that aren't equipped with position-tuned fuel nozzles. When the engine is being operated on the rich side of peak EGT (ROP), the maldistribution causes very little roughness because the slope of the HP-vs-FF curve is so flat that all the cylinders are producing the same power. However, when the engine is operated LOP, there's significant roughness because the HP-vs-FF curve is steeply sloped and so the leanest cylinder is producing significantly less power than the richest cylinder. When different cylinders produce different amounts of power, we perceive the result as roughness.

What's the solution? If the engine is fuel-injected, the fix is an easy one—install position-tuned fuel nozzles to reduce the GAMI spread to as close to zero as possible. A spread of ½ GPH or 3 PPH is quite decent; less is even better.

If the engine is carbureted, we don't have the option of tweaking fuel nozzles. But it's usually possible to improve mixture distribution significantly by means of some operational tricks. One thing that usually helps quite a bit is to avoid operating the engine at wide-open throttle. By retarding the throttle from the wide-open position just enough to obtain a slight decrease in manifold pressure (if the prop is constant speed) or a slight decrease in RPM (if the prop is fixed-pitch), the carburetor's "enrichment circuit" is disabled and turbulence is introduced into the airflow through the carburetor throat by the slightly cocked throttle plate, improving fuel atomization. Another thing that often helps—especially when OATs are cold—is the application of partial carburetor heat, which also improves atomization.

These operational tricks are especially useful with engines like the Continental O-470-series engines used in the Cessna 182 that are notorious for having terrible mixture distribution. Most carbureted Lycomings—especially 4-cylinder Lycomings—have fairly decent mixture distribution right out of the box.

The better the mixture distribution—i.e., the lower the GAMI spread—the leaner the engine can be operated before the onset of objectionable roughness.

Cycle-to-cycle variation

Another source of roughness is cycle-to-cycle variation (CCV). This refers to the fact that each successive combustion event in a particular cylinder is a bit different from the last one. As you can see clearly in the graphics on the facing page, these cycle-to-cycle variations increase rather dramatically at lean mixtures.

A tremendous amount of research has been done into the cause of CCV, particularly now that there's so much emphasis on using very lean mixtures in automotive engines to minimize emissions. It turns out that there are a number of different factors that contribute to the CCV phenomenon.

One principal cause of CCV seems to be the chaotic fashion in which the spark that jumps across the spark plug electrodes ignites the local air-fuel mixture in the immediate vicinity of the spark plug to create an "early flame kernel" that ultimately orga-

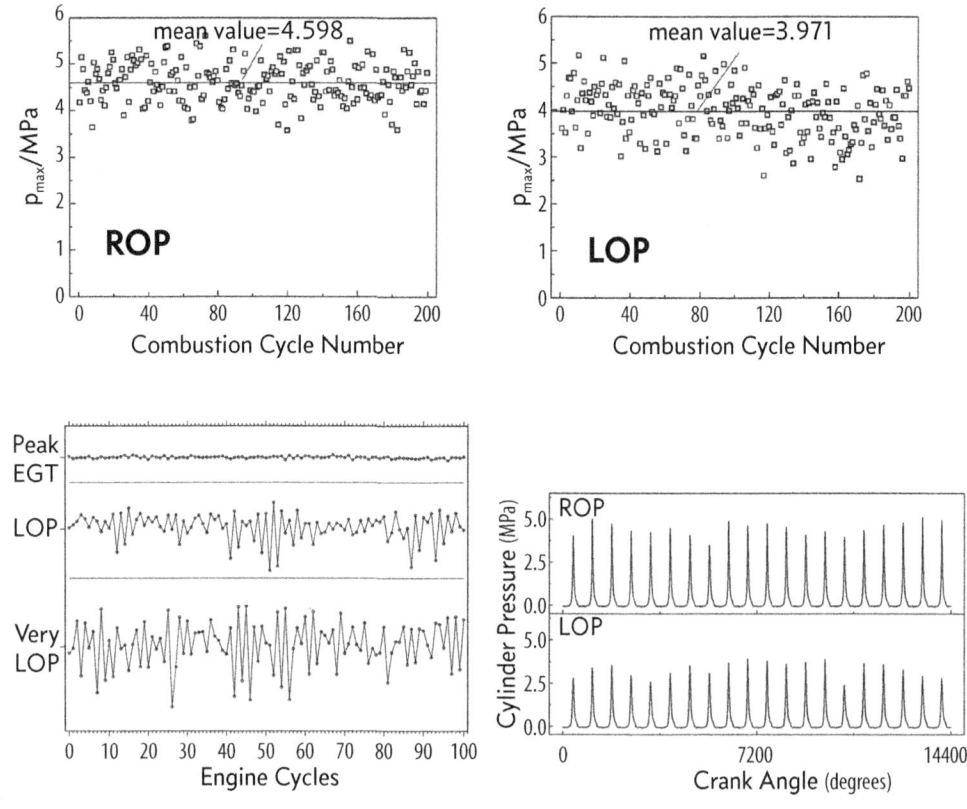

Especially at lean mixtures, there is considerable cycle-to-cycle variation from one combustion event to the next.

nizes itself into a flame front. The local gas flow in the vicinity of the spark plug and local non-uniformities in the air-fuel ratio in the immediate vicinity at the moment of ignition contribute to highly variable and unpredictable combustion in first few milliseconds after the ignition event. The images on the next page offer a good illustration of just how chaotic and unpredictable this process is.

The shape and location of the early flame kernel has a profound effect on the flame front propagation during the remainder of the combustion event. An elongated flame kernel that extends away from the spark plug electrodes will result in faster flame front development than a compact flame kernel that remains close to the spark plug electrodes. The leaner the air-fuel mixture, the greater these variations tend to be.

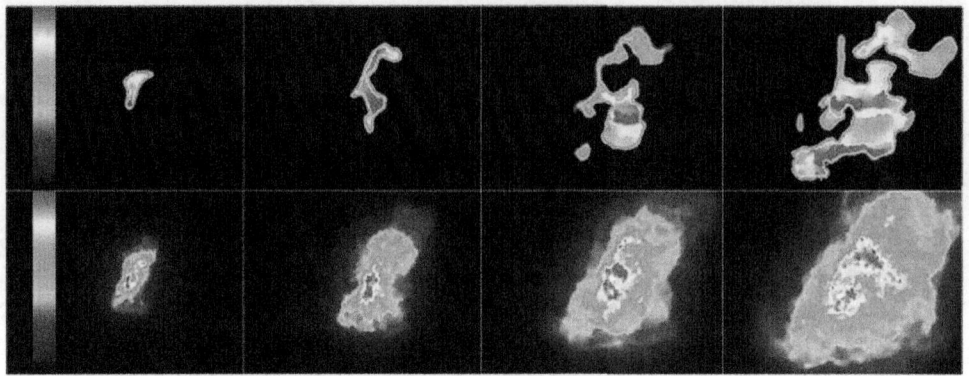

Actual photos of the rather chaotic development of the early flame kernel in a cylinder immediately after the spark plug fires. These remarkable photos were taken using a special piston with a quartz viewing window.

Another significant contributor to CCV is the fact that it is never possible for the engine to expel 100% of the combustion residue out the exhaust valve at the end of a combustion event. There is always some small amount of "end gas" left in the cylinder when the next cycle begins. If the flame front gets a slow start, combustion will be incomplete and there will be some unburned fuel left over at the start of the next cycle. That can cause the next combustion event to start with a richer-than-normal mixture, resulting in a faster burn time and a lower-than-usual amount of unburned fuel at the end of the event. This "residual fraction effect" can cause alternating increases and decreases in peak combustion chamber pressure on successive cycles, particularly with lean mixtures.

CCV helps explain why engines always run a bit rougher LOP than ROP, even if the GAMI spread is close to zero. For the most part, we just have to live with this phenomenon, and accept the fact that the compelling advantages of LOP operation—cleaner, cooler, and much more efficient operation—are more than sufficient to accept a bit more engine roughness.

One thing that can help a bit is using fine-wire spark plugs, which tend to provide slightly more consistency in early flame kernels. Personally, I use massive-electrode plugs and just tolerate the slight additional roughness during LOP cruise. I just can't bring myself to spend close to $100 for a fine-wire plug. Then again, I fly a twin that has 24 spark plugs…and I'm a world-class cheapskate. Your mileage may vary.

36
Diagnosing a Temperamental Ignition

If it hadn't been for the engine monitor in my Cessna 310, I wouldn't have even known I had a problem, much less been able to diagnose it.

I was afflicted with a temperamental ignition problem while on a coast-to-coast trip from California to the East Coast in my Cessna 310. I was on a business trip that would take me from my home base in Santa Maria, California., to Frederick, Maryland, and then to Atlanta, Georgia, and then home again.

I first became aware of the problem as I was climbing out of Santa Maria on the very first leg of what I expected to be a routine 30-hour round-trip. I had reduced to my usual 75% cruise-climb power, was climbing at my usual 130 KIAS cruise-climb airspeed, with the usual 110 pounds/hour fuel flow on each engine. This was all standard routine that I'd performed hundreds of times before.

The engines felt and sounded normal. The airplane was climbing nicely at about 1,000 FPM despite being loaded right at max gross. The air was smooth. My satellite weather display indicated no significant weather all the way to Tulsa, Oklahoma, where I planned to make an overnight stop before continuing on to Frederick. The

SiriusXM audio was tuned to the classical music channel, piping one of my favorite Bach Brandenburg Concertos into my stereo ANR headset. Everything seemed right with the world.

My reverie was interrupted by a flashing amber annunciator light that told me my digital engine monitor was trying to get my attention. Sure enough, when I looked over at the instrument on the right side of the panel, its display was flashing a high TIT alarm on the left engine, and displaying the TIT as 1620°F. I knew from experience that normal TIT in this configuration is around 1570°F and I'd programmed the engine monitor to alarm any time the TIT exceeded 1600°F.

In-flight troubleshooting

Looking carefully at the engine monitor, I noticed that the EGT on the left engine's #5 cylinder was noticeably higher than the other cylinders, and definitely higher than what I was used to seeing. The high #5 EGT suggested to me one of two possible problems: (1) a partially clogged #5 fuel nozzle; or (2) a #5 spark plug that wasn't firing.

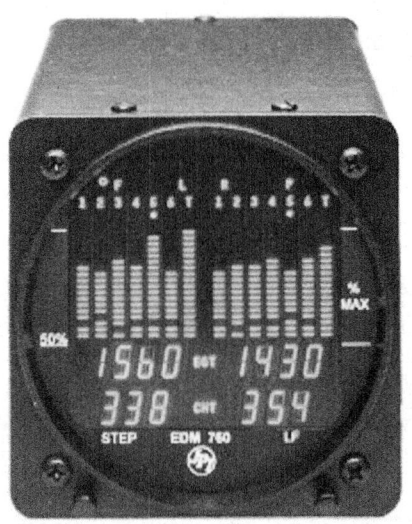

On climbout, the left engine showed excessive TIT and excessive EGT on the #5 cylinder. However, #5 CHT was normal. This suggested that one spark plug might not be firing in the #5 cylinder. An in-flight mag check confirmed that indeed the bottom plug was not firing.

If a clogged nozzle was causing cylinder #5 to run too lean, I would have expected that cylinder also to have elevated CHT while operating ROP during climb. However, the engine monitor did not indicate that the #5 CHT was elevated; if anything, it seemed to be a bit lower than usual.

That suggested to me that a non-firing spark plug was the most likely cause of the elevated #5 EGT. To confirm this theory, I performed an in-flight mag check. When I shut off the left engine's right magneto, the left engine started running quite rough and the #5 EGT bar on the display dropped out of sight.

Bingo! I'd just proved beyond the shadow of a doubt that a non-firing spark plug in the #5 cylinder was the culprit.

But which of the #5 plugs wasn't firing? Because the cylinder went cold when I shut off the right magneto, the non-firing plug had to be the one connected to the left magneto. On my engines (as with most big-bore Continental engines), each magneto fires the top plugs on its side of the engine and the bottom plugs on the opposite side of the engine. Since cylinder #5 is in the right bank of cylinders, its top plug is fired by the right mag and its bottom plug is fired by the left mag. Therefore, I reasoned, my non-firing spark plug had to be the bottom plug on cylinder #5.

(Bottom plugs tend to misfire much more often than top plugs, in my experience, because the bottom ones are so vulnerable to oil-fouling and contamination with debris.)

That's odd, I thought. I had done a thorough runup prior to takeoff, including the usual preflight mag check at 1700 RPM. All 24 spark plugs appeared to be working just fine. Why would one decide to quit firing now? Definitely odd.

I leveled off at my cruising altitude of 13,000 feet and did the "big mixture pull" to transition to LOP. The engine monitor continued to show elevated #5 EGT on the left engine. During the next couple of hours, I repeated the in-flight mag check a couple of more times and got the same result: The bottom plug on cylinder #5 was definitely kaput. Sometimes fouled plugs clear themselves spontaneously. But not this time.

Not to worry, that's why I carry a couple of spare spark plugs in the emergency toolkit I always carry in my airplane, together with the necessary tools to change out a plug on the ramp if necessary. So I knew what I had to do.

Spark plug transplant

I landed at my first planned refueling stop, Saint Johns, Arizona. KSJN is a frequent fuel stop for me going eastbound because it consistently has among the lowest 100LL prices west of the Mississippi. Also, SJN has a field elevation of 5,736 feet MSL, which shortens the descent for landing and the subsequent climb back to altitude. All in all, it's one of my favorite places to refuel.

After topping the tanks, I retrieved my emergency toolkit and proceeded to remove the bottom spark plug from cylinder #5 of the left engine.

The plug had been in service for about 100 hours, and it looked fine to me. But since it clearly wasn't firing, I decided to swap it out anyway. I installed a brand-new spark plug in the bottom plug boss of left engine cylinder #5, torqued it to 360 in.-lbs. using the torque wrench I carry in my emergency toolkit, and reattached the ignition lead.

After closing the left engine nacelle and stashing my emergency toolkit back in the wing locker, I fired up and taxied out for departure. At the runup area, I performed an extra diligent runup and mag check to verify that all plugs were firing properly; they were. I then took off and turned eastbound toward Tulsa.

Climbing out of KSJN, I tuned the SiriusXM audio to the '60s oldies channel and was just getting into the groove when it happened again: The amber light started flashing and the engine monitor started complaining about high TIT on the left engine. A quick cycle of the instrument and a quick in-flight mag check confirmed that the bottom plug on #5 was once again not firing. Yes, the very same brand-new spark plug that I'd just installed!

Plug transplant, part deux

I continued to Tulsa, taxied to the FBO, and broke out the emergency toolkit once again. This time, I removed the newly-installed plug and installed my one remaining spare. I wasn't sure it would solve my problem, but figured it was worth a shot.

The next day, climbing eastbound out of Tulsa, I actively monitored the engines looking for signs of trouble. Everything seemed to be working fine. After leveling off in cruise and switching to LOP, I tried another in-flight mag check. The left engine continued to run smoothly on each magneto individually, and the engine monitor confirmed that everything was operating normally now. Whoopie!

I flew nonstop to Frederick at FL210 (to stay above a bunch of rather nasty frontal weather). High-altitude LOP operation is pretty demanding on the ignition system, but the engines didn't miss a beat and another in-flight mag check at altitude confirmed that all was well.

After completing my business in Frederick, I flew to Charlotte, North Carolina to spend a few days with my in-laws who live there, then proceeded on to Atlanta for another business meeting. After that, I headed home to the west coast with stops in

Memphis and Denver. The engines continued to run perfectly, and I pretty much forgot about the earlier ignition problem.

It's baaaack!

After returning home to Santa Maria and resting up a bit, I decided it was time to do some preventive maintenance on the airplane. I changed the oil, sent oil samples to the lab for analysis, replaced the oil filters, and cut open the old filters for inspection.

Since the spark plugs had over 100 hours on them, I pulled them and sent them out for cleaning, gapping, and bomb testing. All my cleaned/gapped spark plugs passed with flying colors and came back a week later, whereupon I reinstalled them in the engines.

After closing the engine nacelles, I took the airplane out for a post-maintenance test flight. A thorough pre-flight runup indicated that everything was working fine. But the test flight once again revealed elevated #5 EGT on the left engine, and an in-flight mag check showed the bottom #5 spark plug was once again not firing. Are you kidding?

It slowly started to dawn on me that the ignition problem must be something other than a bad spark plug. It had to be either a problem with the magneto itself or a problem with the ignition harness.

I tried replacing the insulator ("cigarette") and contact spring on the bottom #5 ignition lead, but another test flight showed that this did not solve the problem. I pulled the left mag and opened it up but couldn't find anything wrong. The distributor cap was clean inside, the contact springs looked good, the point gap was correct and the internal and external mag timing was spot-on.

Harness transplant

By the process of elimination, that left the ignition harness. I examined the #5 bottom ignition lead and couldn't spot any visual anomalies. But since I was running out of ideas, and since a brand new full harness (for both mags) cost less than $500.00, I decided to order one and install it. Even though the existing harness looked fine, it did have nearly 2,000 hours on it, so I figured it was fully depreciated.

There are a variety of ignition harnesses that are PMA approved for my engines. I have always preferred the Slick harnesses because of their superior construction and flexibility, so I ordered a new Slick M1740 harness to mate with my Continental/Bendix S-1200 magnetos.

Removing the old harness and installing the new one was more time-consuming that I expected. Doing the job correctly involves considerable Adel clamping, grommeting, and tie-wrapping to ensure that the ignition leads cannot vibrate or chafe on anything and have no tight bends. It took me about six hours to complete the job, including retiming both mags. (I am, of course, the world's slowest mechanic. I imagine a professional A&P could have done it much more quickly.)

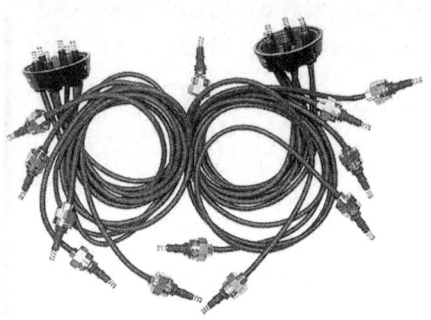

A new full harness (for both magnetos) costs only about $500. (A harness for just one magneto is called a "half harness.") Figure on four hours of labor to install. I prefer the Slick-brand harnesses (shown above) because of their superior construction and flexibility.

Finally, it was time to do yet another post-maintenance test flight. This time, I was overjoyed to find that everything was perfect. The engine monitor readings were just as they should be, and a high-power in-flight mag check showed all systems go. Success at last!

Lessons learned

I learned some important lessons from this experience. One is that the usual pre-flight mag check is a laughably inadequate test of ignition system performance. While trying to track down my problem with a non-firing #5 bottom plug, the ignition system repeatedly showed no problems whatsoever during the pre-flight mag check, only to fail immediately and repeatably as soon as the aircraft was in flight.

Clearly, the pre-flight mag check is not a very demanding test of the ignition system, and won't detect anything but the grossest ignition anomalies. An in-flight mag check is a far more demanding and revealing test. The most demanding ignition system test is a high-power in-flight mag check with the engine leaned aggressively (preferably LOP).

Many pilots have never done an in-flight mag check, and many are afraid to perform one. I've even known some experienced A&P mechanics that discourage pilots from shutting off a magneto in flight. Obviously, I don't agree with that advice. In fact, in the wake of my experience, I now make a point of performing an in-flight mag check on almost every flight, and I heartily recommend that you consider adopting the same practice.

Another lesson I learned here is the tremendous diagnostic value of a modern digital probe-per-cylinder engine monitor. If I hadn't installed one in my airplane, I'd never have known that my #5 bottom plug was not firing. It's quite possible that this situation could have gone on for months and hundreds of hours without being detected. Once again, my engine monitor proved that it is worth its weight in gold.

Finally, I learned that ignition harnesses have a finite useful life. They may look perfect upon visual inspection yet develop internal electrical leaks that seriously compromise ignition system performance. Since a new harness is relatively inexpensive (as aircraft parts go), it probably wouldn't be a bad idea to replace the ignition harness every 1,000 hours or so just on general principles. In fact, I decided to order another new harness and installed it on my right engine, so both engines got new harnesses.

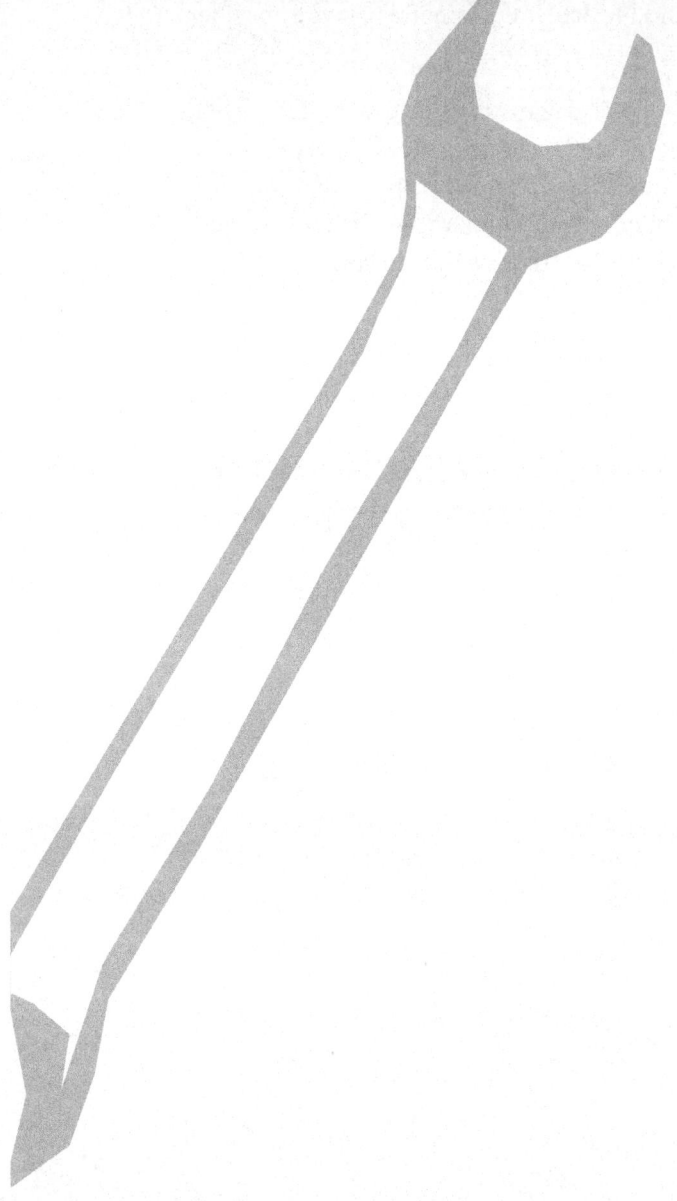

348 ENGINES

37
High Oil Consumption

If your engine is using too much oil, it's got to be going somewhere.

A fellow named Ted phoned me to say that his 1984 Cessna T210 was in the shop for its annual inspection, and his mechanic was suggesting a $14,000 top overhaul. "Mike, I've read a lot of your stuff and I know you're not a big fan of top overhauls," Ted told me, "so I thought I'd ask your opinion before I tell my mechanic to proceed."

"What's the problem with your engine?" I asked.

Ted told me his oil consumption had suddenly increased dramatically to a quart every 3 or 4 hours, and that the compressions on his 1200-hour engine were mediocre (mostly 60s, a couple of high 50s). The mechanic said his cylinders were "tired," that this was typical for a mid-time Continental TSIO-520-R, and that turbocharged Continentals seldom make TBO without cylinder replacement.

I counseled Ted to slow down and take a thoughtful approach. While his increased oil consumption was certainly a matter of concern, it wasn't yet an airworthiness or safety-of-flight issue. Continental says maximum permissible oil consumption for this 310

The Cessna T210's oil consumption had suddenly increased and the owner's mechanic recommended a $14,000 top overhaul.

hp engine is about one quart per hour (and Ted's oil consumption wasn't anywhere close to that). No airplane has ever fallen out of the sky because of high oil consumption... unless the oil consumption was so high that the engine ran out of oil before it ran out of fuel (which is always embarrassing). With a sump capacity of 12 quarts (8 useable) and a maximum IFR range of 5 hours, it would be hard to do that in a T210. So before throwing a five-digit sum of money at the problem, I suggested to Ted that some systematic troubleshooting was in order.

Where's the oil going?

When troubleshooting an oil consumption problem, the first question to ask is always "where's the oil going?" There are only three possibilities:

1. The oil is exiting the engine through its breather line.

2. The oil is leaking out of the engine somewhere else.

3. The oil is getting into the combustion chamber of one or more cylinders and being "consumed" in combustion.

It's usually not hard to figure out the answer. If the oil is departing through the breather, then there will be excessive oil on the belly of the aircraft, especially aft of the tip of the breather tube. If it's leaking from the engine, then there will be excessive oil inside the engine compartment, probably a real mess. If it's being consumed in combustion, then the inside of the tailpipe(s) will be dark and oily instead of being coated with the normal light powdery residue.

I questioned Ted about these possible symptoms. He responded that he hadn't noticed any more oil on the belly than usual, and that the engine compartment seemed relatively dry and oil-free—but now that I mentioned it, the residue on the inside of the tailpipe did strike him as being darker and moister than what he was accustomed to seeing.

This suggested that the oil was being consumed in combustion. If so, the next obvious questions in the diagnostic process are "how is the oil getting into the combustion chamber(s)?" and "which cylinder(s) is/are involved?"

One way that oil can wind up in a cylinder's combustion chamber is a bad intake valve oil seal that permits oil from the cylinder's rocker box area to be sucked into the cylinder during the intake stroke through the annulus between the intake valve stem and intake valve guide. This problem is usually easy to localize and diagnose by using a borescope inserted into the combustion chamber through the top spark plug hole to inspect the intake valve stem when the intake valve is open. If the intake valve stem appears oily, then you've found the problem. If it's a Continental, the fix is quick and easy—simply replace the intake valve oil seal, a task that can be done without removing the cylinder. If it's a Lycoming (which does not use seals), the fit of the valve stem in the guide should be checked for proper clearance according to Lycoming Service Bulletin 388C. If clearance is excessive, you will have no recourse other than to remove the cylinder and fit new guides. Of course, if it's that bad, you may have already been experiencing sticking valves (a.k.a. Morning Sickness) upon engine start.

In a turbocharged engine like Ted's, there's another possibility. If the turbocharger has a bad oil seal that allows oil to escape from the center section into the compressor section, that oil will be "inhaled" by all cylinders, consumed in combustion, and leave a dark oily residue inside the tailpipe. In discussing this possibility with Ted, he admitted that he'd noticed a small quantity of oil dripping from the clamp where the induction air duct connects to the turbocharger compressor discharge. This struck me as being a possible smoking gun.

After a bit of Q&A with the owner, I concluded that the T210's oil consumption problem would most likely be cured by a $1,400 turbocharger swap rather than a $14,000 top overhaul.

I advised Ted to have his mechanic disconnect the air duct from the turbocharger and inspect the compressor, which should be bone dry. If oil is found in the compressor, then my presumptive diagnosis would be confirmed, and the cure for Ted's oil consumption problem would be a $1,400 turbocharger overhaul/exchange rather than a $14,000 top overhaul.

Oil-on-the-belly syndrome

Had Ted told me that he'd noticed a big increase in oil discharge from the breather onto the belly of the aircraft, our conversation would have focused on an entirely different branch of the troubleshooting tree. Excessive breather discharge is the most common reason for elevated oil consumption. It manifests itself as excessive oil on the belly, and it's usually—but not always—caused by excessive blow-by past the compression rings in one or more cylinders.

Excessive discharge of oil on the belly is often caused by excessive blow-by in one or more cylinders—but not always. Unless the oil is turning dark and opaque very quickly after each oil change, blow-by probably isn't the cause.

But just because you have elevated oil consumption and oil on the belly doesn't necessarily mean that it's appropriate to do a top overhaul. Usually it isn't. For one thing, the problem might involve only one or perhaps two cylinders, not all of them. For another, excessive blow-by is not the only cause of oil-on-the-belly syndrome; there are several other possibilities. Once again, a thoughtful approach and systematic troubleshooting is called for before doing anything invasive or expensive.

A first step that is often useful is to measure crankcase pressure and find out whether it's actually excessive. Continental has an old service bulletin (M89-9, "Excessive Crankcase Pressures") that explains how to do this. The procedure involves hooking up a length of tubing to the crankcase—via the dipstick tube, a modified oil cap, or a modified timing plug, depending on engine model—and hooking up the other end of the tubing to an old airspeed indicator that is used as a sensitive pressure gauge. The engine is then run up at full power, and the crankcase pressure is read from the airspeed

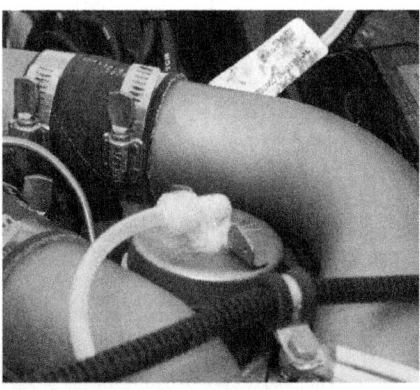

One way to check for excessive crankcase pressure is to run up the engine with an old airspeed indicator plumbed to a modified oil filler cap.

indicator. Continental's service bulletin specifies the maximum acceptable pressure reading on the airspeed indicator; for small-bore engines (A- and C-series, O-200, O-300), the maximum is 44 mph (38 knots), and for big-bore engines (360-, 470-, 520-, 550-series), the maximum is 90 mph (78 knots). A reading above these thresholds indicates excessive blow-by. (The same approach would work for Lycomings, although Lycoming doesn't have a Service Bulletin about it.)

If crankcase pressure is high, then the problem is likely excessive blow-by in one or more cylinders. A confirming symptom of excessive blow-by would be oil that turns dark and opaque very quickly after each oil change.

The best way to determine which cylinder(s) is/are at fault is to perform a borescope inspection with emphasis on the condition of the cylinder barrels. Any cylinders exhibiting excessive barrel wear (shiny spots with no crosshatch), substantial barrel damage (vertical scoring), or excessive corrosion pitting are candidates for coming off for rework or replacement. Low compression with air leakage past the rings audible through the oil filler would tend to confirm the high blow-by diagnosis for a particular cylinder, although a compression test alone is not reliable enough to use as a primary diagnostic tool.

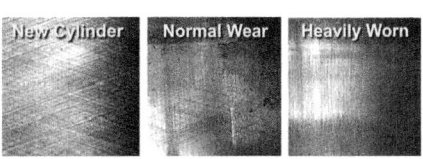

A simple borescope inspection is the best way to tell which cylinder(s) is/are worn to the point of causing excessive blow-by. A compression test is much less reliable.

Yanking all cylinders ("top overhaul") rather than taking the time to determine which one(s) is/are at fault is the worst kind of "shotgunning" in my opinion. What's the probability of all six cylinders going bad simultaneously? Probably about the same as drawing a royal straight flush in poker. Think about it…

If crankcase pressure is NOT excessive, then the problem isn't high blow-by and the problem isn't cylinder-related. So what could be causing oil-on-the-belly syndrome? One of two things: Either something is pressurizing the crankcase in the air but not on the ground, or else something is sucking oil out the breather in flight. Pressurization of the crankcase in flight could be caused by something as simple as a leaky gasket or O-ring in the oil filler cap or dipstick, or something else that permits ram air to enter the crankcase in flight. Suction could be caused by an improperly positioned breather outlet that's located in a low-pressure area in flight.

Air-oil separators: A "Band-Aid" solution

Many aircraft owners have chosen to address the oil-on-the-belly "problem" by installing an aftermarket air-oil separator in the crankcase breather plumbing. The separator is essentially a swirl chamber that uses centrifugal force to separate liquid from gas, returning the liquid to the crankcase while allowing gas to pass out the breather tube.

Air-oil separators such as this one can conceal the symptom (oil on the belly) rather than address the root cause (crankcase pressurization). I consider this a Band-Aid solution.

Several such aftermarket air-oil separators have STCs for installation on most engines. The most common such devices are the M-20 Model 300 (manufactured by M-20 Oil Separators LLC) and the Walker AirSep (manufactured by Airwolf Filter Corp).

I call such devices "Band-Aid solutions" because their purpose is to conceal the symptom (oil on the belly) rather than to address the root cause (crankcase pressurization). I have long discouraged the installation of such devices, and many world-class engine experts have been quite outspoken in opposition to their use.

If you think about it, the quantity of oil discharge from the engine breather is an important indicator of engine health, and installation of an air-oil separator deprives the owner and his mechanic from using this tool. Small air-oil separators can also freeze up in cold weather, blocking the breather and possibly causing engine failure.

In addition to relieving crankcase pressure, the crankcase breather allows the engine to purge moisture from the oil and vent it overboard. This is extremely important, because corrosion is the number-one reason that engines fail to make TBO, and internal moisture build-up is one of the two factors that cause internal engine corrosion. (The other factor is loss of protective oil film due to disuse.) To the extent that an aftermarket air-oil separator returns moisture-laden oil to the crankcase rather than allowing it to be discharged overboard, it can contribute to the corrosion problem and wind up costing you truly big bucks.

If you have excessive oil discharge from the crankcase breather, I would suggest that you do some troubleshooting to identify and eliminate the problem at its source, or alternatively you learn to live with a little oil on the belly.

Oil leaks

While an engine oil leak is a third possible escape route that could account for increased oil consumption, it's the least likely in my experience. While engine oil leaks are quite commonplace, it would take a massive leak to cause a noticeable increase in oil consumption. Most engine oil leaks are tiny, although even a tiny leak can make a major mess in the engine compartment. A little bit of leaking oil can cover a lot of real estate rather quickly when exposed to a 100-knot breeze.

Most common sources of oil leaks—rocker cover gaskets, pushrod housing seals, magneto gaskets, garlock seals, oil filler cap gaskets, dispstick O-rings, etc.—are quick and easy fixes, and are almost always worth fixing to make sure they don't cover up something more serious like a crankcase crack, chafed oil pan, or leaky oil quick-drain.

Chasing down the source of an oil leak is often quite frustrating because the oil spreads so rapidly throughout the engine compartment. The key is to spray down the engine with solvent until it's scrupulously clean of oil, then run the engine on the ground (absent the 100-knot breeze) and inspect it. The application of some white aerosol powder like dye-penetrant developer (or jock-itch spray) can sometimes be helpful in localizing the source of the leak.

None of my recommendations to Ted or the other things I've discussed here involve rocket science, only simple logic. My advice is simply that when faced with elevated oil consumption, resist the temptation to throw money at the problem. Better to throw neurons at it instead.

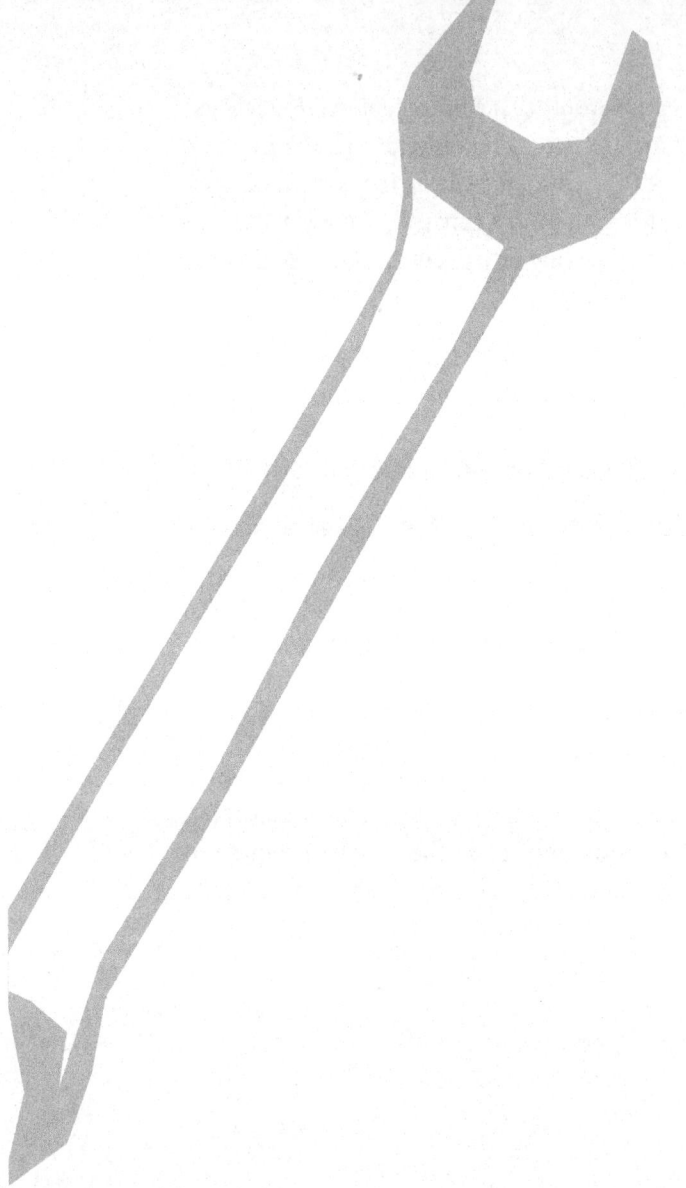

38
Troubleshooting Turbosystems

Turbocharging system problems can be elusive to find and fix because they occur only in flight.

One of the most frequent problem areas for owners of turbocharged aircraft is the turbocharging system. Often these problems are "fixed" by overhauling or replacing costly turbosystem components without having resolved the trouble. As always, it's almost never a good idea to "throw money at the problem" until sufficient troubleshooting has been done to identify the actual cause of the problem.

There are very good reasons that turbocharging problems tend to be difficult for mechanics to troubleshoot. Most aren't reproducible on the ground, often occur only at quite high altitudes, and are sometimes quite erratic or intermittent. More often than not, the mechanic has no choice but to rely entirely on a description of the symptoms provided by the owner or pilot. That description is often incomplete or misleading because the owner or pilot doesn't really understand what the mechanic needs to know to diagnose the problem correctly.

Turbocharging system instrumentation consists mainly of the manifold pressure gauge.

Furthermore, there's precious little cockpit instrumentation that offers any direct measurement of what the turbocharging system is doing. We don't have a turbocharger spindle speed gauge or a wastegate position indicator on the panel. Turbo problems generally show up on the manifold pressure gauge, and that instrument provides only a very indirect indication of what's going on with the turbocharging system.

To make matters even worse, the fundamental design of automatically-controlled turbocharging systems tends to compensate for—and conceal—problems with the engine and turbosystem, often leaving pilots blissfully ignorant that mechanical problems are developing until those problems get quite serious.

Nevertheless, if the owner or pilot understands how the system works and knows what to look for, and if the mechanic employs a logical procedure for troubleshooting the system, it's usually possible to isolate turbocharging problems without having to resort to "shotgunning"—replacing one component after another until the problem disappears.

A wide variety of mechanical ailments can interfere with the proper operation of the system. These include:

- Induction leaks
- Exhaust leaks
- Internal engine problems
- Wastegate problems
- Controller problems
- Turbocharger problems

All of these problems can result in improper operation of the turbocharging system, but each one tends to produce symptoms that are subtly different in character. Therefore, a careful analysis of the symptoms can often help pinpoint the cause of the problem, or at least rule out some of the possibilities and help narrow the search. But since many turbosystem problems show up only at high altitudes and not on the ground, it's often up to the pilot (rather than the mechanic) to make critical observations of the symptoms and decipher what they mean.

Induction leaks

One of the most common causes of turbosystem problems are leaks in the induction system. I recall, for example, helping an aircraft owner troubleshoot his turbocharged twin and it turned out that an engine control cable had been chafing against the engine's induction manifold in a hard-to-see location. Eventually the steel cable wore a slot all the way through the wall of the cast aluminum induction pipe, creating a fairly significant induction system leak.

The pilot squawked the problem when he noticed a significant manifold pressure "split" between the two engines while cruising at the Flight Levels. The engines appeared to be operating normally on the ground, during takeoff, and when operating at low and middle altitudes. The problem only showed up when the airplane was flying up high. If you think about the consequences of an induction leak, that's not surprising.

Consider what happens during a full-power takeoff at or near sea level. Manifold pressure inside the induction manifold is at red-line (say 32 inches), but that's only a trifle greater than outside ambient pressure (around 30 inches). So relatively little induction air escapes through the leak. What little loss there is will be sensed by the turbocontroller as a loss of upper deck pressure (UDP), and the controller will cause the wastegate to close a bit, compensating for the small loss and effectively concealing the problem. From the cockpit point of view, both manifold pressure indications are right where they should be and everything appears nominal.

Climbing through 1,000 feet we throttle back to 29 inches of manifold pressure for cruise-climb. What's the outside ambient pressure at 1,000 feet? About 29 inches! So now, the induction leak becomes a total irrelevancy, since the pressure inside and outside the induction manifold are virtually identical, so there's no loss of pressure through the leak at all.

As we continue to climb at 29 inches of manifold pressure, outside ambient pressure decreases by about 1 inch per 1,000 feet so the pressure differential between inside and outside the induction manifold increases with altitude. More and more induction air escapes through the leak. However, the turbocontroller senses this loss and keeps closing the wastegate and cranking up the turbo output to compensate for it. In the cockpit, the manifold pressure indication never wavers from 29 inches and so the pilot remains blissfully unaware of the problem. The wastegate on the leaky engine is closed more than it should be, and the turbocharger on that engine is spinning faster than it should be, but the engine is running just fine and there's no cockpit instrumentation to provide the slightest clue that something's awry.

> Classic symptoms of an induction leak are normal operation at takeoff and low altitude, and premature onset of bootstrapping at higher altitudes.

The pilot's first indication of a problem comes as the airplane climbs through 15,000 feet and the manifold pressure indication on the troubled engine starts to fall and become erratic, while the indication on the other engine remains rock-solid at 29 inches. What's happened, of course, is that the pressure differential across the induction leak has become so great (about 15 inches now) that even the maximum output of the turbocharger can no longer keep up with the loss. The controller, it its now-futile attempt to compensate for the leak, has commanded the wastegate to go fully closed, and the engine has started to "bootstrap"—something that should normally not happen until the airplane climbs well into the Flight Levels.

These are the classic symptoms of an induction leak problem: normal operation at takeoff and low altitude, and the premature onset of bootstrapping (i.e., loss of manifold pressure regulation) at higher altitudes. However, there are other kinds of problems (e.g., exhaust leaks) that can produce the same symptoms and confuse the diagnosis.

So how can you be sure?

It turns out that there's another symptom that can often be used to distinguish an induction leak problem from various other kinds of turbo-related problems. Best of all, this symptom is one that can be checked without having to take the airplane up to high altitude, or even leaving the ground at all! The tip-off is *higher-than-normal manifold pressure when the engine is throttled back to idle.*

Consider an engine idling on the ground. The engine is "trying to breathe" but the throttle is retarded to idle, closing the throttle butterfly and choking off most of the available induction air. (It's called a *throttle* because it chokes off the engine's airway!) The result is a significant *vacuum* in the induction manifold, as the engine consumes the air in the induction manifold but the closed throttle butterfly blocks the inflow of air to take its place. In the cockpit, this shows up as a very low manifold pressure indication (typically, something on the order of 12 to 15 inches), far below outside ambient (around 30 inches at sea level).

But suppose there's a substantial leak in the induction plumbing somewhere between the throttle and the cylinders. What happens? Ambient air rushes in through the leak because of the vacuum in the induction manifold. In the cockpit, this shows up as a higher-than-normal manifold pressure indication at idle—perhaps three or four inches higher than it should be. The engine will also be idling leaner than usual— since the leak lets in more air but not more fuel—so the engine may tend to stumble a bit when you throttle-up for taxi (at least if the leak is big enough).

High manifold pressure at idle isn't a perfect tool for diagnosing induction leaks. Some induction leaks won't produce this symptom (e.g., leaks in the upper deck portion of the system prior to the throttle body). Also, the symptom can be produced by other things besides an induction leak (e.g., a non-firing cylinder or a badly misadjusted idle mixture).

But certainly if you see both abnormal bootstrapping at altitude and high idle manifold pressure, the odds favor an induction leak, and that's probably the first place you should look for trouble.

If you suspect an induction leak (based on the observed symptoms), the first step should be to confirm the diagnosis by performing a *critical altitude check*. The procedure will be described in detail in your aircraft service manual, and consists of a test flight at altitude in which certain power settings are established at certain altitudes, and the manifold readings are recorded. The service manual has tables that establish how much manifold pressure you should be able to obtain at these benchmark altitudes under specified conditions of RPM, fuel flow and temperature. If your engine falls significantly short, then you can be sure you have a problem—and odds are that it's an induction leak (although there are other possibilities that we will discuss later on).

If you can't find anything obviously wrong after careful visual inspection of the induction system, a simple pressure check may be in order. All that's required is to pressurize the induction system with a few PSI of air—one good way is simply to pump air into a cylinder as if you were doing a compression check, but rotate the prop so that the cylinder's intake valve is open—then close the throttle and go over the entire induction system with a soapy water spray, looking for leaks that reveal themselves by blowing bubbles. Some leaks are expected at the induction system drains and, to a lesser extent, around the throttle shaft, but the rest of the induction system should be airtight.

Exhaust leaks

An exhaust leak can produce similar symptoms to an induction leak—the onset of bootstrapping at a lower-than-normal altitude—because any exhaust that escapes through a leak bypasses the turbocharger just as if it escaped through an open wastegate. Just like with an induction leak, the turbocontroller will try to compensate for (and thereby cover up) the problem by commanding the wastegate to close, so the symptoms generally won't show up until the airplane is at high altitude. (Unlike a lower-deck induction system leak, an exhaust leak will not affect manifold pressure at idle.)

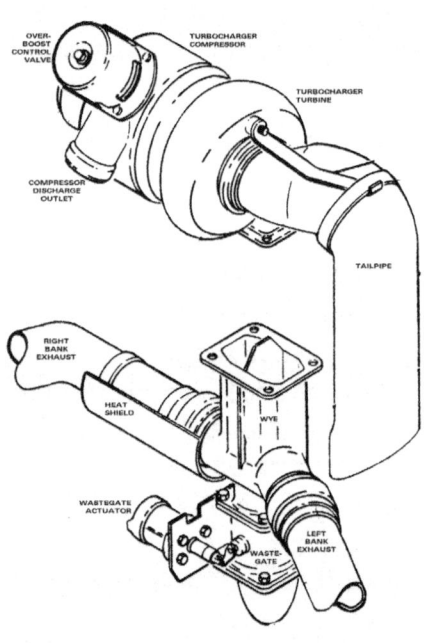

Typical turbocharger, wastegate, and exhaust.

Exhaust leaks are inherently much more dangerous than induction leaks, because of the very serious threat of in-flight fire. Fortunately, exhaust leaks are usually a lot easier to detect because they typically leave brightly-colored exhaust stains—and sometimes also obvious heat damage—that can be detected visually during an engine compartment inspection. All turbocharged aircraft should have their exhaust systems meticulously inspected for leaks every 50 hours, something that is mandated by AD for some models.

Because the exhaust system operates under extreme heat and pressure, and because exhaust gas is so very corrosive, exhaust leaks can sometimes develop suddenly (a "blowout") rather than gradually. The pilot of a turbocharged aircraft who experiences a sudden unexplained loss of manifold pressure in-flight should assume that an exhaust failure may have occurred, and should put the airplane on the ground at the earliest possible moment. If the aircraft is a twin, the pilot should consider the possibility of shutting down and securing the affected engine to minimize the threat of in-flight fire and collateral damage.

I don't mean to frighten you with this statement. The risk of an in-flight exhaust failure (particularly one of the "blowout" variety) is extremely remote on aircraft whose exhaust systems have been properly maintained and inspected. In my experience, the vast majority of in-flight blowouts and exhaust-related fires involved exhaust components with very high time and usually ones with poor-quality weld repairs that failed.

Internal engine problems

A third possible cause of bootstrapping at a lower-than-normal altitude is an internal engine problem that prevents one or more cylinders from firing. This would most likely be something that reduces the compression of a cylinder to near-zero (such as a valve that's badly burned or stuck open), or something that prevents both spark plugs in the cylinder from firing (such as severe lead fouling of both plugs). A non-firing cylinder reduces the exhaust output of the engine, and this means less flow through the turbocharger. As usual, the controller will try to compensate (and cover the problem up) by closing the wastegate, but this means that the wastegate will go full-closed at a lower-than-normal altitude.

A non-firing cylinder can cause bootstrapping at a lower-than-normal altitude.

You'd think that an engine that wasn't firing on all cylinders would be very obvious to the pilot, wouldn't you? That's usually true in a single-engine airplane, but not so much in a twin where any roughness or loss of power is masked by the other engine. Don't ask me how I know this…

So, if you notice bootstrapping at unusually low altitudes but can't seem to find any leaks in the induction and exhaust, it's definitely worth doing a compression check and having a look at the plugs to see whether one cylinder is not firing and/or operating at near-zero compression. If so, you've probably found your culprit.

By the way, a zero-compression cylinder will generally cause the same abnormally high manifold pressure at idle as an induction leak. This is a symptom you should watch for before every flight. If you see it, it's usually a tip-off that something significant is wrong.

Wastegate problems

The other most common cause of turbosystem problems, besides induction leaks, are problems with the wastegate and wastegate actuator. It makes perfect sense that the wastegate would be one of the most problematic parts of the turbocharging system, because it performs such an unenviable job: regulating the flow of incredibly hot and corrosive exhaust gases.

Most wastegate problems are of the "sticky wastegate" variety in which the shaft on which the wastegate butterfly pivots gets "coked up" with byproducts of combustion (a nasty concoction of lead, carbon and sulfur) to the point that it no longer opens and closes smoothly when commanded to do so by the wastegate actuator. Another somewhat less common cause of "sticky wastegate syndrome" occurs when the wastegate actuator itself starts to bind as a result of the accumulation of oil-borne deposits, O-ring deterioration, and/or scoring of the actuator cylinder.

Whatever the exact cause of the sticky wastegate, the result is that the constant series adjustments commanded by the turbocontroller—which are normally executed so rapidly and smoothly that they are unnoticeable to the pilot—become jerky and erratic. The result shows up as abnormal manifold pressure fluctuations, especially during periods of constant wastegate movement such as climb, descent, and flight in turbulent air.

It's easy to confuse the erratic manifold pressure fluctuations caused by a sticky wastegate with bootstrapping, but they're really quite different if you know what you're looking for. Bootstrapping (due to a fully-closed wastegate) is a condition that predictably occurs at high altitude and low engine RPM, and which can be made to disappear at will by increasing RPM slightly or descending a bit. On the other hand, erratic fluctuations due to a sticky wastegate generally occur at various altitudes and RPM settings, and are most obvious during changes in altitude, power settings, and airspeed (all times when wastegate adjustments are most likely to be commanded by the controller).

If you suspect you might have a sticky wastegate, it's easy to check in the shop. Simply remove the oil line that runs from the engine oil pump to the wastegate actuator. Hook a source of adjustable regulated air pressure to the oil inlet port of the actuator—an ordinary cylinder compression tester is ideal for this purpose. Now simply watch the wastegate assembly as you slowly and repeatedly vary the air pressure from zero to 50 PSI and back. As air pressure reaches 15 PSI or so, the wastegate should start to close smoothly, reaching its fully-closed position when the pressure reaches around 50 PSI. As you back the pressure down towards zero, the wastegate should open smoothly. Watch for any signs of jerkiness or binding as you exercise the wastegate in this fashion. Any tendency to stick should be obvious during this test. Also make sure the wastegate butterfly opens and closes fully, a total movement of approximately 90 degrees of shaft rotation.

If the wastegate appears to be sticky, it's possible that you might be able to "rescue" it by giving it a good soak overnight in a strong penetrating oil like Mouse Milk or AeroKroil. But don't count on it. If a penetrant soak doesn't result in silky-smooth action, it's time to yank the wastegate and send it out for overhaul. Make sure that the overhauled wastegate is re-installed with the proper high-temp attaching hardware and new gaskets.

Controller problems

When the turbocharging system starts acting up, pilots and mechanics alike have a tendency to name the turbocontroller as prime suspect. In fact, the controller is much less likely to be the culprit than the wastegate assembly, and it's an unfortunate fact that lots of perfectly good controllers are sent out for overhaul in the course of "shotgunning" turbosystem problems.

The turbocontroller is seldom the culprit for two reasons: It has a terribly easy job, and there's not a whole lot that can go wrong with it. Think about it for a minute: The controller never sees hot exhaust gases or high engine temperatures like other turbosystem components. It spends its life in almost air-conditioned luxury, watching upper-deck pressure at one end, and regulating oil flow to the wastegate actuator at the other. If I had to come back as a turbosystem component in my next life, I'd almost surely apply for the job of turbocontroller!

That's not to say that the controller cannot cause turbo problems, only that it's one of the last system components you should suspect. Before you send the controller out for overhaul (which doesn't come cheap), there are a few things you should try first:

- If you're flying a twin, try swapping the left and right controllers and see if the problem changes sides or stays put (or goes away altogether). The swap generally takes less than an hour and could save you many hundreds of bucks and a week or two of downtime.

- If the controller is at fault, it might just be that its poppet valve is sludged up. Disconnecting the oil lines and shooting a few shots of shop air into the oil return port might dislodge the gunk and fix the controller problem.

- The controller's upper deck air reference line and inlet port should contain *nothing but air*. Disconnect the line and inspect for any signs of liquid contamination (fuel or oil). If you see any, purge the line with solvent and shop air, and disassemble the controller's aneroid chamber and clean it out, too. (That's a whole lot less scary than it sounds.)

If the problem follows the controller when you swap sides (twins only) and a simple cleaning doesn't resolve the problem (singles or twins), only then should you consider sending the thing out for overhaul.

Turbocharger problems

Of course, turbocharging problems can also be caused by—ta da!—the turbocharger itself. I'm listing this component last, not because it seldom fails, but because when it does fail, the failure is more or less obvious. While a turbocharger can last a full engine TBO if the engine is operated with sufficient TLC, it's certainly not unusual for a turbo to need a mid-term overhaul.

Turbochargers have several failure modes, most of which are more-or-less self-diagnosing. For example, turbos sometimes fail catastrophically—the engine suddenly goes normally-aspirated (or quits from over-rich mixture) and the aircraft starts trailing black smoke (actually, oil being pumped into the hot exhaust). You put the airplane on the ground fast, the tower rolls the equipment, and your mechanic doesn't have much difficulty figuring out what to do next. A catastrophic failure like this

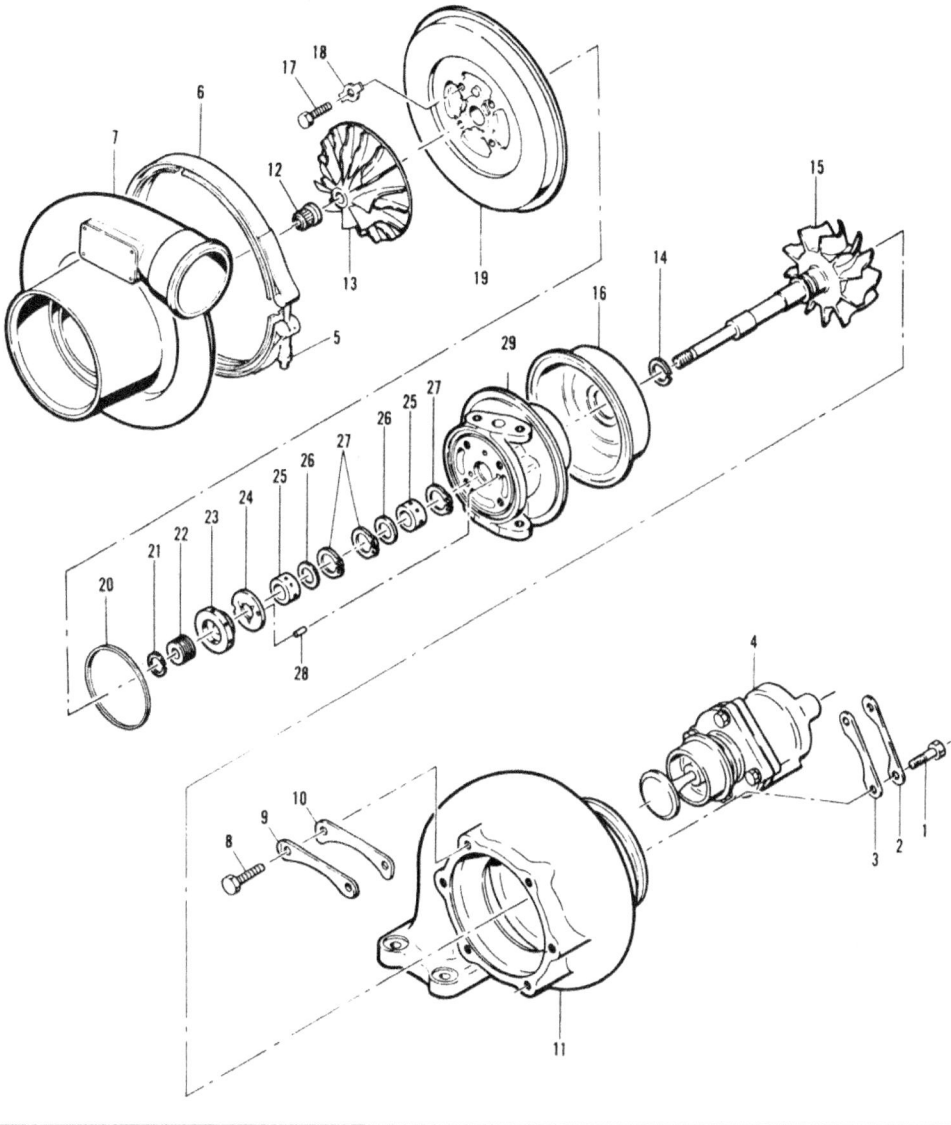

Exploded view of turbocharger.

generally stems from a turbocharger that suddenly develops a serious out-of-balance condition while spinning at 50,000 RPM or so, often because the turbo ingested some foreign object (like a nut, bolt, alternate air door hinge, or chunk of exhaust valve) that damaged the spinning compressor or turbine wheel.

A less spectacular failure mode occurs when the turbocharger center section wears out, generally resulting in engine oil winding up where it doesn't belong: in the compressor and/or turbine portions of the turbo. If oil leaks into the compressor, it will result in oil-soaked induction plumbing (you'll likely see oil dripping from an induction system drain) and oil-fouled spark plugs. If it leaks into the turbine, it will result in an abnormal accumulation of oily deposits on the tailpipe and belly, and sometimes oil dripping from the tailpipe after shutdown. A visual inspection of the turbocharger will confirm oil where it doesn't belong, and often excessive play when the shaft is wiggled. Time to yank and overhaul the beast.

A third turbocharger failure mode is a bit more subtle, and stems from the fact that the turbine wheel operates under tremendous centrifugal forces as it spins at 50,000 to 80,000 RPM, while the metal loses much of its strength at the white-hot temperatures (1600°F to 1800°F) at which the turbine operates. The result is that the hot turbine blades develop a very gradual "stretch" over the life of the turbocharger—particularly if it's run hot and hard—and ultimately they can stretch enough that they actually start to scrape on the turbine housing. The turbo should be inspected for signs of blade scrape at each annual (and any other time the tailpipe is removed). If it is noted, it's overhaul time.

Troubleshooting strategy

If you have a turbosystem problem, the best way to avoid falling victim to the expensive, time-wasting "shotgun" approach is to devise a logical troubleshooting strategy. Your differential diagnosis methodology should seek to find or rule out potential causes in a sequence that starts with the most common failure areas and the quickest, easiest and cheapest troubleshooting steps, and then gradually works towards rarer failures and more difficult and costly actions. It's impossible for me to suggest a one-size-fits-all strategy, but here's a starting point:

The first step in any turbo troubleshooting strategy should be a thorough test flight to document the exact symptoms. By all means do a "critical altitude check" as documented in the service manual for your airplane, to determine if you have a premature bootstrapping problem and to quantify just how serious it is. Be sure to note whether any erratic or abnormal manifold pressure indications occur only at high altitudes and/or low RPMs (indicating bootstrapping), or if they occur primarily during climbs, descents, and/or airspeed changes (indicative of a sticky wastegate). Also note whether manifold pressure at idle is higher than normal (indicative of an induction system leak).

Do a quick compression check and spark plug inspection to determine whether you have a cylinder not firing or with near-zero compression. Of course, you'll know this already if you have a digital engine monitor installed (and you should).

The first step in any turbo troubleshooting strategy should be documenting the exact symptoms in flight.

Visually inspect the induction system looking for any sign of a leak (chafed-through manifold, cracked induction coupling, loose hose clamp, etc.). If nothing suspicious is found, pressurize the induction system with shop air to a few PSI above ambient and use soapy water to search for leaks.

Visually inspect the exhaust system looking for any sign of an exhaust leak—tip-off is usually brightly colored exhaust stains. If nothing suspicious is found, pressurize the exhaust system and use soapy water to search for leaks.

Connect a source of variable regulated shop air (such as a cylinder compression tester) to the oil inlet port of the wastegate actuator, and exercise the actuator by repeatedly varying the air pressure between 0 and 50 PSI, watching for any sign of sticky operations. If the wastegate is sticky, try an overnight penetrant soak. If that doesn't free it up, pull the wastegate and have it overhauled.

Remove the oil lines from the turbocontroller and blow shop air through the poppet valve to dislodge any sludge. Remove the air reference line and inspect for any signs of fluid contamination. In a twin, try exchanging the two controllers to see whether the problem moves with the controller. If it does, and if cleaning the controller doesn't help, send it out for overhaul.

Check the turbocharger for signs of foreign object damage (FOD), oil in the compressor or turbine, excessive center section play, and turbine blade scrape. If any of those problems are noted, pull the turbo for overhaul.

Once you understand the theory of operation behind the turbosystem, troubleshooting it can and should be done in a logical sequence. Avoid the temptation to "shotgun" the system by replacing costly components, hoping to get lucky. Instead, use your system knowledge, don your differential diagnosis hat, and methodically cross off possible failure modes one by one until you catch the culprit redhanded.

PART VIII
Maintenance

39
Oil Changes

Changing the oil is a routine maintenance task, but important enough to be worth thinking about.

I had just returned from my annual July pilgrimage to EAA AirVenture in Oshkosh, Wisconsin. As in past years, I'd headed up to the Great Lakes Region early to participate in a pre-AirVenture floatplane weekend at Cadillac Lake, Michigan. Organized by my friend Rick Durden, the Cadillac Lake shindig was an informal affair where a bunch of pilots spend two days flying Super Cubs on floats and renewing old friendships. It was relaxing and enjoyable and exactly what I needed before the pressure-cooker week of AirVenture.

My 3,500-nm trip from home base (Santa Maria, Calif.) to Cadillac, Oshkosh, and return involved 19 hours of flying time in my Cessna 310. While preparing for the trip, I checked my maintenance records and found that it had been 39 hours since I changed the oil. My normal oil-change interval is 50 hours, so it seemed obvious that I'd better change the oil and filter before leaving on the trip, 11 hours early. Otherwise, by the time I got back from Oshkosh, my oil would be 58 hours old.

As I thought about this a bit more, I considered that once I got back from Oshkosh at the end of July, my schedule didn't call for another trip until September. Consequently, the airplane would likely be sitting idle for the entire month of August. If I changed the oil before the trip, then the engines would be sitting dormant for five weeks with 20-hour-old oil in them. On the other hand, if I delayed the oil change until I got back from Oshkosh, then the engines would be full of fresh, clean oil during the dormant period. I realized that 20-hour-old oil is considerably more corrosive than clean fresh oil, because it's contaminated with combustion byproducts, fuel, moisture and acids.

Maybe the decision wasn't so obvious after all…

In the end, I decided to postpone my oil change until after the trip, putting 58 hours on the oil before draining it—eight hours over my usual limit. I started the oil draining immediately upon my return home and went down to the hangar the next day to finish the job by changing the oil filters, cutting open the old ones, and servicing the engines with fresh Aeroshell W100. Then I took the airplane up for 20 minutes to make sure that all the internal engine parts were thoroughly coated with fresh oil.

While at AirVenture, I ran into Ed Kollin at the ASL CamGuard booth. Ed is a lubrication research chemist by trade—formerly with the Exxon Engine Research Laboratory, Advanced Fuels and Lubricant Group—and knows more about the chemistry of engine oil than anyone I know. I described my predicament to Ed and asked him whether he thought I made the correct decision to change the oil after the trip rather than before. Ed agreed that after was better. He added that perhaps an even better choice would have been to do two oil changes, once before the trip and one immediately afterward. Ed was quite emphatic that letting the engines sit unflown for a month while full of 20-hour-old oil was not such a hot idea because 20-hour-old oil is pretty nasty stuff.

Why we change the oil

We don't change oil regularly because the oil breaks down in service and its lubricating qualities degrade. The fact is that conventional petroleum-based oils retain their lubricating properties for a very long time, and synthetic oils retain them nearly forever.

Consider, for example, that most auto manufacturers now recommend a 7,500-mile oil-change interval for most cars and trucks, which is the equivalent of about 250 hours of driving. In fact, oil analysis studies have shown that a synthetic automotive oil like Mobil 1 or Amsoil can go 18,000 miles without appreciable degradation, and that's the equivalent of about 600 hours.

No, the reason we change oil in our aircraft engines every 25 to 50 hours is not because it breaks down. It's because it gets contaminated after 25 to 50 hours in an aircraft engine. In fact, it gets downright filthy and nasty.

Compared with automotive engines, our piston aircraft engines permit a far greater quantity of combustion byproducts—notably carbon, sulfur, oxides of nitrogen, raw fuel, partially burned fuel, plus massive quantities of the corrosive solvent dihydrogen monoxide or DHMO—to leak past the piston rings and contaminate the crankcase. This yucky stuff is collectively referred to as "blow-by" and it's quite corrosive and harmful when it builds up in the oil and comes in contact with expensive bottom-end engine parts like crankshafts and camshafts and lifters and gears.

To make matters worse, avgas is heavily laced with the octane improver tetraethyl lead (TEL), which also does truly ugly things when it blows by the rings and gets into the crankcase. If you're as old as I am, you may recall that back before mogas was

There's more to changing the oil than most aircraft owners realize.

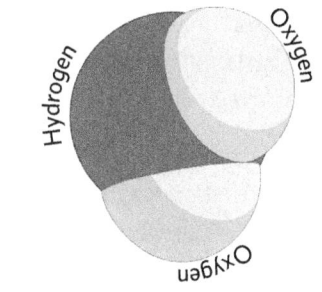

Dihydrogen monoxide (DHMO) is a highly corrosive chemical that is produced in copious quantities during combustion and can cause great harm to costly engine components when it blows by the piston rings and contaminates the engine oil. You may be more familiar with DHMO's common chemical formula: H_2O.

unleaded, typical oil-change intervals were only 3,000 miles. It was largely the switch to unleaded fuel that made it possible to go 7,500 miles between oil changes.

So, one of the most important reasons that we need to change the oil regularly in our Continentals and Lycomings is to get rid of these blow-by contaminants before they build up to levels that are harmful to the engine's health.

How often to change

Another reason we need to change the oil regularly—arguably even more important than disposing of contaminants—is to replenish the oil's additive package, particularly its acid neutralizers. When sulfur and oxides of nitrogen mix with DHMO, they form sulfuric acid and nitric acid. If you remember these dangerous corrosives from your high school chemistry class, then you'll certainly appreciate why you don't want them attacking your expensive engine parts.

To prevent such acid attack, aviation oils are blended with acid neutralizer additives. These are alkaline substances that neutralize these corrosive acids, much as we might use baking soda to neutralize battery acid. These acid neutralizers are consumed by the process of neutralizing acids, so it's imperative that we replenish them before they get used up to an extent that might jeopardize our hardware. The way we replenish them is to change the oil.

How can we tell when the acid neutralizers in the oil have been used up? It turns out that there's a laboratory test that measures the level of acid neutralizers remaining in the oil. This is known as the "total base number" or "TBN" test. Many oil analysis laboratories can perform this test on your oil samples. However, it's not routinely done as part of the normal oil analysis report, so you need to specially request a TBN test when you send in your oil sample (and be prepared to pay extra).

Fresh virgin aircraft oil typically has a TBN around 10. By the time the TBN decreases to 2 or 3, it's time to retire the oil because its ability to neutralize acids is just about gone. If you want to determine scientifically how long you can safely go between oil changes in your aircraft engine (or car, motorcycle or tractor, for that matter), the TBN test lets you do that.

However, most owners don't bother with the hassle and expense of TBN testing. They simply change their oil at a conservative interval that's guaranteed to get the junk out and fresh additives in before anything untoward is likely to occur. As a rule-of-thumb, I generally like to change my oil at 50 hours or four calendar months, whichever comes first. The effect of this rule is that operators who fly at least 150 hours a year can go 50 hours between oil changes, but operators who fly less will use a proportionately reduced oil-change interval.

Another important result that does appear in the normal oil analysis report provided by some labs is the "insolubles" test. This test is performed by placing the oil sample in a centrifuge to separate out all solids and liquids in the sample that are not oil-soluble.

Virgin oil normally contains no insolubles. The insolubles found in drained used engine oil come from three sources: (1) oxidized oil that breaks down due to excessive heat; (2) contaminants from blow-by of combustion byproducts; and (3) particulate contamination caused by poor oil filtration. If your oil analysis report reveals above-normal insolubles, it might be indicative of an engine problem—high oil temperature, excessive blow-by, inadequate filtration—and almost certainly means you should be changing your oil more frequently.

Did I mention that I'm a big fan of laboratory oil analysis? I use it religiously, recommend it strongly to all piston aircraft owners, and believe that it's one of the most important tools we have—along with oil filter inspection and borescope inspection—for monitoring the condition of our engines and determining when maintenance is necessary.

Choosing oil and additives

Much has been written about the pros and cons of various types and brands of engine oil. Exxon, Shell and Phillips each spend immense sums of money each year for advertising that explains why you should choose their brand of aviation oil over their competitors'.

The fact is, however, that our piston aircraft engines—by virtue of their low RPMs, low operating temperatures, and wide dimensional clearances—have extremely modest lubrication requirements (compared to high-revving automotive engines, for example) that can easily be satisfied by virtually any brand and type of aviation oil.

Blackstone Labs, my favorite oil analysis outfit, maintains an extensive database of wear metals produced by almost all types of piston aircraft engines. They conducted a study that compared the wear metals generated by 571 Lycoming IO-360 engines using four types of oil: Aeroshell W100, Aeroshell 15W-50, Exxon Elite

20W-50, and Phillips 20W-50. They found no significant difference in wear metals between the four types of oil.

Another important oil-related issue is preventing corrosion (rust) when an engine is dormant for a period of weeks or months—something that seldom happens to "working airplanes" (charter, rental, flight school, flying club, etc.), but is unfortunately common among owner-flown airplanes. Rust is the #1 reason that engines fail to make manufacturer's recommended TBO, so preventing it is extremely important.

This is where controversy abounds. Both Aeroshell and Exxon go to great lengths in their advertisements to proclaim the superior corrosion-preventive capabilities of their flagship multigrade oils (Aeroshell 15W-50 and Exxon Elite 20W-50) and offer all sorts of laboratory test results to substantiate their claims. Similarly, an independent study performed by and published in *The Aviation Consumer* confirmed that these two oils were superior at preventing rust on steel plates in a salt-water-cabinet test.

Although any approved aviation oil will easily meet your engine's lubrication requirements, I generally recommend single-weight oil (like this Aeroshell W100 Plus) for owner-flown airplanes that fly irregularly and sometimes sit unflown for weeks at a time.

However, I've discussed this issue with many experienced engine builders at most of the leading aircraft engine overhaul shops, and they have a very different view. Almost without exception, the engine builders recommend that owner-flown aircraft that fly irregularly (as so many do) and are therefore at high risk for rust damage are better off using a thick single-weight oil (such as Aeroshell W100 or W100 Plus) and avoiding the use of multigrade oils except when necessary due to cold-weather operations. I personally agree with this recommendation. For the past 50 years, I've used nothing but single-weight Aeroshell W100 in my aircraft and have enjoyed engine longevity that is nothing short of phenomenal. I add a pint of ASL CamGuard at each oil change for additional anti-corrosion and anti-wear protection.

Do it yourself?

Changing the engine oil is a procedure approved for holders of private pilot certificates on aircraft they own and operate. FAR 43 Appendix A lists this and other tasks designated as "preventive maintenance" that a pilot-rated aircraft owner who isn't an A&P or certificated repairman may perform on his or her own recognizance without supervision. To do your own oil changes, you must have the proper maintenance manuals and proper tools, and you must make a FAR 43.9 logbook entry documenting your work in the aircraft maintenance records. Advisory Circular (AC) 43-12a provides additional information about owner-performed preventive maintenance. It's wise to ask an A&P to show you how to do it properly before you attempt it unsupervised.

An oil change is one of the simplest but most important forms of preventive maintenance an aircraft owner can perform. It's a terrific opportunity to get a close look at your engine compartment, become familiar with what's under the hood, and ensure there's nothing leaking, burning or chafing that might come back to bite you in flight. By doing this yourself, you can give your engine some much-needed TLC, save yourself some money, and learn a lot about your airplane. Highly recommended!

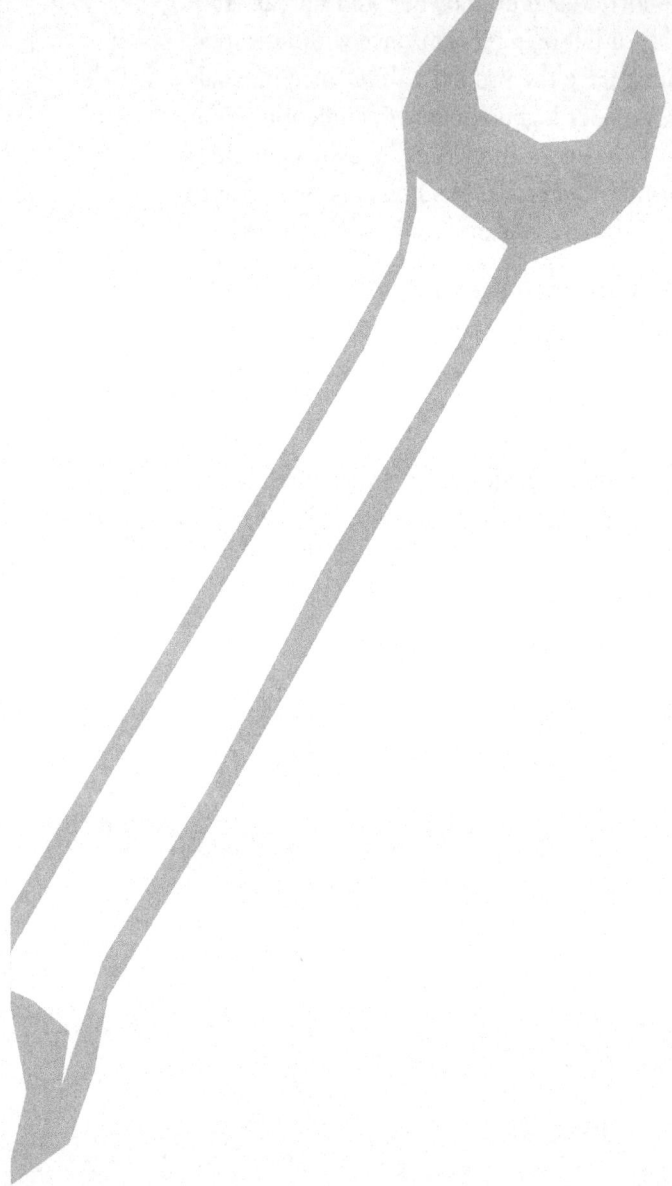

380 ENGINES

40
Spark Plugs

They start the fire going, but there's more to them than that.

We all know what aircraft spark plugs do: They accept high-voltage pulses from a magneto or electronic ignition unit and produce an electric spark inside the cylinder's combustion chamber to ignite the air/fuel mixture and initiate a flame front. They do this about 20 times per second under hostile conditions of extreme temperature and pressure.

A spark plug has three coaxial components: a center electrode, a ceramic insulator, and a barrel. The steel barrel is threaded on both ends, one end that screws into the cylinder head and the other that mates with the ignition lead. The insulator is made from aluminum oxide ceramic to provide good strength, thermal conductivity and dielectric properties. The center electrode comes in two flavors: massive and fine-wire.

Aviation spark plugs use a part numbering scheme that specifies various important parameters about the plug's type, design and specifications. Both manufacturers of aviation spark plugs—Champion Aerospace and Tempest Plus—use

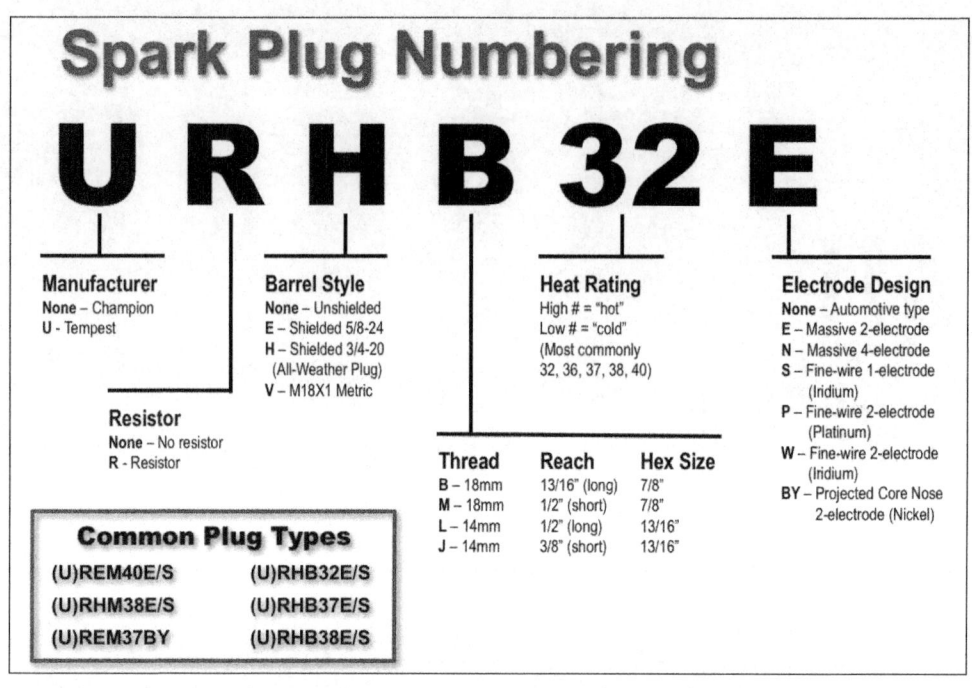

The part number specifies various parameters concerning the spark plug's type, design, and specifications.

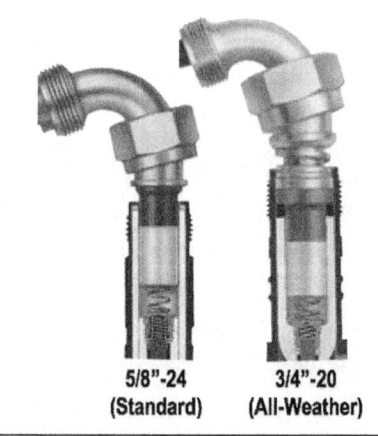

Different barrel designs require different ignition harnesses.

the same numbering scheme, except that Tempest part numbers have a "U" prefix and Champion part numbers don't. So, for example, an RHB32E plug from Champion is interchangeable with a URHB32E plug from Tempest.

The "R" in the part number designates that this is a resistor plug (as almost all aviation plugs are nowadays). The resistor serves as a current limiter that results in a longer-duration lower-current spark that reduces electrode erosion and prolongs the service life of the plug.

The "H" specifies that the ignition-lead end of the barrel has 3/4-20 threads to mate with an "all-weather" harness. An "E" in this position would denote 5/8-24 threads, and an "M" would mean 18mm metric threads. The barrel design of the plug must conform to the thread size of the ignition harness caps.

The "B" specifies that the cylinder-head end of the barrel has 18mm threads that are 13/16 inch long—a so-called "long-reach" plug. An "M" in this position would denote a "short-reach" plug with 18mm threads that are only 1/2 inch long. ("L" and "J" denote 14mm threads.) The reach of the spark plug must correspond to the length of the threaded Heli-Coil insert in the cylinder head spark plug boss; using the wrong-reach plug can be a catastrophic mistake.

The "32" specifies the "heat rating" of the plug, which is a measure of its ability to transfer the heat of combustion from the nose core to the cylinder head. Hot plugs have a long insulator nose that creates a long heat dissipation path; cold plugs have a shorter insulator nose and shorter heat dissipation path. The plug's nose core must operate hot enough to minimize carbon and oil deposits that could foul the plug, but cool enough to prevent preignition. Ideal temperature is somewhere between 1,000°F and 1,300°F. Generally, turbocharged and high-compression engines need cooler plugs (29-32) to avoid preignition, while low-compression engines need hotter plugs (36-40) to minimize fouling.

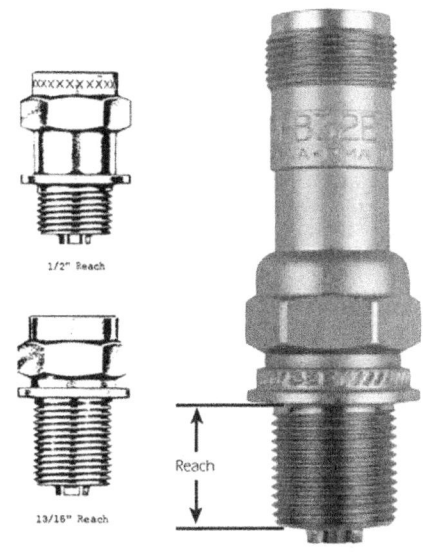

The reach of a spark plus is the length (in inches) of the threaded end that goes into the cylinder. Installing a plug with the incorrect reach can be catastrophic!

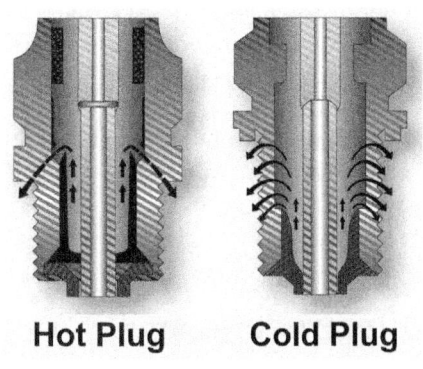

Cooler plugs are generally used in high compression and turbocharged engines to protect against preignition. Hotter plugs are typically used in low compression engines to protect against carbon and lead fouling.

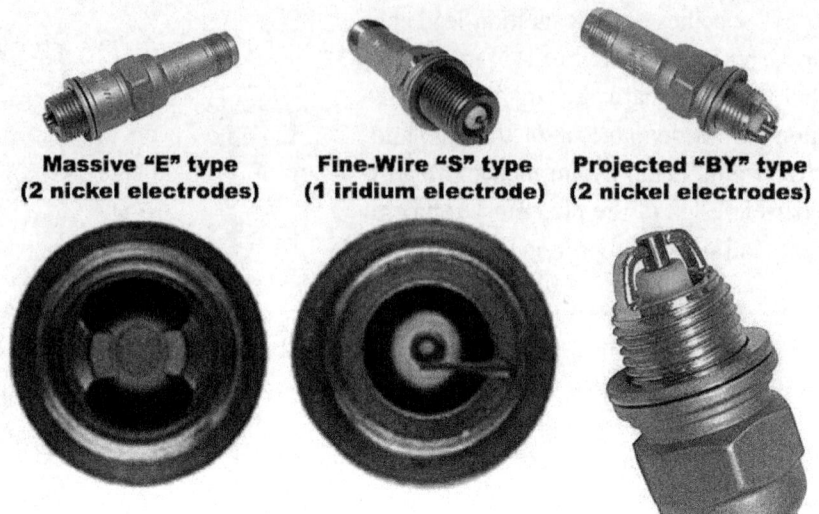

Spark plug electrodes can be massive, fine wire, or projected BY types.

Finally, the "E" specifies a massive center electrode with two ground electrodes, while an "S" would denote a fine-wire center electrode with one iridium ground electrode and a "BY" would denote a projected-core center electrode with two ground electrodes. E-style plugs are most common. BY-style plugs are designed to reduce fouling problems, while the exotic S-style fine-wire iridium plugs are both resistant to fouling and much longer-lasting than conventional massive-electrode plugs (but also much more expensive).

Common aviation spark plugs include (U)REM40E/S, (U)RHM38E/S and (U)REM37BY short-reach plugs and (U)RHB32E/S, (U)RHB37E/S and (U)RHB38E/S long-reach plugs.

Electrode wear

Whenever an electric spark jumps the gap between the spark plug's center electrode and ground electrode, one of the electrodes loses a trace amount of metal due to ionization. Since this occurs roughly 75,000 times per hour when the engine is running, this loss of metal from electrodes gradually increases the gap size and eventually erodes the electrode to the point that the plug becomes unserviceable and should be replaced.

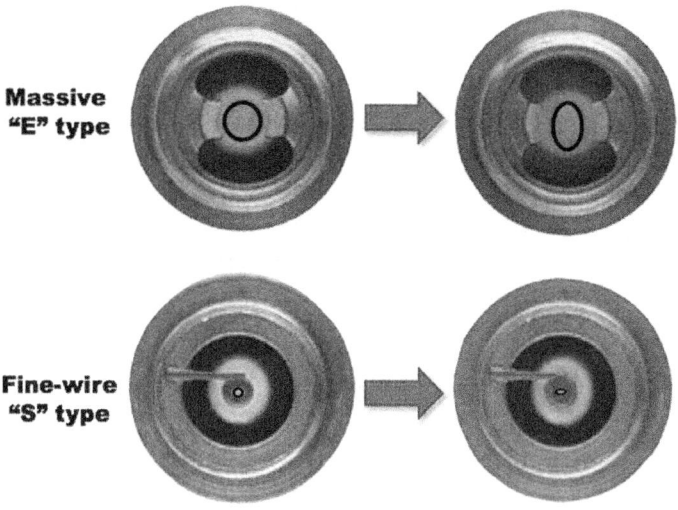

Plugs should be scrapped if either the center or ground electrode reaches one half its original dimension.

Center electrodes start off with a circular cross-section, and gradually become football-shaped in service due to erosion. When the short dimension of the electrode reaches one-half the long dimension, it's time to retire the plug. Similarly, when ground electrodes wear to half their original dimension, the plug should be retired. Simple go/no-go gauges are available to facilitate deciding when a plug has reached end of life.

Magnetos generate high-voltage pulses of alternating electrical polarity. When a spark plug is fired with a negative pulse, metal is lost from its center electrode. When fired with a positive pulse, metal is lost from the ground electrode(s). Consequently, in a horizontally-opposed engine with an even number of cylinders, half the plugs in the engine suffer erosion primarily of the center electrode, while the other half of the plugs suffer erosion primarily of the ground electrode(s). By rotating the spark plugs regularly to positions of opposite polarity, wear of the center and ground electrodes can be balanced, and the useful life of the plug can be extended significantly.

Also, in horizontally-opposed engines, bottom spark plugs are much more prone to contamination and fouling than are top spark plugs. During plug rotation, we usually move plugs from bottom to top (and vice versa) to equalize the time that each plug spends doing latrine duty over the course of its useful life.

Note that radial engines have an odd number of cylinders, and their spark plugs fire with alternating polarity, so plug rotation is not usually necessary on these engines.

Preventive maintenance

Traditionally, spark plugs are removed for cleaning, gapping and rotation every 100 hours of engine operation. An alternative procedure that has worked well for me is to do regular in-flight LOP mag checks (as described in Chapter 34) and to do spark plug maintenance only when the in-flight checks reveal less-than-stellar ignition performance. I've found that this "on-condition" approach reduces how often I do spark plug preventive maintenance.

When it's time to do preventive maintenance on spark plugs, the first step is to remove the plugs from the engine, place them in a spark plug tray (so you won't forget which plug came from which position in which cylinder), and inspect them carefully under a bright light. Look for things like abnormal electrode wear, cracked nose core insulators, oil, carbon and lead deposits, electrode bridging, etc.

Inspect the plugs under a bright light. The condition of the electrodes and nose core insulator can reveal a lot about how the cylinder has been operating.

If plugs suffer from persistent fouling, more often than not the problem can be solved by changing how the engine is operated. The use of aggressively lean mixture settings—particularly during idle, taxi, and other ground operations—is usually sufficient to eliminate fouling problems. If fouling persists, the use of fine-wire (S-suffix) or projected-core (BY-suffix) plugs can help. In the case of lead fouling, using a lead-scavenging fuel additive like Alcor TCP may also help.

Once the plugs have been inspected, the next step is to clean them. You can do this using an inexpensive grit blaster powered by compressed air. If you do a lot of spark plugs, more elaborate and expensive cleaners are available. Grit blasting removes oil and carbon deposits from the plugs, but sometimes small globules of metallic lead (from the TEL in avgas) adhere stubbornly to plug electrodes and nose core insulators and can't be dislodged by grit blasting. Such lead globules may need to be removed using a sharp pick or an electric vibrator.

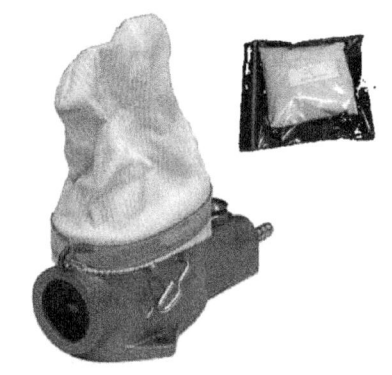

Spark plugs can be blasted clean with an inexpensive grit blaster run from an air compressor.

Once the plugs have been cleaned, you'll want to measure the electrode gap(s) and adjust them as necessary. A special round-wire feeler gauge is used to measure each gap. Don't try to use a flat-blade feeler gauge, because you'll get inaccurate measurements if you do.

The proper gap for most aviation spark plugs is between 0.016 and 0.021 inch. I recommend gapping spark plugs "tight" at 0.016 inch to provide maximum "headroom" for the gap to grow in service before re-gapping is necessary. If you fly a turbocharged aircraft, such "tight" gapping also reduces the risk of high-altitude misfire.

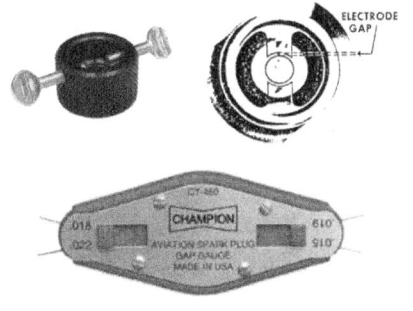

Spark plug gap should be carefully checked to ensure proper operation. The typical plug gap is 0.016 inches.

Massive-electrode plugs (E-suffix) will typically need to have their gaps adjusted every 100 hours or at each on-condition preventive maintenance cycle. Fine-wire plugs (S-suffix) hold their gaps much longer; they should be measured every cycle and re-gapped only when gap is excessive.

Once they have been re-gapped, massive-electrode plugs can be checked with a simple go/no-go gauge to determine whether they should remain in service or be replaced. Fine-wire plugs can't be checked this way and must be evaluated using the eyeball method. My experience is that massive-electrode plugs generally last 400 to 500 hours, while fine-wires are good for at least 1,500 hours and often make it to engine TBO without having to be replaced.

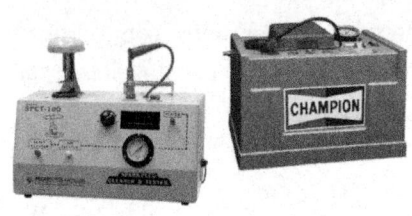

Spark plug "bomb testers" allow the spark to be viewed when the plug is operated under pressures similar to actual combustion pressures.

Plugs that pass electrode dimensional checks may be tested for proper electrical performance using a contraption called a "bomb tester" which fires the plug in a chamber pressurized with compressed air and equipped with a mirror and viewing window allowing you to tell if the plug is yielding a strong and steady spark across its electrodes. Note that this "bomb test" is not nearly as demanding or revealing as the in-flight LOP mag check described in Chapter 34, but it's something that can be done in the maintenance hangar.

Installing plugs

Whether you're reinstalling plugs after cleaning and gapping or installing brand new plugs, it's crucial to use the right tools and procedures to ensure correct installation. The plugs to be installed should start off in a spark plug tray so that you know what cylinder and position they occupied when they were removed.

Remove one spark plug from the tray and remove and discard its used copper gasket. Install a new copper gasket with the flat side facing the spark plug flange. (For 18mm plugs, use Champion M674 or Tempest U674 gaskets.) While it's possible to reuse the old, work-hardened copper gasket if you anneal (soften) it by heating red-hot with a torch, I always install new gaskets because they're so cheap (less than 50 cents each) that I don't think it's worth the hassle of annealing and reusing old ones.

Carefully apply a small amount of aircraft spark plug thread lubricant and anti-seize to the second and third threads of the plug barrel. Use either Champion 2812 or Tempest T556. Apply to the threads sparingly, being extremely careful not to allow any of the thread lube to drip onto the electrodes or nose core insulator, otherwise the highly conductive lube might short out and ruin the plug.

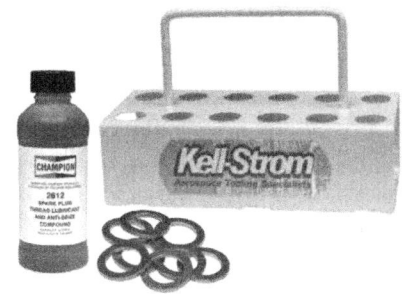

When reinstalling spark plugs, use anti-seize compound sparingly and torque them to manufacturer's specifications with fresh copper gaskets.

Install the spark plug into its new rotated location following the rotation diagram. This ensures that each plug moves from bottom to top (or vice versa) and from positive to negative polarity (or vice versa). Start the lube-prepared end of the plug into the threaded cylinder head boss using only your fingers. If the plug resists being screwed into the head with finger pressure, you may have cross-threaded it—so remove it and try again. Never force it with a wrench because you could ruin the plug or (even worse) the Heli-Coil insert in the cylinder head. In stubborn cases, you might want to use an aircraft spark plug "thread chaser" tool to clean up the Heli-Coil threads.

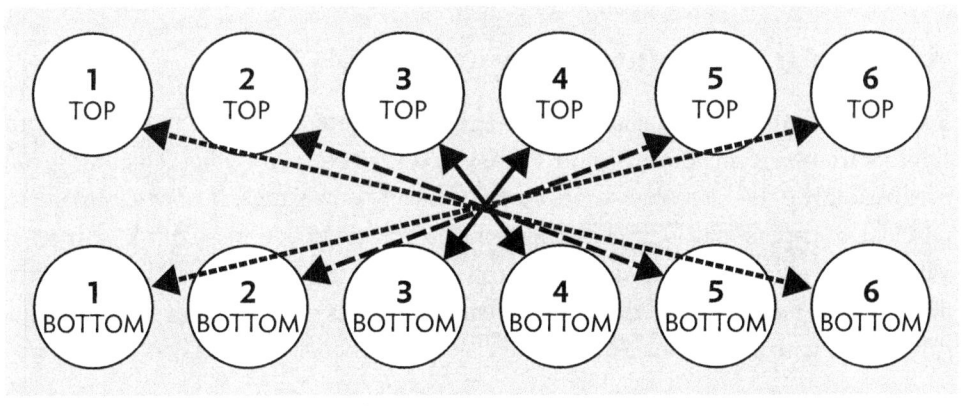

This rotation sequence assures that plugs move from top to bottom and from negative to positive polarity. (The sequence for a six-cylinder engine is shown here, but the sequence for a four-cylinder engine is similar.)

Once the plug has been screwed into the cylinder head boss finger-tight, use a special-purpose deep aircraft socket and a calibrated torque wrench to tighten the spark plug to its final torque. Continental calls for 25-30 foot pounds (300-360 inch-pounds) of torque, while Lycoming calls for 30-35 foot-pounds (360-420 inch-pounds). Never install aircraft spark plugs using the TFAR (that feels about right) method; always use a torque wrench.

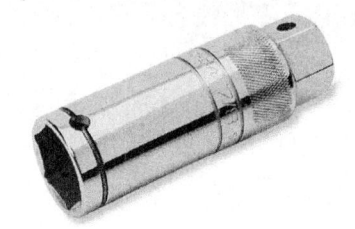

Always install spark plugs with a special-purpose deep aircraft spark plug socket and a calibrated torque wrench.

Prepare the ignition lead by wiping its contact spring and "cigarette" insulator with a lint-free rag moistened with MEK or acetone. Insert the cigarette into the spark plug and screw on the cap finger-tight, then use a 7/8-inch wrench to tighten the cap while holding the lead with a 7/16-inch end wrench to prevent it from twisting. It's extremely important not to overtighten the cap (a rookie mistake), because doing so will make it nearly impossible to remove next time. While most A&Ps use the TFAR method here, I prefer to use a torque wrench to tighten the ignition lead cap to the recommended 10 foot-pounds (120 inch-pounds).

Another do-it-yourself opportunity

Spark plug maintenance is a task that aircraft owners are permitted to do themselves without A&P supervision. It's easy to do, not terribly time-consuming, and requires only a few inexpensive tools. It's also a lovely excuse for owners to get an up-close-and-personal look at their engines on a regular basis—think of it as an advanced preflight inspection. I love to see owners get involved with the maintenance of their aircraft, and doing your own oil changes and spark plug maintenance is a great way to get involved.

If you choose to do it yourself, there are various rookie mistakes to avoid. I've already mentioned several of these: Over-applying thread lube, overtightening ignition lead caps, allowing a lead to twist while tightening the cap, and reusing a work-hardened copper gasket. Don't do these things.

Another common mistake is dropping a spark plug during removal or installation. Never reuse a spark plug that has been dropped; toss it in the trash. Dropping an aviation spark plug just a few inches onto a hard surface can fracture its fragile nose core insulator. Even an invisible hairline insulator crack can lead to destructive preignition that could trash your engine and ruin your day. Enough said.

Always use the right tools. Never attempt to adjust the gap of a spark plug using needle-nose pliers or a vise or an automotive gapping tool; that's a sure way to ruin the plug. Always use a proper aircraft spark plug gapping tool. There are inexpensive gapping tools available for both massive-electrode and fine-wire plugs, and you should always use the proper tool for the type of aircraft plug you use. Be cautious when adjusting the gap, because if you get the gap too small you can't enlarge it without wrecking the plug. Creep up on the proper gap slowly, checking frequently with your feeler gauge and making sure not to overshoot your goal.

Aircraft Spruce and Specialty Co. sells a nice "aircraft spark plug maintenance kit" for $120 that contains almost everything you need to maintain your spark plugs: spark plug tray, spark plug cleaner, feeler gauge, gap setting tool for massive-electrode plugs, aircraft spark plug socket, some copper gaskets and a bottle of thread lubricant/anti-seize. The only additional thing you'll need is a suitable torque wrench, which you can pick up for about $20 from Harbor Freight. That's a total investment of $140 that'll more than pay for itself the very first time you do your own spark plug maintenance instead of paying your shop $100/hour to do it.

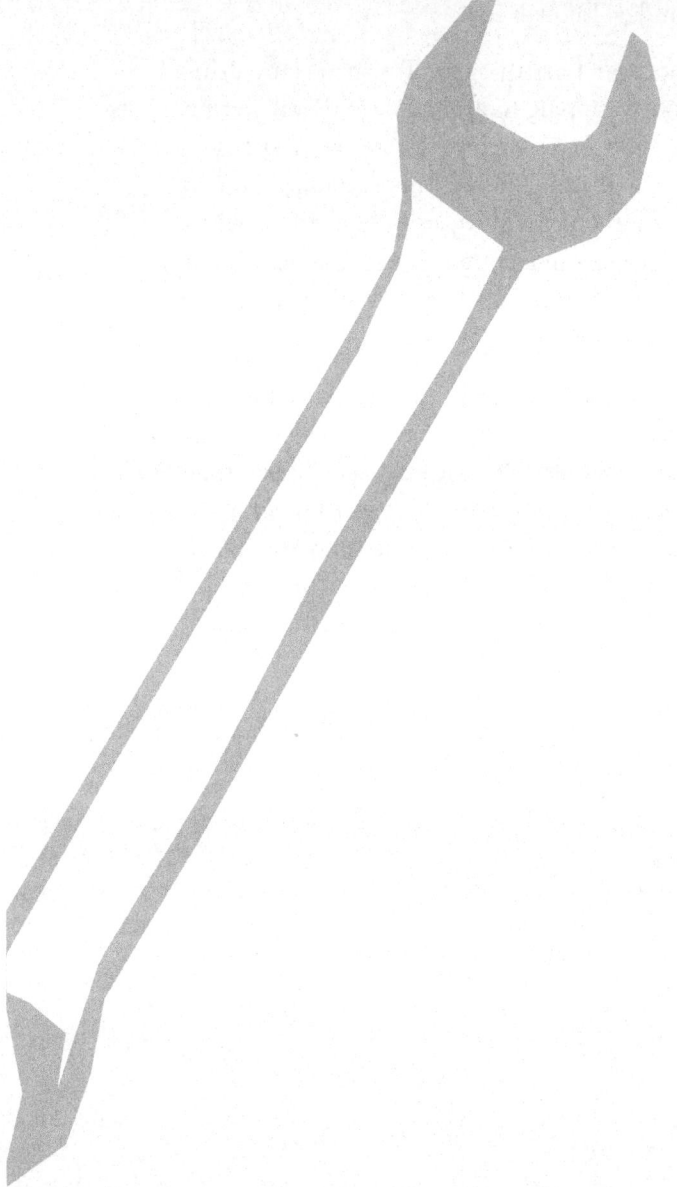

41
The Dark Side of Maintenance

Maintenance has a dirty little secret: It sometimes hurts more than it helps.

How often have you had the experience of putting your airplane in the shop—perhaps for an annual inspection, to correct some squawk, or even for a routine oil change or spark plug servicing—only to discover when you get the airplane back and take it aloft for the first time after maintenance that something that used to work fine no longer does?

As I discussed in my previous book *Manifesto*, maintenance has a dark side that we don't often hear discussed, especially by mechanics: Although maintenance is supposed to make our aircraft safer and more reliable, the fact is that it sometimes accomplishes exactly the opposite.

When something in an aircraft fails due to something that a mechanic did—or neglected to do—I refer to it as a "Maintenance-Induced Failure," or "MIF" for short. Such MIFs occur a lot more often than anyone cares to admit.

How often do MIFs occur?

MIFs seem to occur quite frequently. All too often, I receive an email or read an online post in which a frustrated aircraft owner is complaining about some aircraft problem that is obviously a MIF.

For example, I was contacted by the owner of a 1974 Cessna 182P. He explained that he'd put the plane in the shop for a routine oil change and installation of an STC'd exhaust fairing. A couple of months later, he decided to have a digital engine monitor installed.

The new engine monitor revealed that the right bank of cylinders (#1, #3 and #5) all had very high CHTs, well above 400°F. This had not shown up on the standard factory CHT gauge because its probe was installed on cylinder #2. (Another good reason that every piston-powered aircraft should have a digital engine monitor.)

At the next annual inspection (performed by a different A&P), the inspector discovered some induction airbox seals missing, which the owner was convinced were left off when the exhaust fairing was installed. The missing seals were installed during the annual, and CHTs returned to normal.

Sure sounds MIFfy, doesn't it?

Unfortunately, the problem was not caught and corrected early enough to prevent serious heat-related damage to the right-bank cylinders. All three jugs had compressions down in the 30s with leakage past the rings, and a borescope inspection revealed visible damage to the cylinder bores. Oil consumption increased from one quart in 12 hours to one quart in 2 hours, and the oil in the sump started turning jet black within 10 hours after oil change. The owner now faced replacing three cylinders, and since he had no way of proving that the first A&P left out the airbox seals, he was on the hook for the cost of the three jugs—probably around $6,000 including labor.

Maintenance is invasive!

Most owners and mechanics don't think enough about the fact that maintenance is inherently invasive. Any time a mechanic takes something apart and puts it back together, there's a risk that something won't go back together quite right, and the result will be a MIF. Some maintenance operations are more invasive than others; the more invasive the maintenance, the greater the risk of a MIF.

In the world of medicine, invasiveness is taken quite seriously. If you develop gallstones, for example, the traditional treatment had long been cholecystectomy (gall bladder removal), which is major abdominal surgery in which the surgeon removes the gall bladder through a 5- to 8-inch incision. Recovery typically involves a week of post-surgical hospitalization, followed by several weeks of recovery at home. This standard treatment is extremely invasive, and so not surprisingly the incidence of complications and even death is significant. (My dad nearly died as the result of complications following an open cholecystectomy operation).

So, the medical community developed a less invasive procedure called laproscopic cholecystectomy in which the traditional large incision is replaced by several tiny incisions, and the surgery is performed using a tiny video camera inserted through one of the incisions and various microsurgery instruments inserted through the others. This procedure is far, far less invasive than the traditional open procedure, and recovery usually involves only one night in the hospital and a few days at home. More important, the risk of complications is substantially reduced. Consequently, the laproscopic procedure has now replaced the open procedure as the first-choice treatment for gallstones (although about 5% of the time, the laproscopic procedure proves infeasible and the surgeon must switch to the more invasive open procedure).

> Most owners and mechanics don't think enough about the fact that maintenance is inherently invasive.

Even less invasive treatments for gallstones exist. In some cases, the stones may be dissolved slowly by taking a long course of oral medication (Actigall or Chenix), or quickly by direct injection of a drug (methyl tert butyl) into the gall bladder. Extracorporeal shockwave lithotripsy (ESWL) has also been used to break up gallstones with shock waves, although the success rate hasn't been very high.

Likewise, some aircraft maintenance procedures are more invasive than others. The more invasive a procedure is, the greater the risk of a MIF. Therefore, when considering any maintenance task, we should always think carefully about how invasive it is, whether the benefit of performing the procedure is worth the risk, and whether less invasive alternatives are available.

Awhile back, for example, I received an email from an aircraft owner who said that he'd recently received an oil analysis report showing an alarming increase in iron. The oil filter, however, showed no visible metal. The lab report suggested flying another 25 hours and then submitting another oil sample for analysis.

The owner showed the oil analysis report to his A&P, who expressed real concern that the elevated iron levels might indicate that one or more cam lobes on his Continental engine were coming apart. The mechanic suggested pulling one or two cylinders and inspecting the camshaft. The owner wisely decided to seek a second opinion before authorizing something as invasive as cylinder removal, so he emailed me to ask for my recommendation.

Unless you see something like this in your oil filter, chances are that your cam and lifters are okay.

I advised the owner that in my opinion, the elevated iron was almost certainly NOT due to cam lobe spalling. I explained that a disintegrating cam lobe throws off fairly large particles or whiskers of steel that are usually clearly visible during oil filter inspection. The fact that the oil filter was clean suggested that the elevated iron was coming from microscopic metal particles less than 50 microns in diameter, too small to be detectable in a filter inspection, but easily detectable via spectrographic oil analysis. Such tiny particles were probably coming either from light rust on the cylinder walls (if the aircraft had been inactive for awhile), or from some very slow wear process.

I suggested to the owner that a borescope inspection of the cylinder barrels (a very non-invasive procedure) would be a good idea to see whether the cylinder bores showed evidence of rust. I also advised that no invasive maintenance procedure should ever be undertaken solely based on a single oil analysis report. I thought the oil lab was spot-on by recommending that the aircraft should be flown another 25 hours and another oil sample submitted.

I went on to explain that even if a cam inspection was warranted (and I didn't think it was), there was a far less invasive method of accomplishing it. Instead of a 10-hour cylinder removal, the mechanic could do a 1-hour removal of the intake and exhaust lifters for inspection, and then determine the condition of the cam by inserting a pick into the lifter boss, rotating the propeller, and determine whether the cam lobe had any pits sufficient to grab the tip of the pick. Not only would this procedure involve about 10% of the labor of cylinder removal, but the risk of a consequential MIF would be almost nil.

Sometimes, less is more

Many owners seem to believe—and many mechanics seem to preach—that preventive maintenance is inherently a good thing, and the more of it you do the better. I consider this a wrongheaded view.

To the contrary, I believe we often do far more preventive maintenance than necessary, and we often do it using excessively invasive procedures. By doing this, we increase the likelihood that our maintenance efforts will cause failures rather than preventing them.

In my earlier book *Manifesto*, I wrote at length about Reliability-Centered Maintenance (RCM)—a concept developed at United Airlines in the late 1960s, and universally adopted by the airlines and the military during the 1970s and 1980s. One of the major findings of RCM researchers was that preventive maintenance often does more harm than good, and that safety and dispatch reliability can often be improved substantially by reducing the amount of preventive maintenance we do, and using the least invasive methods possible.

Sadly, this way of thinking hasn't seemed to trickle down to piston GA maintenance, and is considered absolute heresy by most mechanics because it contradicts everything they were taught in A&P school. The long-term solution is that GA mechanics need to be educated about RCM principles, but that isn't likely to happen any time soon.

In the meantime, aircraft owners can improve the situation by exercising caution before authorizing any invasive maintenance procedure on their aircraft—and doing what the above-mentioned owner did: get a second opinion.

Finally, aircraft owners need to fully appreciate that the most likely time for a mechanical failure to occur is the first flight after maintenance, and that the risk of such MIFs is very substantial. It's therefore imperative that owners conduct a post-maintenance test flight—without passengers, in VMC, preferably in the vicinity of the airport just in case something bad happens—before launching into the clag or putting passengers at risk. In my judgment, even the most innocuous maintenance task—e.g., a routine oil change—deserves such a post-maintenance test flight.

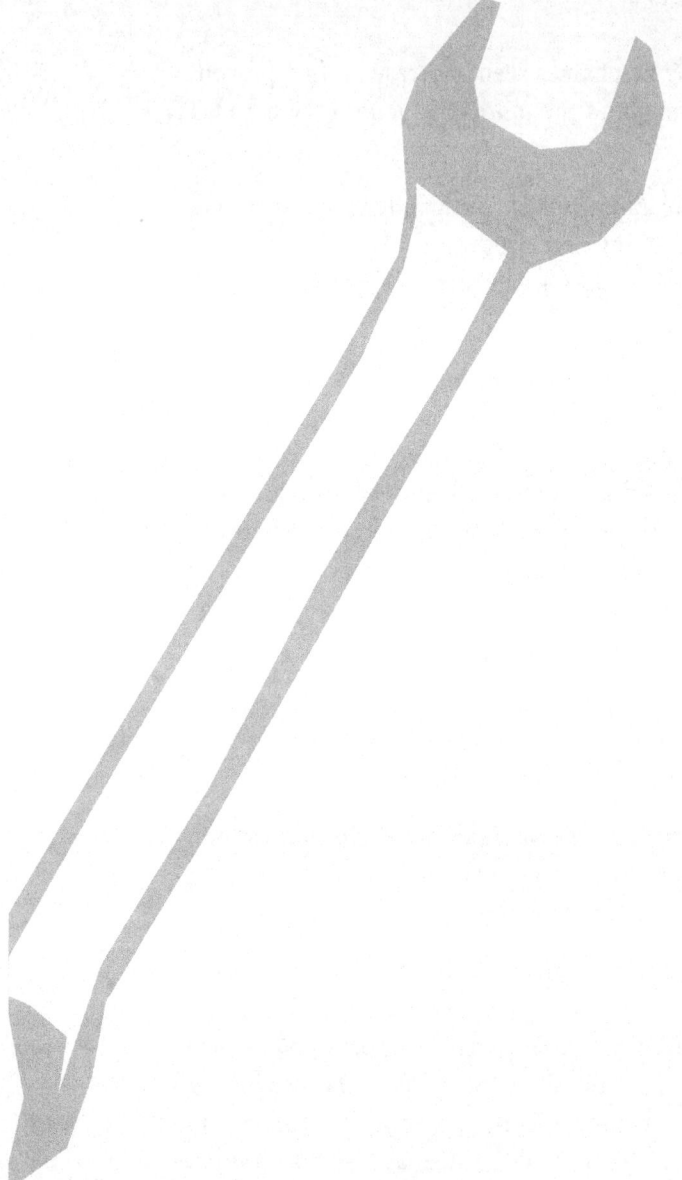

42
Watch Your Language!

When requesting maintenance, the words you use can be very important.

The voice on the phone identified himself as a Cessna 182 owner—let's call him Jim—who said he was considering overhauling his O-470-R engine and could use some advice. I asked Jim why he was considering overhauling his engine, thinking maybe it was approaching TBO and hoping to convince him this wasn't a valid reason to tear down a perfectly healthy engine. As it turned out, this engine wasn't yet close to TBO.

Jim said his A&P recently pulled a cylinder because it had a leaking exhaust valve. With the jug removed, the mechanic discovered a badly worn cam lobe and showed it to Jim, explaining that the engine would need to be removed from the airplane and have its crankcase split to replace the damaged cam. The mechanic offered to do the work himself in-house (rather than send the engine out), and estimated the engine repair would come to roughly $10,000, including parts and labor. Jim agreed and authorized the mechanic to proceed.

In a follow-up conversation, Jim asked his A&P whether it might make sense to overhaul the engine, given that it was already going to be opened up for a cam transplant. The mechanic replied that he'd be glad to do a full overhaul on the engine, but that it would be a lot more expensive by at least a factor of three.

An engine overhaul can cost up to three times more than a simple engine repair which requires the case to be split.

Jim wasn't expecting that, and inquired why overhauling the engine would be so expensive. The mechanic pulled out Continental Standard Practice Maintenance Manual M-0 and showed Jim all the parts that would need to be replaced at overhaul. These included, among other things, a pair of new gears (crankshaft and camshaft) that alone would add $5,000 to the tab. There were 24 additional categories of parts that must be replaced with new ones at overhaul. All the engine accessories—starter motor, starter adapter, alternator, magnetos, carburetor—would have to be sent out for overhaul. All the cylinder assemblies would have to be sent out for overhaul or replaced with new ones. The list was rather daunting.

"Do I really have to do all that stuff?" Jim asked me. He pointed out that Continental's list of "mandatory replacement parts" was contained in the M-0 manual. "I thought Part 91 operators like me were not required to comply with manufacturer's maintenance instructions," he added, "but my mechanic says he's required by regulation to replace all those parts at engine overhaul. Who's right?"

"You're both right," I replied, and paused for a moment while the irony of that apparent contradiction sunk in before starting my explanation.

"As a Part 91 operator, you aren't required by regulation to comply with any manufacturer's instructions unless compelled to do so by an Airworthiness Directive," I explained. "Nor are you required by regulation to overhaul your engine...now or ever! Legally, you could run that engine for 20,000 hours—repairing it whenever it breaks—and never have the word 'overhaul' appear in your engine logbook."

"However," I continued, "if you ask your mechanic to work on your engine and sign that work off as an 'overhaul,' he is required by regulation to follow Continental's guidance to the letter, and that includes complying with the "must replace at overhaul" guidance in Continental's Manual M-0. So if you don't want him to replace those $5,000 gears and you're happy with your fuel pump and alternator, just ask him to log his engine work as a 'repair' instead of as an 'overhaul' and then he can do—and not do—whatever you ask."

Watch your language!

When an aircraft component (such as an engine, propeller, instrument, accessory or other appliance) becomes inoperative, unairworthy, or just plain sick or worn out, we usually have several options: We can replace the bad component with a new one or a rebuilt one, or have our defective component overhauled or repaired.

These four words—new, rebuilt, overhauled and repaired—are terms of art that have distinct, specific meanings in the context of aircraft maintenance. Those meanings are defined in the FARs. It's crucial for owners to understand precisely what they mean and how they differ.

"New" means never used. Dimensionally, a new component meets new fits and limits (obviously).

"Rebuilt" means a previously used component that has been overhauled to new fits and limits (possibly using approved oversize or undersize parts) by the original

"New", "rebuilt", "overhauled", and "repaired" each have unique meanings defined in the FARs.

manufacturer. For example, only the Continental factory in Mobile, Alabama, can "rebuild" a Continental engine, although any A&P mechanic or engine shop may "overhaul" one. When the factory rebuilds a component, it receives a new data plate with a new serial number; if it's an engine it receives a new zero-time logbook with no record of previous operating hours or history.

"Overhauled" means disassembled, cleaned, inspected, repaired as necessary, reassembled and tested in accordance with the manufacturer's approved technical data—

normally the overhaul manual as supplemented by service bulletins. The word "overhaul" implies conformance to service limits—not necessarily new limits—so if you want new limits you must specify a "new-limits overhaul." A new-limits overhaul is essentially the same as "rebuilt" except that it doesn't have to be performed by the original manufacturer and doesn't receive a new serial number or a zero-time logbook.

"Repaired" means inspected and repaired as necessary ("IRAN") to restore the inoperative component to proper working condition. This term implies nothing about fits and limits, because there is no requirement to measure anything when performing a repair. One could, for example, remove a cylinder, replace a burned exhaust valve and guide, then reinstall the cylinder back on the engine without measuring anything, and call it a "repair."

A repair differs from an overhaul primarily in that there's no obligation to follow the fits, limits, mandatory parts replacements, and other procedures in the manufacturer's overhaul manual. The FAA's late great Bill O'Brien—who long served as the agency's top maintenance guru—used to drive this point home to mechanics who attended his IA renewal seminars by using this catchphrase: "If you used a micrometer, then it's an overhaul; if you didn't, then it's a repair."

The O-word is one of the most expensive and overused words in aviation maintenance.

When an owner asks a mechanic to "overhaul" something instead of "repair" it, he virtually ties the mechanic's hands. The mechanic is no longer able to use discretion as to which parts are worn out and need to be replaced and which parts look fine and can be retained. He must slavishly replace 100% of the parts on the manufacturer's mandatory replacement list (even if they look to be in perfect condition), and he must measure all the other parts and replace them if they fail to meet the manufacturer's service limits.

The words "overhauled" and "rebuilt" are defined in FAR 43.2, and have these very specific regulatory meanings. If a mechanic documents that something is "overhauled" and hasn't complied with every jot and tittle of the overhaul manual and other manufacturer's guidance, he can lose his A&P certificate. However, if he documents that something is "repaired," he can do as much or little as he sees fit to do, so long as he is satisfied that his repair work was done properly in accordance with acceptable methods, techniques and practices.

In short, if you ask for a "repair" you give the mechanic or technician considerable discretion to do only as much work as he believes needs to be done…and that's usually a good thing, assuming the mechanic is competent and you trust his judgment. On the other hand, if you ask for an "overhaul" you eliminate the mechanic's discretion and require him to do everything precisely "by the book."

Watch your wallet…

Repair is almost always the lowest-cost method to get a mechanical problem resolved. Often, the cost of having something overhauled will be much higher than the cost of having it repaired. For example, a propeller "overhaul" is typically at least twice as expensive as an IRAN ("reseal repair"). I've seen cases where having a malfunctioning instrument "overhauled" can cost ten times as much as having it "repaired."

The "O-word" is one of the most expensive and overused words in aviation maintenance. It is invariably a waste of money to have something overhauled if a simple repair will suffice—often a lot of money! Of course, the word "rebuilt" is even more expensive, and "new" is the most expensive word of all.

From my perspective, the object of the game is to use the least expensive word that will make the aircraft safe and get it back in the air. If you hate to waste money on maintenance as much as I do, make a point of using these four words carefully when requesting or approving maintenance work on your aircraft.

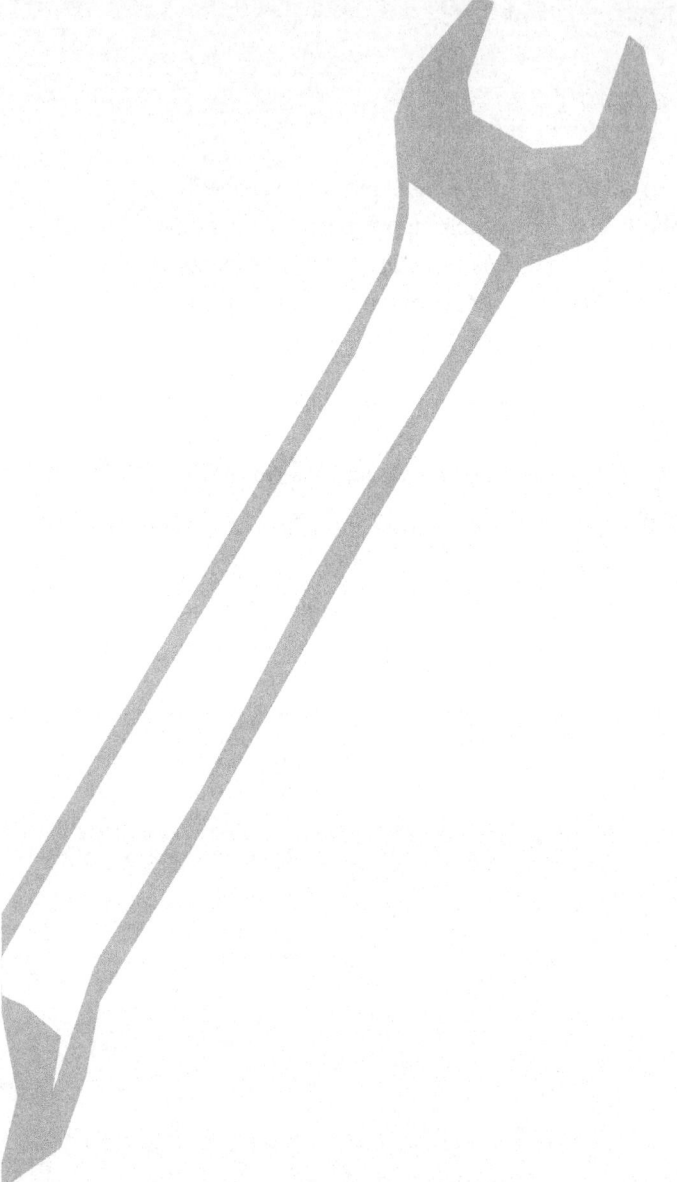

404 ENGINES

43
A Little Dab'll Do Ya... In

Even a seemingly trivial mistake by a well-intentioned mechanic can have dire consequences.

On April 19, 2005, a 1966 Beechcraft Debonair was cruising at 8,500' on a VFR flight from Van Nuys to San Jose, Calif., when the pilot heard a loud noise and the Continental IO-550 engine started running rough. The pilot checked the oil pressure and oil temperature gauges and found both had normal readings in the green. He throttled back, advised Oakland Center that he was diverting to Paso Robles, and began descending.

Then there was another loud bang from the engine compartment, engine RPM dropped dramatically, and the cockpit started to fill with white smoke. The PIC—a 32-year-old 500-hour commercial pilot who had recently earned his CFI—throttled back to idle, made a mayday call to Oakland Center, and established an 80-knot glide. The smoke cleared from the cockpit and the pilot determined that he was within gliding distance of Paso Robles Airport.

Just as it seemed like the story might have a happy ending, the Debonair's lefthand engine cowling door popped open and stayed open. That increased drag and caused

The Debonair made a forced landing in a vineyard two miles short of the Paso Robles (Calif.) airport. The aircraft was not equipped with shoulder harnesses, and the pilot and passenger both sustained serious head injuries.

the rate of descent to increase to nearly 1,000 feet per minute. The pilot now realized he wasn't going to be able to make the airport.

He set up for an off-airport landing in what he thought was a green field about two miles short of the airport. At the last minute he realized that it was a vineyard and his flight path was at a 45-degree angle to the rows. The pilot extended the landing gear in an attempt to mitigate the impact, but the left main gear snagged a grapevine row pole and wires. The aircraft veered left and stopped abruptly.

Both the pilot and his right-seat passenger were wearing lap belts, but the aircraft was not equipped with any upper-body restraints. (Shoulder belts were not offered by Beech when the airplane was built in 1966, and no regulation required that they be retrofitted.) Both pilot and passenger sustained serious injuries when their heads struck the instrument panel. The 44-year-old passenger was knocked unconscious, and his head trauma tragically resulted in permanent brain injury.

NTSB investigation

The NTSB investigated the accident and issued its probable-cause report about a year later. The investigation included a detailed examination of the engine by the Safety Board investigator-in-charge and a technical representative of Continental Motors.

The NTSB examined the Debonair's engine and found that the #4 cylinder and piston had departed the engine in-flight.

The post-accident engine examination revealed that the #4 cylinder and piston had departed the engine in-flight; they were never recovered. The two righthand cylinder hold-down studs appeared relatively undamaged, but their hold-down nuts had backed off. The four remaining hold-down studs were sheared, and both through-bolts were sheared as well. The #4 connecting rod punctured a 3-inch x 7-inch hole in the top of the crankcase. The #4 connecting rod, cap and rod bolts and nuts were recovered from inside the engine cowling, as were most of the sheared-off studs, through-bolts and nuts.

The engine logbook entry for the June 2004 annual inspection conducted 10 months prior to the accident indicated that cylinders 2, 4, 5 and 6 had been removed and replaced with four overhauled cylinders. The engine had subsequently undergone a 100-hour inspection and then a 50-hour oil change. No defects were noted.

The NTSB concluded the obvious:

> The National Transportation Safety Board determines the probable cause(s) of this accident to be the separation of the #4 engine cylinder due to improperly torqued cylinder hold-down nuts. A finding in this accident was the lack of a shoulder restraint system in the airplane, which contributed to the occupants' injuries.

Ah, but the story didn't end there...not by a long shot. As so often happens these days, it ended up in court. The ensuing litigation revealed that there was a lot more to this accident than what the NTSB found.

The smoking gun

The brain-injured passenger sued. Defendants included the PIC, the corporation that owned the Debonair, the repair station that maintained it (and performed the cylinder replacement), and engine manufacturer Continental Motors. The litigation went on for five years and culminated in a jury trial in 2010 in California Superior Court.

The lead attorney for the plaintiff was himself a GA pilot and aircraft owner who had enough experience doing owner-assisted annuals on his Grumman American Tiger that he knew his way around a toolbox, a maintenance manual, and FAR Part 43. In reviewing the photographs taken during the NTSB's examination of the engine, the attorney discovered something crucial that the NTSB missed: several dabs of light brown sealant on the cylinder deck area of the crankcase where cylinders 2, 4, 5 and 6 were attached. This was literally the "smoking gun" that explained precisely why the #4 cylinder departed the engine in-flight.

The "smoking gun"—sealant on the cylinder deck. But how did it get there?

The mating surfaces of the crankcase's cylinder deck and the cylinder's mounting flange need to be perfectly clean when the cylinder is mounted and its hold-down nuts are torqued up. Any sort of contamination of those surfaces—even a thin coat of paint—is enough to cause the hold-down studs to lose their pre-load and ultimately to fail. The presence of even a tiny dab of flexible sealant on the mating surfaces could have disastrous consequences. Any competent A&P mechanic who works on piston aircraft engines knows this (or at least should know it).

So how the heck did the sealant get on the cylinder deck?

Discovery

In a videotaped deposition, the young mechanic who had installed the cylinders in June 2004 readily admitted that he had applied a sealant called "Gasket Maker"

(Continental part number 646942) to the cylinder base O-rings when he installed the four cylinders. He had installed cylinders on Continental engines on several prior occasions and was quite familiar with the process. However, this was the first time he'd ever applied Gasket Maker to the cylinder base O-rings.

The young mechanic also acknowledged that he did not hold an FAA mechanic certificate at the time he did the cylinder work on the accident airplane, although he had earned his A&P subsequently. As an uncertificated apprentice mechanic, he had been working under the supervision of a senior A&P who signed off his work. (The use of uncertified apprentice mechanics is quite common in FAA-certified repair stations and is explicitly permitted by FAR Part 43.)

So why did the apprentice mechanic decide to use Gasket Maker on the cylinder base O-rings this time, a departure from the procedure he'd been taught and used on previous occasions? The answer to that question was interesting.

It turns out that shortly before the Debonair's annual commenced, the repair station had hosted an American Bonanza Society service clinic. A Continental technical representative was in attendance at the service clinic and was touting a new Continental product called Gasket Maker as the ultimate solution to preventing and correcting oil leaks in Continental engines. In his deposition, the apprentice mechanic stated that he asked the Continental tech rep where Gasket Maker could be used and was told "anywhere you have a leak." Immediately after the service clinic ended, the apprentice mechanic ordered a tube of Continental Gasket Maker for his toolbox and was on the lookout for opportunities to use it.

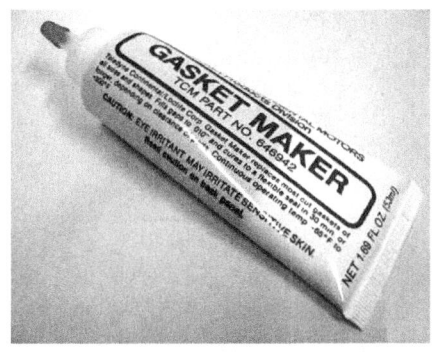

The Continental maintenance manual said that Gasket Maker could be used on "all uncoated oil seals." That turned out to be misleading.

When it came time to install the four cylinders on the Debonair's engine, the apprentice mechanic consulted with a senior A&P to determine whether using Gasket Maker on the cylinder base O-rings would be appropriate. The A&P had no familiarity with the Continental Gasket Maker product, so he consulted the section of the Continental IO-550 maintenance manual that listed approved lubricants, sealants and adhesives. In that section, he found a reference to Gasket Maker that listed several specific applications, followed by a catch-all item indicating that Gasket Maker was appropriate for use on "all uncoated oil seals."

The A&P then looked up the cylinder base O-ring in the IO-550 illustrated parts catalog and found that it was listed there as a "seal." After some discussion, the apprentice mechanic and the senior A&P agreed that a cylinder base O-ring was indeed an "uncoated oil seal" and therefore the use of Gasket Maker was appropriate.

So, the apprentice mechanic applied Gasket Maker to the cylinder base O-rings before installing the four cylinders on the Debonair's engine. When he torqued down the cylinder hold-down nuts, the sealant on the O-rings was extruded onto the mating surfaces of the cylinder base flange and the crankcase cylinder deck. About 150 hours later, the #4 cylinder departed the engine and the Debonair fell out of the sky.

Trial and verdict

When the case finally came to trial in 2010, the plaintiff's attorney knew that he had to convince the jury that Continental was at least partially to blame for the accident, because none of the other defendants had any significant assets or insurance. Although the apprentice mechanic clearly screwed up by using sealant on the cylinder base O-rings, the attorney made sure that the jury understood that the mechanics were trying to do the right thing after consulting the appropriate Continental manuals. The manuals weren't particularly helpful, stating that Gasket Maker could be used on "all uncoated oil seals."

The jury might have let Continental off the hook altogether but for one thing: Continental's expert conceded under cross examination that Continental knew of other incidents where their engines came apart because mechanics had applied Gasket Maker to cylinder base O-rings, yet had not revised its manuals to make it clear that this was not an acceptable use for the product.

In the end, the jury awarded $15 million in damages to the brain-injured plaintiff. They assessed comparative fault to the various defendants as follows: 55% for the repair station that worked on the engine for its inappropriate use of sealant, 35% for Continental because it failed to warn against doing so, and 10% for the PIC because his forced landing in the vineyard left a lot to be desired. However, under California law (as in many other states), each defendant is liable for 100% of the economic damages (e.g., medical expenses, loss of future income), but only for its proportionate share of non-economic damages (e.g., pain and suffering). Since the other defendants had little ability to pay, Continental wound up being on the hook for the lion's share of the $15 million award.

Takeaways

Aircraft owners need to understand that invasive piston engine maintenance—particularly cylinder removal—is fraught with peril. There are countless little details that must be done exactly right. Even a seemingly insignificant error or omission can have catastrophic consequences.

Owners also need to understand just how commonplace it is for critical work like this to be performed by young, inexperienced, and often uncertificated mechanics (affectionately known as "nuggets"), even at large FAA-certified repair stations. It takes 30 months of full-time aircraft maintenance experience to earn an A&P certificate. How do you suppose most mechanics obtain that experience, so they can qualify to take their A&P exam? By swinging wrenches on airplanes—quite possibly on your airplane. In this case, the apprentice mechanic was a bright, thoughtful, conscientious fellow who was trying his best to do the right thing. But a little dab of Gasket Maker no larger than a pea brought down the Debonair and seriously injured its two occupants.

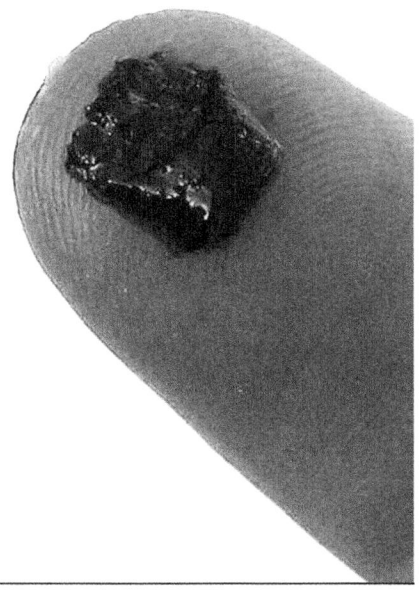

Here's what a $15,000,000 dab of Gasket Maker looks like.

Finally, owners need to understand the extent to which the threat of litigation casts a chilling shadow over every person and company involved in building, maintaining and operating aircraft. If you wonder why aircraft, engines and parts are so expensive, consider the cost of product liability that everyone involved in the industry must bear.

Aftermath

As a result of this lawsuit, Continental has added the following warning to its technical publications:

> ***WARNING***
>
> *Apply Gasket Maker only as directed. The improper use of sealants may cause engine malfunction or failure. Do not apply any form of sealant to the crankcase cylinder deck, chamfer, cylinder mounting flange, cylinder base O-ring, or cylinder fastener threads. The use of RTV, silicone, Gasket Maker or any other sealant on the areas listed above during engine assembly will cause a loss of cylinder deck stud or through-bolt torque. Subsequent loss of cylinder attachment load, loss of main bearing crush and/or fretting of the crankcase parting surfaces will occur. The result will be cylinder separation, main bearing movement, oil starvation and catastrophic engine failure. USE ONLY CLEAN 50 WEIGHT AVIATION ENGINE OIL ON SURFACES LISTED.*

44
Fear and Balderdash

Maintenance decisions need to be fact- and evidence-based.

The owner of a vintage J-model Bonanza emailed me for advice. He'd purchased the airplane just four months earlier with a fresh annual inspection and was already stressing out about what to do with his engine at the next annual eight months hence.

According to the owner, the airplane's Continental IO-470 engine was the original one that had been in the airplane when it rolled out of the factory in 1958. It had been overhauled seven years later in 1965 at its published TBO of 1400 hours, and then overhauled again nine years later in 1974 when it hit TBO again. Now the engine was once again at 1400 hours SMOH, but this time it had taken more than 40 years to get there.

The owner explained that the engine was using about a quart of oil every three hours but was otherwise running strong and smooth with decent compression readings and not making metal. He wanted my opinion as to whether he should

The owner's J35 Bonanza came with the original Continental IO-470 engine that rolled out of the factory in 1958. Was it time to overhaul after 40 years of operation?

be considering a major overhaul at the forthcoming annual, a top overhaul, or no overhaul. "Considering this engine is 57 years old and more than 40 years since the last overhaul, am I tempting Lady Luck?" he asked.

Calendar age

The original manufacture date of the engine was irrelevant, I told the owner. The fact that more than 40 years had elapsed since the prior overhaul was relevant only insofar as it meant that the aircraft wasn't very active during the past four decades, having averaged only about 35 hours a year. This naturally raised a concern about the possibility of internal corrosion.

I explained that the level of concern here was primarily a function of where the aircraft lived during those four decades. "If the airplane lived in Houston or Tampa, I'd be very worried," I told him, "but if it lived in Denver or Tucson I wouldn't be worried at all." Turns out that the plane now lived in Denver, but for most of the 40 years it lived in Redlands, California. Redlands is in the "Inland Empire" of Southern California, and has a climate that is hot and dry, with about the same annual rainfall as Denver and Tucson. Corrosion was probably not an issue.

Oil consumption

I told the owner that oil consumption of a quart in three hours was not intrinsically an airworthiness issue according to Continental's guidance. He could continue to live with it, but it was on the threshold of concern.

Before deciding whether to do a top overhaul, I suggested that some basic troubleshooting was in order to determine the reason for the elevated oil consumption. The problem might be confined to one or two cylinders, in which case it would be a shame to replace all six. Or it might be something unrelated to cylinders—perhaps a leaky oil filler cap gasket or a mis-positioned breather line that was pressurizing the crankcase in flight—in which case the owner would probably be upset if he did a top overhaul and it didn't cure the problem. On the other hand, perhaps the cylinders were badly worn, or the oil control rings were badly fouled with lead sludge. In that case, the cylinders would need to come off.

"Seems to me some detective work is in order to diagnose the cause of the oil consumption before throwing money at it," I told the owner. "Given that this is not yet an airworthiness issue, you've got time to troubleshoot. There's no gun to your head to do anything at the annual if you don't want to."

Fear and balderdash

The owner sounded very relieved. "Thanks, Mike, it's tough out there," the owner emailed me back. "I spoke to a rep at Continental Motors and he just about told me that I was nuts to fly with my engine!! That can rattle you."

He went on to describe his conversation with the Continental rep, who told him that the crankshaft in his engine is three generations old, and that the IO-470 crankcase has been upgraded three times since his was manufactured. The rep went on to say that his crankshaft was considered unairworthy "by today's standards."

I told the owner that the Continental rep was technically accurate when he said that the crankshaft was unairworthy "by today's standards." I explained that Airworthiness Directive AD 97-26-17 required that if his engine was overhauled or if the crankcase was split for any reason, his old "airmelt" crank would have to be scrapped and replaced by newer-design "vacuum arc remelt" (VAR) crankshaft at the cost of mucho kilobucks.

"But what the Continental rep didn't tell you," I continued, "was that the new VAR crankshafts have suffered far more in-flight failures than the old airmelt ones ever did." In fact, there was a rash of failures of new VAR crankshafts in 1999, and another in 2000, both prompting massive recalls by Continental. "That AD was a complete boondoggle that cost owners hundreds of millions of dollars for no valid reason," I opined.

The old IO-470 engine is probably the most reliable engine that Continental ever built.

The Continental rep was also technically accurate when he stated that the engine has an old-design crankcase that had been upgraded three times since 1958. "But what he didn't tell you is that the old-design crankcase worked just fine in 470-series engines, and only started cracking in the higher-horsepower 520- and 550-series engines, particularly the turbocharged ones."

I concluded my discussion by telling the owner that "the old IO-470 is probably the most reliable engine that Continental ever built. So, relax!"

If it ain't broke, don't fix it

Bottom line is that it's quite likely that there was nothing at all wrong with the engine in this owner's Bonanza. At worst, perhaps it had a couple of worn cylinders requiring eventual replacement. (Even that's not clear, since the owner didn't mention low compressions.)

Both Lycoming and Continental cleverly designed their engines so that the cylinders were bolt-on accessories that can be repaired or replaced without removing the engine from the airframe.

Even if a borescope inspection revealed that the engine had a worn-out jug or two, so what? Both Continental and Lycoming cleverly designed their engines so that the cylinders were bolt-on accessories that can be repaired or replaced without removing the engine from the airframe or splitting the case. If the engine actually did have badly worn cylinders, that's a reason to repair or replace the jugs, not to tear down the whole engine.

Think about this for a moment. If some other bolt-on engine accessory went bad—say an alternator or vacuum pump or magneto or prop governor—would you let your mechanic remove the engine and have it major overhauled? Of course you wouldn't. If you had a hangnail, would you let your doctor perform an amputation and hand transplant? No, I didn't think so!

Why would an aircraft owner even consider major overhaul or engine replacement just because one or two cylinders might be worn out? To my way of thinking, it doesn't matter whether an engine is at 100 hours since new or 100 hours past TBO—a sick cylinder calls for cylinder replacement, not engine replacement.

The Bonanza owner's experience is hardly an isolated case. I frequently see owners making major, costly maintenance decisions based on fear, myths, half-truths, poppycock and balderdash. Such decisions really need to be made based on facts and evidence, not emotions. Fact and evidence make one thing crystal clear: If it ain't broke, don't fix it.

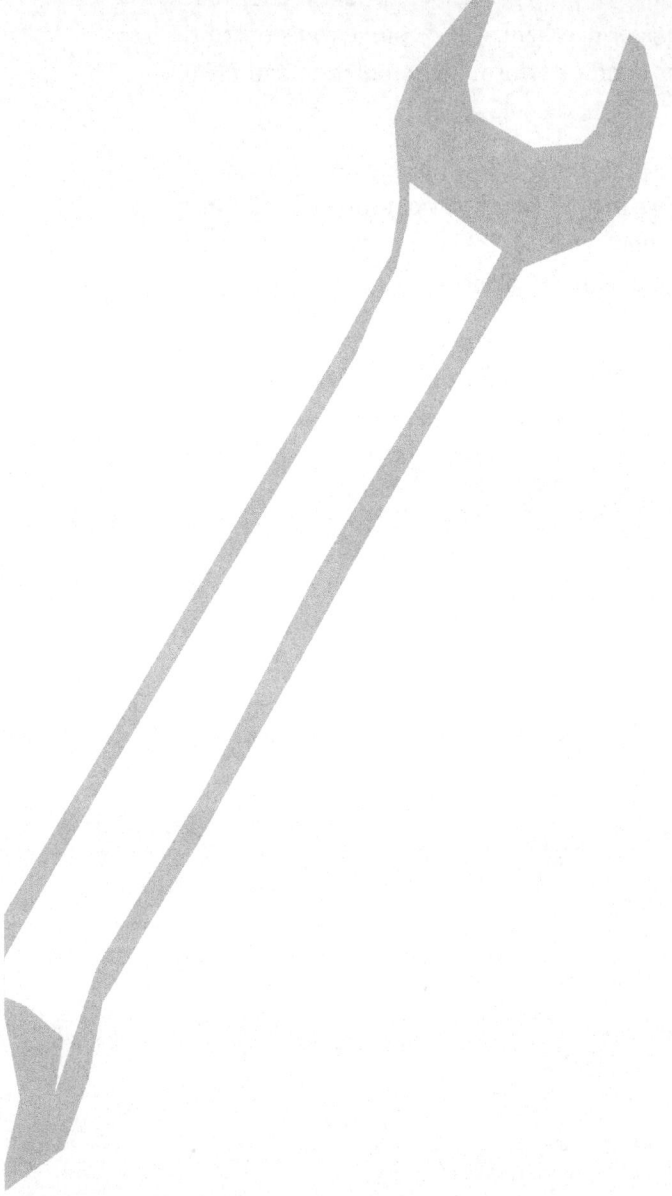

418 ENGINES

45
The Perils of Cylinder Work

It's a daunting task to install a cylinder properly when the engine is on the airplane.

I suppose it comes as no surprise that I'm not a big fan of top overhauls. I never like to see any cylinder removed from any piston aircraft engine unless there's a compelling reason. I truly hate to see multiple cylinders removed simultaneously. Removing all cylinders at once—the so-called "top overhaul"—is something I try to avoid at all costs, except in truly extraordinary circumstances.

One of those truly extraordinary circumstances had my phone ringing off the hook in early 2014. The calls were from frantic owners of Continental 520- and 550-series engines whose Superior Millennium-brand investment-cast cylinders were being forcibly euthanized by Airworthiness Directive 2014-05-29 and its predecessor 2009-16-03 that legislated thousands of these magnificent cylinders out of existence the moment they reached the calendar age of 12 years. We were successful in convincing the FAA to extend this to 17 years by means of a Global Alternative Means of Compliance (AMOC), but at that point there was really nothing aircraft owners could do but comply.

A similar fate befell ECi cylinders in 2016 after a handful of cylinder head fatigue failures, none of which resulted in accident or injury; AD 2016-16-12 became effective in September of that year, and required premature removal from service of up to 30,000 ECi cylinders based on time-in-service. The fallout from this misguided AD ultimately drove ECi out of business, a tragic loss for the GA industry.

Risky business

Cylinder replacement is a highly invasive and risky procedure with a long history of precipitating catastrophic in-flight engine failures that cause airplanes to fall out of the sky and sometimes hurt or kill people. I have been personally involved with at least a half-dozen of these maintenance-induced catastrophic engine failures—either as expert witness or investigator—where the engine either "threw a rod" through the crankcase or suffered the complete separation of a cylinder from the engine, resulting in a total loss of power. In some cases, the pilot made a successful forced landing; in others, the outcome was serious injury or death.

This Continental IO-550 threw the #2 rod through the top of the case less than 10 hours after the #2 and #4 cylinders were replaced at annual.

Cylinder replacement—and especially replacement of multiple cylinders at once—is a procedure that needs to be executed perfectly. If it isn't, there can be dire consequences. Yet it's a procedure that most career general aviation A&Ps perform routinely without any apparent fear or concern, seemingly oblivious to the fact that even a tiny mistake could result in an engine failure that could destroy an aircraft and injure someone.

Why aren't these mechanics nervous? Undoubtedly because they are convinced that they always perform the cylinder transplant procedure properly, and that only careless or incompetent mechanics screw it up. Of course, this is a dangerous attitude. Even the most experienced, careful and cautious mechanics screw up occasionally.

But wait. It gets worse!

This Continental IO-360 spun a main bearing after cylinder work. Fortuitously, the situation was caught at a routine oil change (where the filter was found to have thousands of flakes of bearing metal) literally an hour or two before the engine would have self-destructed in flight.

Is "properly" impossible?

I discussed this issue with Roger D. Fuchs—veteran A&P/IA, aircraft engine overhauler, accident investigator, expert witness, FAA-Designated Engineering Representative (DER), and really smart guy—who specializes in research on fastener torque and engine assembly practices. Roger chastised me for blaming these maintenance-induced engine failures primarily on mechanic screw-ups.

> It cannot be established that any maintenance action should be withheld because of the vague possibility that it might be performed improperly," Fuchs wrote. "My concern comes rather out of decades of experience with Continental engines, particularly the design of crankcases, main bearings, fasteners and assembly practices. All my experience indicates that there is

a major risk to safety and airworthiness when performing [top overhauls] on mid- to high-time Continental engines, not as a result of improperly performed maintenance actions but rather **as the result of maintenance actions that experienced mechanics are attempting to perform properly according to the manufacturer's recommendations as published.**"

In short, Fuchs contends that there's a significant risk that an engine might come apart after cylinder work (especially a top overhaul) even when the work is performed exactly as directed by the manufacturer. How can that be? According to Fuchs, the root cause of spun bearings, thrown rods and separated cylinders is simply "failure to achieve sufficient preload in the assembled fasteners."

It's all about preload

"Preload" is the technical term for the clamping force created by tightening a fastener (typically a threaded bolt or stud) that holds assembled parts together. Having sufficient preload is the key to a strong and reliable bolted joint that will not loosen, break or shift under load. In order for a bolted joint to be stable under cyclic repetitive stress, the preload on the fasteners must be greater than the maximum stress that is trying to pull the joint apart. If this condition is met, the joint will not separate and the fasteners won't "feel" the repetitive stress cycles. But if it isn't, the joint will shift under load and the fasteners will ultimately fail from repetitive stress fatigue.

Failure to achieve sufficient preload in the assembled engine fasteners has ruined many engines.

Consider a cylinder on a Continental 520- or 550-series engine. During the peak pressure point in each combustion cycle, the pressure in the combustion chamber will reach 800 to 1,000 PSI. The piston diameter is 5.25 inches, so its surface area is 21.6 square inches and the peak force trying to pull the cylinder off the engine is on the order of 20,000 pounds. Each cylinder is bolted to the engine with eight fasteners—six 5/16 inch-diameter "deck studs" threaded into the crankcase and two 1/2 inch-diameter "throughbolts" that pass all the way through the crankcase and do triple duty of clamping the main bearing supports in the crankcase halves together plus holding down a pair of opposing cylinders. In a perfect world, the 20,000 pounds of force trying to rip the cylinder off the engine 20 times every second would be equally divided among the eight fasteners, so each one would bear 2,500 pounds of force. (In the real world, of course, things are never that simple.) To be on the safe side, we'd want each of those eight fasteners to be tightened to a preload of 3,000 or 4,000 pounds.

How do we obtain the desired preload? In a perfect world we'd tighten the cylinder base nuts so that the deck studs stretched by about .005 inch and the through-bolts stretched by about .035 inch. In the real world, unfortunately, mechanics have no practical way of measuring the stretch of the deck studs and through-bolts, so they are forced to rely on using a calibrated torque wrench to tighten the nuts to manufacturer-specified torque values in an attempt to establish fastener preloads that are in the desired ballpark. This turns out not to be a very reliable method.

A pair of new through-bolts for Lycoming (top) and Continental (bottom) engines. Both have 1/2 inch cad-plated steel threads. The Continental through-bolts are stiffer (larger shank diameter) and have grooves for O-rings.

The trouble with torque

The problem with the torque-wrench method is that the amount of fastener preload generated by torqueing a nut to a specified torque value can vary quite a bit. That's because the lion's share of the applied torque is dissipated overcoming friction—both friction under the nut face and friction of the threads—leaving only a small and rather unpredictable portion of the applied torque to generate preload on the fastener.

If the fastener is torqued dry, then 85% to 90% of the applied torque is dissipated overcoming friction, leaving only 10% to 15% to generate preload. That's why both Continental and Lycoming specify that cylinder deck studs and through-bolts should

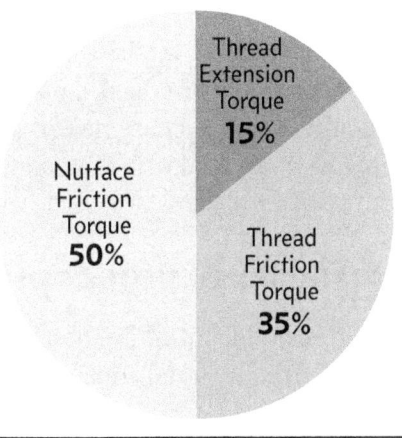

When using a torque wrench to tighten a dry steel fastener, at least 50% of the tightening torque is dissipated overcoming friction under the nut face, and another 35% overcoming friction of the threads, leaving only 10% to 15% generating preload on the fastener. Lubrication of the nut face and threads can have a dramatic effect on this.

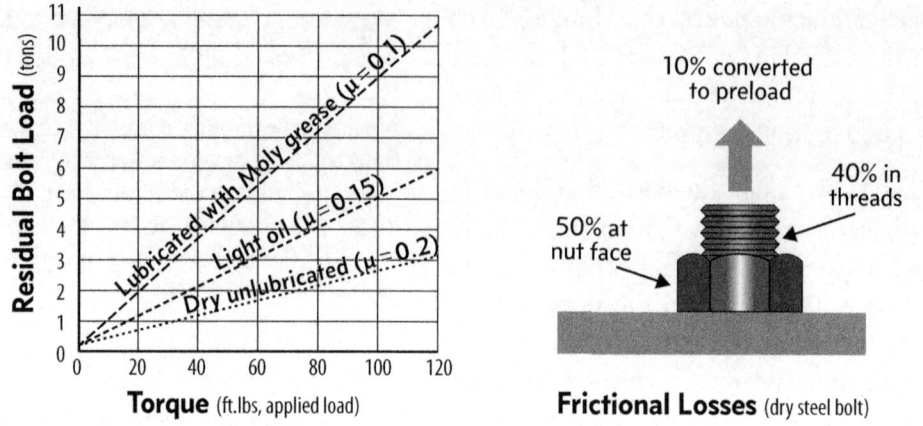

Lubricating fasteners before torqueing can have a profound effect on the amount of fastener preload by reducing frictional losses.

be torqued "wet" (by applying lubricant to the fasteners) to reduce these frictional losses and increase the preload achieved by torqueing to a specified value.

The "wet" torque method works adequately during initial engine assembly at the factory or engine overhaul shop when the engine is mounted on an assembly stand and all the fasteners are new. But it works far less well when cylinders are being replaced in the field with the engine still mounted in the airplane.

Obstacles to proper preload

Roger Fuchs identified five obstacles to achieving proper fastener preload when performing cylinder installation on an engine mounted in the airplane:

1. The fasteners aren't new. When an engine is initially assembled at the factory (or by a first-rate overhaul shop), the through-bolts, deck studs and cylinder hold-down nuts are all new components with cadmium-plated threads in perfect condition. The cad-plating is very slippery (helping to reduce friction) but very thin (typically 8 microns thick, one tenth the thickness of a human hair) and relatively soft, making it easy to damage.

Many field-overhauled engines are assembled with repaired crankcases in which the deck studs are not replaced and may have been torqued numerous times with most or

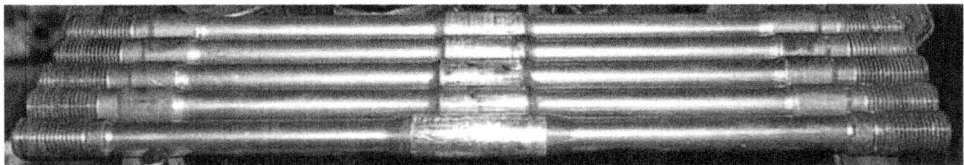

Through-bolts in typical "as removed" condition. The damage to threads and loss of cad-plating are obvious.

all cad-plating removed by wear or cleaning. Through-bolts and hold-down nuts may also be reused (although they shouldn't be).

Things get even worse when cylinders are replaced with the engine on the airplane. Through-bolts and deck studs are never replaced, and hold-down nuts may be reused at the mechanic's discretion. The threads of through-bolts, deck studs and hold-down nuts may be damaged but it's almost impossible for a mechanic in the field to evaluate this.

2. Fasteners may not be adequately lubricated. Both Continental and Lycoming specify that cylinder fasteners are to be torqued "wet." Continental calls for using 50-weight engine oil, while Lycoming suggests using a 90/10 mixture of engine oil and STP.

To achieve proper preload, the lubricant needs to be slathered onto both the fastener threads and the nut face area. But according to Fuchs, "mechanics are generally reluctant to use much oil on threads and nut faces during cylinder assembly" because it "conflicts with their innate desire for tidy-looking engines." Fuchs goes on to say "I find use of too little lubricant rather common in the maintenance industry; it's a very serious issue when assembling used fasteners."

3. The lubricant itself is rather poor. Fuchs also points out that plain 50-weight engine oil (e.g., Aeroshell W100) is a lousy thread lubricant because it lacks synthetics like PAO and anti-friction additives that would make it much slipperier. Add this to the likely loss of slippery cad-plating from the worn fastener threads, not to mention the possibility of thread damage, and it's anybody's guess whether proper torque will result in proper preload.

4. Wrench access is limited. Both Continental and Lycoming call for a two-phase tightening procedure where all the cylinder hold-down nuts are first torqued to 50% of their final torque in a specified tightening sequence, and then they are torqued

Achieving final torque needs to be accomplished with one smooth motion of the torque wrench. This is usually possible when the engine is out of the airplane and being assembled on a stand, but it's almost impossible when the engine is in the airplane.

to 100% of their final torque following the same sequence. An important reason for doing it this way is that consistent results can only be obtained if the final tightening sequence is performed using a single continuous motion of the torque wrench. If the movement of the wrench is interrupted, the "click" from the wrench that signifies that the specified torque has been achieved can occur too early, because breakaway torque is significantly higher than running torque.

While it's usually easy to do this properly when the engine is out of the airplane and sitting on an engine stand with unobstructed access, it's almost impossible to do when the engine is mounted in the airplane and various components restrict wrench movement. Frequently, two or three "bites" of the wrench are needed before final torque is achieved, and each adds uncertainty to the final result. "This is particularly true when wrench rotation must be stopped as the nut is approaching the desired 'click' of the wrench—but not there yet," says Fuchs. "Time after time when the wrench is removed before it clicked, the wrench will do so upon attempting to tighten the nut further with no additional rotation of the nut." This deceives the mechanic into believing that proper preload has been achieved when it almost certainly hasn't.

5. Manufacturer instructions are incomplete. The published guidance from Continental and Lycoming leaves a lot to be desired. To cite one glaring example, Continental's overhaul manuals and torque charts emphasize that when a cylinder is replaced, the nuts on both ends of each through-bolt must be torqued. In my experience, many mechanics don't bother with torqueing the nut on the opposite side of the engine. But even if they do, there's nothing in Continental's guidance suggesting that the opposite-side through-bolt nut should be removed and the nut and threads be lubricated, nor that the opposite-side threads should be cleaned and the opposite-side nut be replaced with a new one. It's dubious that following the published guidance will accomplish anything with respect to torqueing the opposite-side nut…which is probably why so many mechanics don't even bother with it unless they're also replacing the opposite-side cylinder.

Be afraid

Cylinder changes are sometimes necessary. If you have a cylinder repaired or replaced, your conscientious A&P probably employed what would generally be considered proper maintenance practices. His torque wrench may have been calibrated recently, set to the proper torque value, and given the reassuring "click" indicating that the desired torque value was achieved. And yet it's entirely possible that most of the fastener preloads achieved may well be below the design minimum required for safety and reliability of the engine. There are a number of things that can be done to mitigate this risk.

Never change a cylinder unless it's absolutely necessary. Try hard to avoid changing more than one cylinder at a time. When cylinder changes are unavoidable, extreme care must be exercised during cylinder installation. Clean through-bolt and hold-down stud threads meticulously and inspect them carefully for damage. Slather all threads liberally with lubricant to minimize friction. Never reuse hold-down nuts, always install new ones. Use a recently-calibrated torque wrench to tighten the nuts, following the manufacturer's torque values and torque sequence to the letter. Take great care to achieve the final torque value with one smooth motion of the torque wrench, rather than taking multiple "bites."

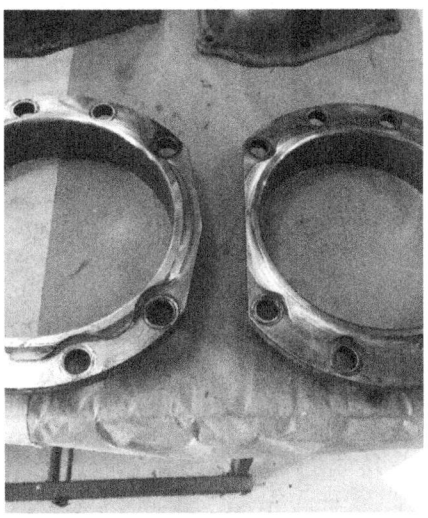

If multiple cylinders must be removed, install torque plates and retighten through-bolts to keep the case halves together.

If multiple cylinders must be removed, install torque plates and retighten the through-bolts to keep the case halves clamped together. Never allow more than one pair of through-bolts to be torque-relieved at one time. It is also critical to never rotate the crankshaft when any through-bolts are torque-relieved; the main bearings may shift in their saddles, causing misalignment of the oil holes in the bearing with the oil passages in the crankcase and cutting off the oil supply to the bearing. The subsequent reduction in bearing lubrication results in increased heat and friction, causing the bearing to eventually spin and leading to catastrophic engine failure.

Even if all these procedures are followed, there's no guarantee that the fastener preload will be correct or consistent. But it'll help a lot.

Of course, any time multiple cylinders are replaced, the probability of failure increases with the number of fasteners that are messed with. Food for thought next time your mechanic suggests that it might be a good idea to do a top overhaul.

PART IX
Overhaul

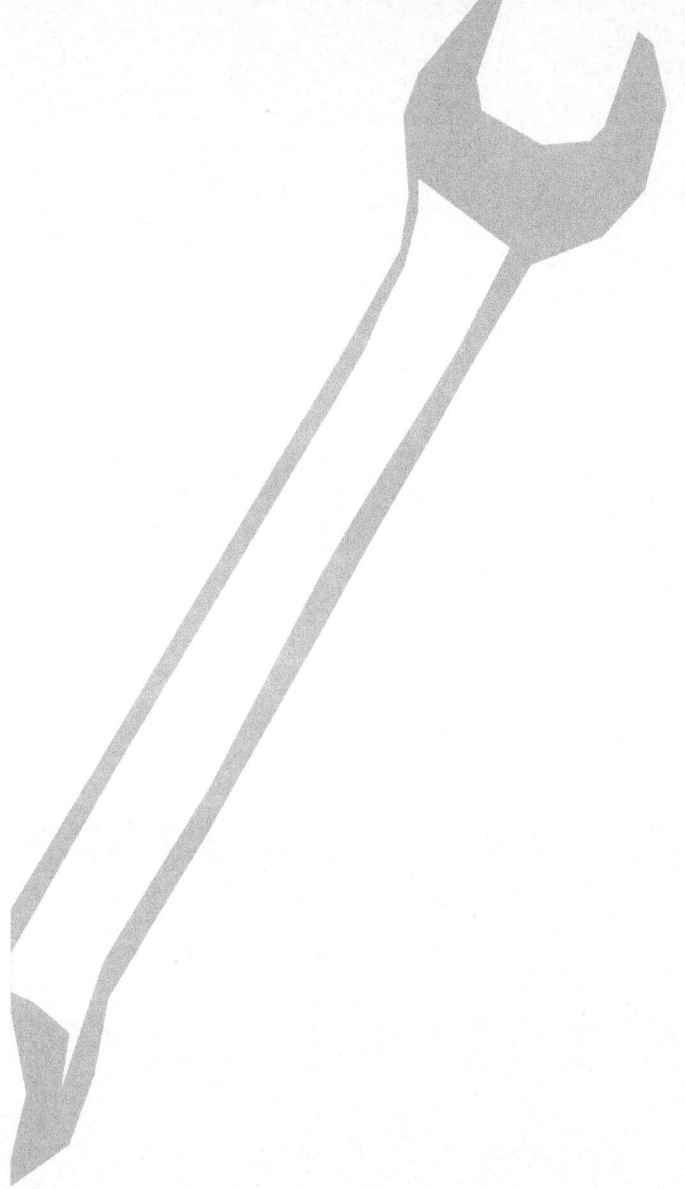

46
Maintenance Intervals

When are manufacturer-specified inspection, overhaul and replacement intervals mandatory for a Part 91 operator?

"It has been six years since your propeller was last overhauled, so we're going to have to overhaul it this year as required by Hartzell."

"Your magnetos are past due; Continental requires that they be overhauled every four years."

"We need to clean your fuel nozzles and adjust your fuel injection system annually as specified by the engine manufacturer."

"The trim tab actuators need to be disassembled, cleaned and lubricated—the Cessna maintenance manual says this must be done every 200 hours."

"We need to pull the wing bolts on your Bonanza and send them out for non-destructive testing—Beech requires this be done every five years."

"The regulator on your STC'd oxygen system needs to be sent out for overhaul every five years according to the manufacturer's ICA."

"The Instructions for Continued Airworthiness (ICA) for your Garmin autopilot requires that the servo clutches be checked for proper breakaway torque at every annual inspection."

My company Savvy Aviation works with hundreds of maintenance shops across the U.S., and we see mechanics telling our aircraft-owner clients these sorts of things every day. An important part of our job as maintenance managers and consultants is to advise our clients to decline most of these things because (1) they're not required (even if the manufacturer says they are), and (2) doing them is often at best a total waste of money and at worst a good way to create a problem where none exists.

Most aircraft owners find it quite uncomfortable to say "no" to their mechanics, so our clients look to Savvy's technical team to do that on their behalf. No problem: We're very comfortable saying "no" to A&Ps. We do it dozens of times a day.

Most of the time when we say "no" to something that a mechanic proposes to do based on manufacturer's guidance, our direction is accepted without debate. Occasionally, however, we get some resistance. That's because most A&P mechanics have been trained to follow manufacturer's guidance to the letter without question. Many even believe that they are required by regulation to do so (even though they generally aren't).

We sometimes find it necessary to give these mechanics a bit of remedial training on exactly what the FARs do and do not require. Occasionally these discussions get interesting.

Hartzell Prop TBO

For example, I found myself in an interesting dialogue with the Director of Maintenance (DOM) of a shop that was performing an annual inspection on a client's aircraft. The shop told my client that they had to send his Hartzell propeller out for overhaul because it was six years old. The overhaul would have cost my client about $3,000, and because this was an always-hangared, actively-flown aircraft, I recommended that my client defer the prop overhaul for at least a few more years.

When we declined the overhaul, the shop's DOM pushed back and opined that it was required by regulation. That surprised me, because most mechanics understand that manufacturer-specified TBOs are almost never required by regulation for Part 91 operators.

Many mechanics believe that Hartzell propellers are required to be overhauled at TBO because the propeller TCDS references Hartzell's TBO service bulletin. They aren't.

When I asked the DOM to explain why he felt the overhaul was required, he responded with an interesting argument. He cited the Type Certificate Data Sheet (TCDS) for the Hartzell propeller in question. Like most such documents, this TCDS contains a series of numbered notes at the end. "Note 12" refers to Hartzell Service Letter HC-SL-61-61() for overhaul periods. In turn, Hartzell Service Letter HC-SL-61-61-Y (the current version) states that "Hartzell propellers installed on reciprocating engines MUST be overhauled at the intervals specified in Section 3, Overhaul Periods, Paragraph B." For this particular prop, Paragraph B specified an overhaul period of 2400 hours, or 72 calendar months.

The DOM argued that while Part 91 operators are normally not required to comply with manufacturer's service bulletins, the fact that service letter HC-SL-61-61-Y was referenced by the TCDS made compliance mandatory. This argument gave me pause, because—like most IAs—I had always been taught anything that appears in a TCDS is an airworthiness requirement. At the same time, I found it difficult to believe that the FAA really intended to make Hartzell's propeller TBOs compulsory. So before responding to the DOM, I decided I'd better do some homework.

My homework turned up two interesting documents. One was a Letter of Interpretation sent to Hartzell in January 2008 by the head of the FAA Chicago Aircraft Certification Office (the FAA office that has direct oversight over Hartzell) specifically addressing "Note 12" in the Hartzell TCDS. This letter states in pertinent part:

> The intent of this Note [12] is to provide the end user a means to determine the proper service information for that particular propeller design. The overhaul periods are NOT mandated by this reference as the FAA-approved airworthiness limitations section (e.g., mandatory inspections of life limits) does NOT include overhaul periods. The addition of this Note does NOT mandate overhaul periods for Part 91 operators…

My homework also unearthed FAA Order 8620.2A which specifically addresses the subject of Type Certificate Data Sheets. This Order states in pertinent part:

> A TCDS is part of a product's type certificate (TC). A TCDS is a summary of the product's type design. It is used primarily by authorized persons during initial or recurrent issuance of a Standard Airworthiness Certificate. ***It is neither a regulation, a maintenance requirements document, or a flight manual document***. As such, for aircraft holding a valid and current airworthiness certificate, a TCDS should not be used as a sole source to determine what maintenance is required or what the flight operations requirements are. ***Any language on a TCDS, by itself, is not regulatory and is simply not enforceable***.

On the subject of TCDS Notes, Order 8620.2A states:

> Some OEM's have placed mandatory language such as 'shall,' 'must,' and 'will' on their TCDS that imply that compliance with TCDS notes is mandatory. However, in the absence of regulatory language, or an AD that makes such TCDS notes mandatory, compliance with such notes is NOT mandatory.

I shared these documents with the DOM. He reviewed them and agreed with me that the six-year prop overhaul was not required.

What is required by regulation?

Mechanics are often confused about exactly what maintenance is or is not required by regulation partly because the regulations are not terribly clear, and partly because A&P training tends to be spring-loaded to the "always do it by the book" position. There exists a large body of FAA Orders and FAA Letters of Interpretation that make it very clear what the FAA's position is on this subject, but very few mechanics have ever read any of this stuff or received any training on the subject.

If you spend a little time studying this (as I have), it turns out that things are very simple, clear and unambiguous. Here it is in a nutshell...

Manufacturer's guidance comes in three basic forms: maintenance manuals (MM), instructions for continued airworthiness (ICA), and service bulletins (SB). It also comes in two basic flavors: how-to's and when-to's. How-to's are the responsibility of mechanics and are covered in FAR Part 43, while when-to's are the responsibility of aircraft owners and are covered in FAR Part 91.

The FARs refer to "how-to" guidance as "methods, techniques and practices" (MTPs). The general rule is that maintenance must be done in accordance with the MTPs specified in the manufacturer's MM or ICA or in accordance with other MTPs that the FAA finds to be acceptable. The specific regulatory reference is FAR 43.13(a) which states in pertinent part:

> **§ 43.13 Performance rules (general).**
>
> (a) Each person performing maintenance, alteration, or preventive maintenance on an aircraft, engine, propeller, or appliance shall use the methods, techniques, and practices prescribed in the current manufacturer's maintenance manual or Instructions for Continued Airworthiness prepared by its manufacturer, or other methods, techniques, and practices acceptable to the Administrator, except as noted in §43.16.

The sole exception to this rule is that "how-to" guidance that is set forth in an FAA-approved Airworthiness Limitations Section of a manufacturer's MM or ICA, which must always be followed exactly:

> **§ 43.16 Airworthiness limitations.**
>
> Each person performing an inspection or other maintenance specified in an Airworthiness Limitations section of a manufacturer's maintenance manual

or Instructions for Continued Airworthiness shall perform the inspection or other maintenance in accordance with that section…

Although the FARs provide for using alternative acceptable MTPs (except for Airworthiness Limitations), such alternative methods are rarely available. So 99% of the time, mechanics wind up performing maintenance using the MTPs ("how-to's") set forth in the manufacturer's MM, ICA and/or SBs.

What about when-to's?

"When-to" guidance includes manufacturer-specified inspection, overhaul and replacement intervals, as well as other manufacturer guidance about how frequently various maintenance tasks are to be performed. Virtually every aircraft maintenance manual contains a long list of scheduled maintenance tasks—things to be done every 50 hours, every 100 yours, every 12 months, etc. The maintenance manual for my Cessna 310 contains more than 250 such items.

Manufacturers of engines, propellers and appliances (e.g., magnetos, vacuum pumps, etc.) usually specify times between overhauls (TBO) or times between replacement (TBR) in MM or ICA or SBs. Lycoming, Continental, Hartzell and McCauley all set forth their engine and propeller TBOs in service bulletins.

The general rule is that Part 91 (non-commercial) operators are NEVER required to comply with such manufacturer-specified intervals ("when-to's") simply because there is no regulation in the FARs requiring them to do so. There are two—and only two—exceptions to this general rule: If such intervals are mandated by an FAA AD or if they are set forth in an FAA-approved Airworthiness Limitations Section (ALS) of the manufacturer's MM or ICA, then compliance is required by regulation. Otherwise, it isn't. The regulatory reference that covers these two

Part 91 operators are not required to comply with manufacturer's Instructions for Continued Airworthiness. They are only required to comply with any Airworthiness Limitations contained therein (if any).

exceptions is FAR 91.403, which states in pertinent part:

> **§ 91.403 General.**
>
> (a) The owner ... of an aircraft is primarily responsible for maintaining that aircraft in an airworthy condition, including compliance with Part 39 of this chapter [Airworthiness Directives].
>
> (c) No person may operate an aircraft for which a manufacturer's Maintenance Manual or Instructions for Continued Airworthiness has been issued that contains an Airworthiness Limitations section unless the mandatory replacement times, inspection intervals, and related procedures specified in that section ... have been complied with.

Frequently-asked questions

Q: Are you saying that I can ignore all the scheduled maintenance tasks listed in my aircraft's maintenance manual?

A: If your maintenance manual has a clearly identified FAA-approved "Airworthiness Limitations Section" (ALS), then any inspection, overhaul or replacement intervals prescribed *in that section* must be complied with. Intervals that appear in any other part of the maintenance manual need not be complied with. The maintenance manuals for legacy aircraft certificated under CAR 3 do not contain an ALS; those for newer-design aircraft certificated under FAR 23 typically do contain an ALS. The maintenance manuals for Lycoming and Continental engines do not contain any Airworthiness Limitations.

This doesn't mean you should ignore all manufacturer-prescribed maintenance intervals. Some of them make sense and are worth following, although many of them don't and aren't. Such intervals are simply recommendations, not requirements. You should feel free to accept or reject them as you see fit—except for ADs and Airworthiness Limitations, which are non-negotiable.

It makes no difference if the manufacturer uses "compulsory-sounding" words like "mandatory" or "required" or "must" or "shall." No manufacturer has the authority to compel you to perform any maintenance task that you don't want to do, regardless of what language the manufacturer uses. Only the FAA has that authority.

When in doubt as to whether to follow manufacturer's when-to guidance, it's wise to seek an independent expert second opinion from someone you trust…and maybe even a third.

Q: Are you saying that I can ignore Instructions for Continued Airworthiness?

A: Yes, unless the ICA contains a clearly identified FAA-approved ALS. If it does, then any intervals prescribed *in the ALS* must be complied with. Intervals that appear in any other part of the ICA need *not* be complied with. Most ICA do not contain ALS, but some do.

Q: Are you saying that I can ignore Service Bulletins?

A: That's exactly what I'm saying. A Part 91 operator is *never* required to comply with any manufacturer's service bulletin, even those marked "mandatory" or "critical," unless compliance is mandated by the FAA by AD. Again, I'm not saying that you should blindly ignore all SBs; some of them are quite important. I'm simply saying that whether you choose to comply with any particular SB is totally up to you—compliance is not required by regulation. When in doubt, seek an independent expert second opinion.

But my mechanic says…

So why do mechanics persist in telling their aircraft-owner customers that their engines, propellers, magnetos, trim tab actuators and oxygen regulators need to be overhauled at certain manufacturer-prescribed intervals? Usually because they believe this to be true, even though it isn't. Every A&P is taught that manufacturer's guidance is always to be followed meticulously and to the letter. That's certainly what I was taught when I was studying for my A&P knowledge and practical tests. It wasn't until I started discussing the subtleties of the FARs with the FAA lawyers at the Regulations Division of the Office of General Counsel at FAA headquarters, and studying FAA Orders and Letters of Interpretation that I discovered that most of what A&Ps (including me) have been taught about this subject is just flat wrong.

47
Engine TBOs

There are lots of misconceptions about TBOs, so let's look at the facts.

Ask any aircraft owner what the TBO is for the engine(s) on his aircraft and you'll almost always get the correct answer without hesitation: "My engine has a 1,700-hour TBO." But ask that owner to explain the significance of that TBO figure and you'll get all sorts of answers, most of them flat wrong. Here are a few of the most common misapprehensions about TBO:

- *It's illegal to fly an airplane if the engine is past the TBO established by the manufacturer.* Nonsense. The TBO figures published by Lycoming and Continental are not Airworthiness Limitations. An engine may be long past TBO and still be legally airworthy. (An engine may also become unairworthy long before reaching TBO.)

- *While it's true that manufacturer's TBO isn't compulsory for non-commercial (Part 91) operators, commercial (Part 121/135) operators are required to overhaul an engine when it reaches TBO.* Not so. Both Lycoming and Continental publish engine TBOs in the form of non-mandatory service bulletins. Some Part 121/135 operators have Operations Specifications that require them to comply with all manufacturer's service bulletins (even non-mandatory ones), while others have Op Specs that require compliance only with mandatory service bulletins. Those in the latter group are no more obligated to comply with published TBO than are Part 91 operators. Those in the former group might theoretically be required to overhaul at published TBO, but most such operators request TBO extensions from their FSDO and these are routinely granted, often for as much as 50% over the engine manufacturer's published TBO. So, in actual practice, published TBO is hardly ever compulsory for any operators, commercial or non-commercial.

- *Continuing to fly an engine beyond TBO could void your aircraft insurance.* Poppycock. I've yet to see any aircraft insurance policy that requires compliance with non-mandatory service bulletins as a condition of coverage. Most policies only require that the aircraft be airworthy and in compliance with FAA inspection requirements. I've owned my Cessna 310 for more than three decades, and have logged more over-TBO hours than under-TBO hours. Over the years, I've insured my aircraft with four major insurance underwriters, and none have expressed the slightest concern that my engines are past TBO. It's simply a non-issue.

- *Continuing to fly an engine beyond TBO is dangerous because doing so increases the chance of an in-flight engine failure.* To the contrary, an engine is much more likely to fail during the first few hundred hours after major overhaul than during the first few hundred hours after passing published TBO. If you exclude fuel starvation or exhaustion (i.e., pilot error), most engine stoppages involve mechanical failure of some "top end" engine component like a cylinder, exhaust valve, piston, magneto, turbocharger, exhaust stack, etc. Such bolt-on components are routinely replaced during normal maintenance without any need to overhaul the engine. The purpose of a major engine overhaul is to inspect, recondition and or replace the engine's "bottom end" components—crankshaft, camshaft, crankcase, gears, bearings, etc.—that cannot be accessed without splitting the case. But these "bottom end" components are seldom implicated in catastrophic engine failures. Furthermore, in those rare cases when these components do fail (e.g., crankshaft fracture), the failure is almost never correlated with time since overhaul. (If a crankshaft is going to fail, it's most likely to fail during the first few hundred hours after manufacture, or after a prop strike.)

- ***Continuing to fly an engine beyond TBO is false economy, because doing so just makes the inevitable major overhaul more expensive.*** This old wives tale probably originated back in the days when new cylinders were very expensive and most engines were field overhauled using reconditioned (chromed or oversized) jugs. In those days, if you pushed an engine to the point that its cylinders could not be reconditioned, you'd have to spend more at overhaul to buy new ones. Nowadays, however, the cost of new cylinders has come down to the point where most major overhauls include all new jugs as standard procedure. Consequently, there's no longer any real advantage to overhauling sooner rather than later. The only things that will impact the overhaul cost are an unserviceable crankshaft or a cracked crankcase, and neither of those items are any more probable for an engine operated beyond TBO.

By the way, it's not just owners who hold these misconceptions. Plenty of A&P mechanics believe these things, too.

TBO from the horse's mouth

Back in the old days, Continental published its engine TBOs in Service Bulletin M91-8, which stated in pertinent part:

> The overhaul periods listed are recommendations only. Any operation beyond these periods is at the operator's discretion and should be based on the inspecting mechanic's evaluation of engine condition and operating environment. These recommended overhaul periods in no way alter Continental's warranty policies.

In 1998, M91-8 was superseded by Continental SIL 98-9 which removed this permissive language. Lycoming engine TBOs are set forth in Service Instruction 1009 which says that non-commercial operators may run past Lycoming's 12-year TBO based on inspection by a qualified mechanic, but says nothing about running past the recommended hours-in-service TBO.

Nevertheless, the FAA is crystal clear on this subject. FAA National Policy Notice N8900.410 titled "Clarification of Inspection and Overhaul Requirements Under

Part 91" (issued in 2017) states "the concept of Part 91 operators not having to comply with manufacturer's TBO is well known." This Notice references FAA Advisory Circular 20-105B (issued in 1998), which says that "for Part 91 operations compliance to the TBO time is not a mandatory maintenance requirement."

Bottom line is that if an engine is still going strong when it reaches TBO, there's absolutely no reason to consider removing it from service for major overhaul, and every reason to continue flying until it starts showing signs that overhaul is warranted.

Overhaul on-condition, not on time

I previously discussed the concept of Reliability-Centered Maintenance (RCM) and explained why the piston engine GA community would do well to adopt its principles. For the past 50 years, the airlines and military have been using the RCM approach to slash maintenance cost and improve reliability. Most of these benefits have come from replacing fixed overhaul intervals with on-condition maintenance.

Unfortunately, RCM has not trickled down to the low end of the aviation food chain. Maintenance of piston GA aircraft is mostly done the same way it was 50 years ago: largely time-directed rather than condition-directed. Most GA owners dutifully overhaul their engine at TBO, overhaul their prop every five to seven years, and replace their alternators and vacuum pumps every 500 hours, just as Lycoming, Continental, Rotax, Hartzell, McCauley, and various accessory manufacturers recommend. Bonanza owners have their wing bolts pulled every five years. Cirrus owners replace their batteries every two years. And the beat goes on…

Does any of this make sense?

After analyzing reams of operational data from major air carriers, RCM researchers concluded that fixed-interval overhaul or replacement rarely improves safety or reliability, and more often than not makes things worse.

When *does* TBO make sense?

For fixed TBO to make sense, the component must have a failure pattern that looks like pattern B in the figure on the next page, where the component can be expected to operate reliability for some predictable useful life, beyond which the probability of failure starts to increase significantly to unacceptable levels.

But piston aircraft engines don't exhibit this kind of failure pattern. We know these engines suffer the highest risk of catastrophic failure not when they pass TBO, but rather when they're fresh out of the factory or field overhaul shop. Look at the NTSB data for the five-year period 2001 through 2005 in the graph on the next page.

This NTSB data can't tell us much about the risk of engine failure beyond TBO, because GA's antiquated maintenance culture causes most engines to be voluntarily euthanized at TBO. What the data does show clearly is that engines fail with disturbing frequency during their first few years and few hundred hours in service after manufacture, rebuild or overhaul. Obviously, our engines have a failure pattern more like A or C in the figure above, with a high risk of "infant mortality" failure.

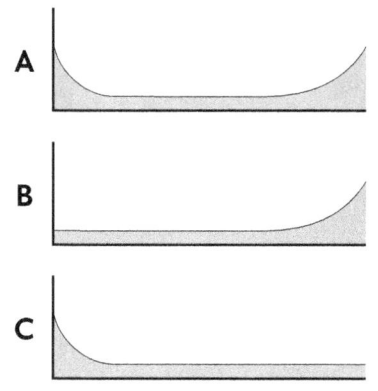

Fixed TBO may make sense for components with failure pattern B, but probably not for those with pattern A and definitely not for those with pattern C.

Would you be comfortable taking your family up in a piston-powered GA airplane with an engine at five hours SMOH? At night? Over rough terrain or water? In IMC? How about at 10 hours SMOH? Or 25 hours SMOH? (These are not easy questions.)

If our engines have a failure pattern like A ("bathtub curve"), then overhauling at a fixed TBO becomes a two-edged sword. On one hand, it keeps us out of the presumptive wear-out zone. On the other hand, it puts us right back inside the infant mortality window where the data shows clearly that engine failure is disturbingly common.

If our piston engines have a failure pattern like C (as turbine engines do), then overhauling at a fixed TBO makes no sense at all, because there's no obvious wear-out zone. That's why the airlines and military overhaul turbine engines strictly on-condition. (There's very little data suggesting that piston engines have an obvious wear-out zone, either.)

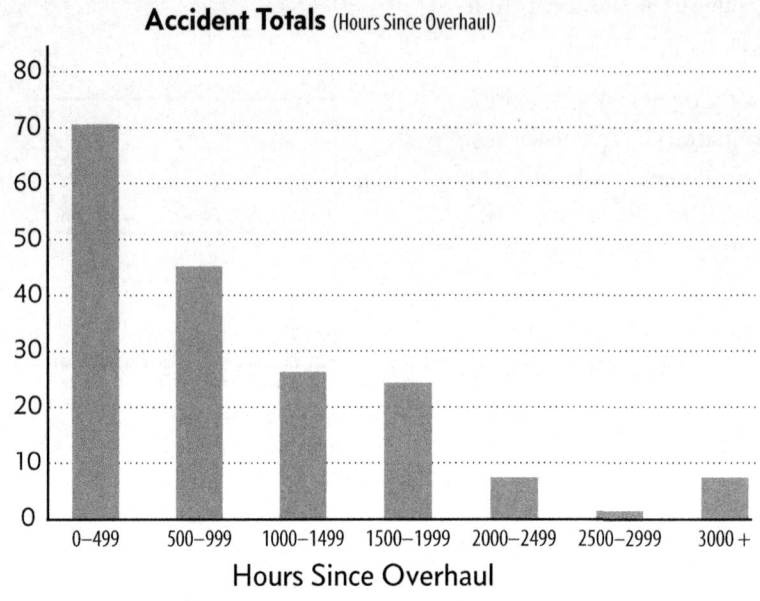

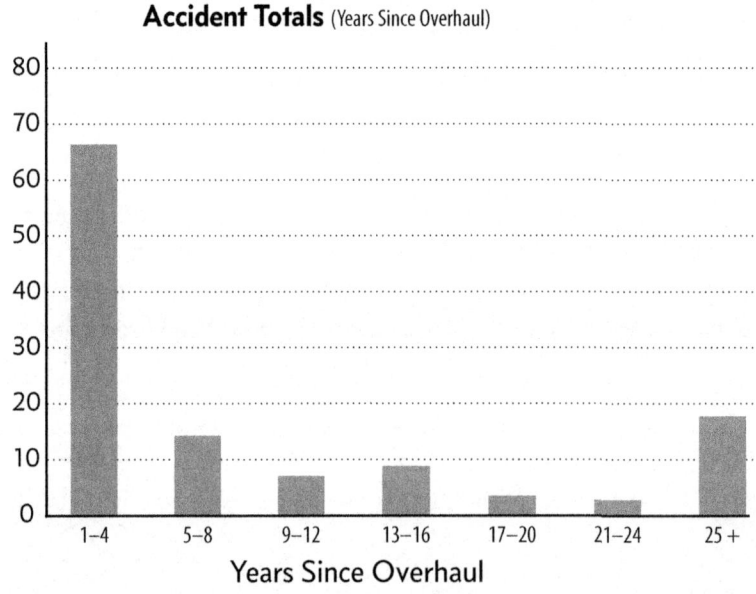

Small piston airplane accidents in 2001 through 2005 attributed by the NTSB to engine failure, by hours and years since engine overhaul. (Thanks to Nathan Ulrich Ph.D for this data.)

Tough sell

Because we have so little data about piston aircraft engines operated beyond TBO, it can be difficult to persuade mechanics and regulators to adopt on-condition maintenance where fixed TBO has been the norm. We can't collect engine failure data unless there are engine failures, and aircraft engine failures are considered unacceptable because they can cause injury and death. So, the maintenance program for an aircraft engine must be designed without the benefit of data about failures which the program is meant to avoid. RCM researchers call this "The Resnikoff Conundrum."

(If you'll permit me a brief detour: The FAA's decades-long opposition to rescinding "the age 60 rule" for airline pilots is a perfect example of The Resnikoff Conundrum. Experts in aviation medicine long agreed that there was absolutely no scientific basis for the FAA's venerable policy of forcing airline pilots to retire at age of 60. The FAA's longstanding argument was that it has no data showing that allowing airline pilots to continue flying beyond age 60 is safe. Well, duh! It took nearly five decades for the FAA to raise the age limit from 60 to 65… and that's still an age limit that has no scientific basis. Isn't it high time that ATPs were retired on-condition?)

We do know that fixed-interval overhaul is counterproductive for turbine engines, because the airlines and military started phasing out such fixed-interval overhauls in favor of on-condition maintenance decades ago (and saved gazillions of dollars by doing so). We have tons of data about high-time turbine engines, and that data makes it crystal clear that fixed-interval overhaul hurts reliability more than it helps—not to mention that it increases maintenance cost and downtime. I firmly believe that the same is true for piston aircraft engines, but sadly I don't have enough failure data on past-TBO engines to prove it beyond a reasonable doubt.

Are we even asking the right question?

Do we really care whether a piston engine has failure curve A or C? Perhaps not. A piston engine isn't a single component with a single dominant failure mode and a well-defined failure pattern.

Engine failures occur for lots of reasons. A piston engine is a complex system made up of hundreds of components—crankcase, crankshaft, camshaft, connecting rods, pistons, piston rings, cylinder barrels, cylinder heads, valves, valve guides, rocker arms, pushrods, gears, bearings, through-bolts, magnetos, spark plugs, etc.—each of which has its own unique failure modes and patterns. An engine failure can be caused by the failure of any of these parts, and each part has distinctively different failure characteristics.

To gain insight into how, when, and how often piston engines fail—and how best to prevent those failures—we need to analyze the failure modes and patterns of each of the engine's critical component parts, rather than try to lump them all into a single failure pattern for the engine as a whole.

Exhaust valves often don't survive to TBO. If we don't catch a potential exhaust valve failure using compression tests, borescope inspections, and engine monitor data, we risk a total failure ("swallowed valve").

Consider exhaust valves, for example. We know from experience that they often don't survive to TBO. When they begin to fail, we're almost always able to catch the potential failure before complete functional failure occurs (i.e., a "swallowed valve") by means of an annual compression test or borescope inspection. If the aircraft is equipped with a digital engine monitor and if the pilot knows how to interpret it, the pilot can detect a potential exhaust valve failure before the valve fails completely. But if the valve fails in-flight, it's usually a mayday situation.

Does this mean that we should reduce engine TBO to something less than typical exhaust valve life? Should we be overhauling our engines every 500 or 1,000 hours to prevent exhaust valve failures? Of course not!

Why not? Because repairing a leaky exhaust valve can be done without an engine teardown; it only involves lapping the valve in place or at worst pulling the cylinder. We've got excellent tools (like borescopes and digital engine monitors) that let us detect potential valve failures before complete failure occurs—provided those tools are used properly and regularly. (Sadly, they often aren't because most A&Ps have no training in borescopy, and most owners have no training in engine monitor data interpretation.)

So what good is TBO?

Does all this mean that the manufacturer's TBO is a worthless figure that should be ignored? No, not at all. In my view, the best way to think about published TBO is the way we think about human life expectancy statistics.

According to the National Vital Statistics Report published by the Centers for Disease Control, the 2015 life expectancy at birth for a white male in the United States is just over 76 years. This statistic might be quite useful in figuring out what premium to charge for a life insurance policy, or how to plan for retirement.

Does this mean that white U.S. males should be euthanized (removed from service) shortly after they reach age 76? (As a 74-year-old white U.S. male, I certainly hope not!) In fact, the same CDC figures show that the current life expectancy for a 76-year-old white male in the U.S. is 11 years. In other words, if you're still kicking at age 76, you can expect on average to live until age 87. Furthermore, if you are still alive at age 87, your life expectancy is five years so on average you can be expected to live until age 92.

Similarly, according to Continental Maintenance Manual M-0, the "life expectancy at birth" (recommended TBO) of TSIO-520-BB engines (like the ones in my 1979 Cessna Turbo 310) is 1,400 hours. This statistic might be quite useful in figuring out a suitable dollar amount for amortizing overhaul expense. Since it costs about $40,000 to overhaul one of these engines, a reasonable "reserve for overhaul" would be $28.57 per hour (i.e., $40,000 divided by 1,400 hours). This figure would also be appropriate for adjusting the "blue book" value of my airplane to account for higher- or lower-than-average engine time.

Does this mean that I should have euthanized my engines when they reached 1,400 hours SMOH, even though they were running great, had excellent compressions, low oil consumption, no metal in the oil filters, and excellent oil analysis reports? No, I don't think so. Although Continental doesn't publish figures for "life expectancy at 1,400 hours" for these engines, it only stands to reason that TSIO-520-BB engines that are in good shape at 1,400 hours surely ought to have a good deal of useful life left in them.

Well, as a matter of fact, those engines made it to 3,200 hours (about 230% of TBO) before being overhauled. My Cessna 310 was down for months while the engines were being overhauled. When those freshly overhauled engines were finally hung on the airplane and it was time for me to get back into the air, THAT's when I was nervous!

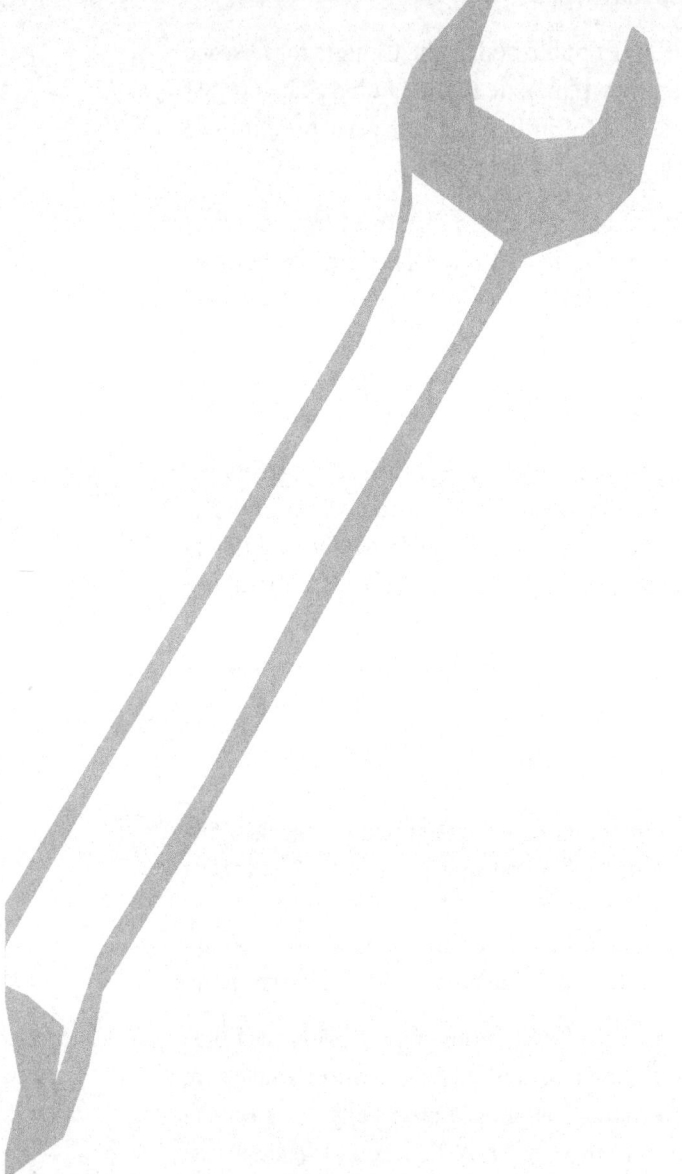

48
Deciding When to Overhaul

It's among the most difficult and agonizing decisions we have to make.

By far one of the most agonizing decisions aircraft owners face involves whether to remove an engine from an airplane and send it to an engine shop for a teardown. Tearing down an engine is the most invasive thing you can do to an engine. It's expensive—typically $15,000 for a minimal teardown inspection and repair, if you include the removal and installation labor, and possibly $40,000 or more if it turns out that a full engine overhaul is warranted. It's inconvenient—usually the airplane is grounded for 60 days or more. And it's risky—NTSB data shows that most catastrophic in-flight engine-failure accidents occur during the first 200 hours or 24 months after the engine is built, rebuilt or overhauled.

As stated in earlier chapters, I'm not a big fan of the idea of overhauling engines at TBO. I consider engine TBO to be a thoroughly discredited concept that has cost aircraft owners hundreds of millions of dollars by causing perfectly healthy engines to be euthanized arbitrarily. The notion of overhauling aircraft engines at a certain number of hours or years was abandoned many decades ago by the airlines and the military. Piston GA is the last segment of aviation that still subscribes to this ridiculous idea.

There are several reasons that TBO is such a flawed concept. One is that engine life has very little to do with engine hours in service. Hours is not what limits the life of our engines. The biggest life-limiting factor is exposure to corrosive environments during periods of disuse. The next biggest is operator abuse, particularly cold starts and improper powerplant management. None of these factors are reflected in the manufacturer's published TBO.

Why euthanize a healthy engine just because the hour meter rolled over to some fixed value?

Does it make any sense at all that the engine on an aircraft tied down outdoors in Tampa would have the same TBO as one hangared in Tucson? Or that one that flies 400 hours a year would have the same TBO as one that flies 40 hours a year? Or that one that flies mostly long-distance cross-country missions would have the same TBO as one used primarily for flight training? Of course not! The whole notion of a one-size-fits-all TBO is inherently nonsensical.

So why are so many owners still removing healthy engines for overhaul just because the hour meter rolled over to some fixed value?

On-condition maintenance

The smart way to deal with engine maintenance—including deciding when to overhaul—is to do it "on-condition" rather than on a fixed timetable. This means that we use all available condition-monitoring tools to monitor the engine's health, and to let the engine itself tell us when maintenance is required. This is how the airlines and military have been doing it for the past 50 years. Technology has finally advanced far enough – and become sufficiently affordable – to allow us to do this in even the simplest GA aircraft.

For our piston aircraft engines, we have a marvelous multiplicity of condition-monitoring tools at our disposal. They include:

- Oil filter visual inspection
- Oil filter scanning electron microscopy
- Spectrographic oil analysis
- Digital engine monitor data analysis
- Borescope inspection
- Differential compression test
- Visual crankcase inspection
- Visual cylinder head inspection
- Oil consumption trend analysis
- Oil pressure trend analysis

If we use all these tools on an appropriately frequent basis and understand how to interpret the results, we can be confident that we know whether the engine is healthy or not—and if not, what kind of maintenance action is necessary to restore it to health.

Use those tools!

The moment you abandon the TBO concept and decide to make your maintenance decisions on-condition, you take on an obligation to use these tools—all of them—and pay close attention to what they're telling you. Unfortunately, many owners and mechanics don't understand how to use these tools appropriately or to interpret the results properly.

For example, I once received an email from the owner of a 2002 Beechcraft A36 Bonanza which said in part:

> My engine has 1,700 hours [it's at TBO] and is running fine. It uses only 2 to 3 quarts of oil per 50-hour oil-change interval. I send in an oil sample to the lab at each oil change, and so far there have been no issues. I don't regularly cut open the filter. My plan is to install a factory rebuilt engine when the time comes…

Yikes! This owner is not cutting open his oil filters on a regular basis. How the heck is he going to know "when the time comes"? Oil filter inspection is by far the single most important tool we use in deciding when to overhaul, and he's not doing it. Oil analysis might or might not warn you when things are starting to go south; oil filter inspection will almost surely tell you when a crisis is imminent.

Oil filter inspection is arguably the most important single tool we use to decide "when it's time" to overhaul or replace an engine.

In a subsequent email exchange, the owner told me that his mechanic felt that there was no need to cut open the oil filters because the engine is on an oil analysis program. Boy oh boy, what terrible advice! Oil analysis is a complement to oil filter inspection, not a substitute. The purpose of oil analysis is to detect microscopic wear metal particles that are too small to see and too small to be caught in the filter.

If the engine develops a serious problem that throws off large quantities of metal chips, flakes, whiskers or fines, chances are that metal will be caught in the oil filter and never make it into the oil sample bottle. Rapid wear events such as cam lobes, lifters or starter adapters coming apart almost never show up in oil analysis because the metal they throw off is too large to pass through the oil filter. If we don't do regular oil filter inspections, we'll never detect problems like this.

Don't panic!

Time after time, I see aircraft owners and their mechanics overreact when elevated wear metals appear in an oil analysis report or visible metal shows up in the oil filter or a cylinder exhibits a drop in compression. You need to resist this urge. Don't panic!

A single bad oil analysis report is essentially meaningless. For Savvy Aviation's managed-maintenance clients (all of whom are on oil analysis), we generally do not recommend taking action until we've seen three consecutive oil reports so we can determine whether the initial bad report was a transient anomaly or the start of a worsening trend. We will often put the engine on a short oil-change interval (e.g., 25 hours) so we can gather data more rapidly. More often than not, the elevated wear metal values wind up self-resolving over the next 50 to 100 hours and no follow-up action is necessary.

Even if the oil analysis results keep getting worse, we never suggest grounding the airplane or taking the engine apart based solely on oil analysis results alone. Keep in mind that what oil analysis is reporting on is microscopic wear metal particles that are too small to be seen or filtered. Metal particles that small certainly do not represent any sort of imminent safety-of-flight hazard. No matter how ugly the oil report looks, we're not going to do anything drastic.

Oil analysis is like an early warning system: A series of deteriorating oil analysis reports simply tells us that we need to start looking more closely using other condition-monitoring tools at our disposal. For example, we might perform a borescope inspection of all the cylinders to see if we can spot anything wrong (scored barrel, burned exhaust valve, etc.) Or we might send the oil filter media out to a lab for scanning electron microscope evaluation, which will tell us the shape and alloy of the metal, and often provide a good indication of where it's coming from. Or we might dump the engine monitor data and analyze it to see if we can spot any combustion anomalies, oil pressure fluctuations, etc.

Similarly, if we cut open an oil filter and spot a significant amount of metal, we won't ground the aircraft unless the amount of metal found is quite large (e.g., a quarter-teaspoon). Most of the time, we will put the engine on a short oil-change interval, and we'll send the filter media out for scanning electron microscopy. If the amount of metal in the filter continues to increase, we might decide to pull some lifters, the prop governor, or perhaps the starter adapter in hopes of finding out exactly where the metal is coming from.

Here's a great example of how a little sleuthing and oil filter analysis saved a perfectly good engine from an untimely teardown. "I've been a happy aircraft owner until this morning," Frank wrote me. "My aircraft is in the shop for its annual inspection, and the mechanic just called to say that they found magnetic chips in the oil filter, and my big-bore Continental engine would have to be taken out and completely disassembled. Do you have any advice how to proceed?

I asked Frank to email me a photo of the filter contents, and he did. I saw a few curved magnetic whiskers and a few small bronze-colored chips. It was enough to get my attention and pique my curiosity about their source, but hardly anywhere close to warranting a $40,000 engine teardown.

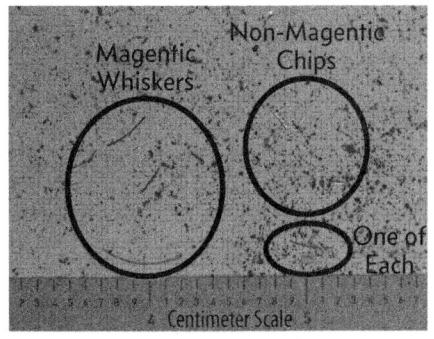

Photo of what was found in Frank's oil filter. Ultimately, it was identified as residual metal from a slipping starter adapter.

I asked Frank a few follow-up questions and learned that his engine's starter drive adapter had recently started slipping and had been replaced. The small quantity of metal in the filter struck me as being entirely consistent with what one would expect to see from a failing starter drive adapter. So, I advised Frank to have his mechanic change the oil and oil filter, then go fly the airplane for 10 hours or so, and then have the new oil filter cut open and inspected. I added that tearing down the engine because of such a small quantity of easily explainable metal in the oil filter struck me as being a grotesque overreaction on the part of his mechanic.

A few days later, Frank reported back that despite his best efforts of persuasion, the mechanic refused to budge. He refused to sign off the annual as airworthy unless Frank agreed to have the engine torn down and inspected. Frank was upset, but felt the mechanic had him over a barrel—a very expensive barrel indeed.

Taking it upstairs

As luck would have it, the next week I would be attending a meeting with a senior executive of Continental Motors, someone I'd known for years. I asked Frank to package up the contents of his oil filter in a clear zip-lock plastic bag and send it to me. I stashed the baggie in my briefcase and brought it to the conference.

At a break in the meeting, I presented the little bag of filter contents to my friend from Continental and asked him, "would you recommend tearing down an engine if you found this in the oil filter?" He shook his head, and then gave me one of those looks that I took to mean "is this a trick question?"

I told him about Frank's predicament, explained that his mechanic was holding Frank's airplane hostage, discussed Frank's recent starter adapter failure and replacement, and asked whether he agreed that the metal most likely came from the now-replaced starter adapter. He nodded in the affirmative and offered to take the sample back to the Continental Motors metallurgy lab for analysis.

A few days later, my friend emailed me to let me know that they'd examined Frank's filter contents and clearly established that the metal was indeed from the starter adapter, just as I suspected. At my request, he then sent an email to Frank's mechanic recommending that the aircraft be flown for 10 hours and the filter cut open and inspected.

After receiving guidance directly from the engine manufacturer, Frank's mechanic relented and agreed to sign off the annual. Some weeks later, Frank brought the airplane back to the shop after flying it for about 10 hours. The mechanic cut open the filter and found it to be free of any metal.

Frank was happy. So was his mechanic. Another engine saved from the chopping block!

Time to overhaul?

It takes something pretty serious before we should conclude it's time to send the engine off to an engine shop for teardown—or to replace it with an exchange engine. Here's a list of the sort of findings that would prompt us to recommend that "the time has come":

- An unacceptably large quantity of visible metal in the oil filter; unless the quantity is very large, we'll often wait until we've seen metal in the filter for several shortened oil-change intervals;

- A crankcase crack that exceeds acceptable limits, particularly if it's leaking oil;

- A serious oil leak (e.g., at the crankcase parting seam) that cannot be corrected without splitting the case;

- An obviously unairworthy condition observed via direct visual inspection (e.g., a bad cam lobe observed during cylinder or lifter removal); or

- A prop strike, serious overspeed, or other similar event that clearly requires a teardown inspection in accordance with engine manufacturer's guidance.

To summarize: Ignore TBO, maintain your engine on-condition, make sure you use all the available condition-monitoring tools, make sure you know how to interpret the results (or consult with someone who does), and don't overreact to a single bad oil report or a little metal in the filter.

Badly damaged cam lobe found during cylinder removal. It's time!

If you adopt this approach, you will be more comfortable with your engine and be able to make decisions based on hard data rather than fear or pressure. Think of it as removing uncertainty—uncertainty in aviation is never a good thing. Using the reliability-centered on-condition approach to engine maintenance, you can obtain the maximum useful life from your engine, saving a great deal of money, downtime and hassle and minimizing risk. It doesn't get any better than that.

49
Choosing an Overhaul Shop

Choose wisely—you'll have to live with your decision for a long time.

Engine major overhaul is the most important (and expensive) maintenance decision most aircraft owners ever face. It's also perhaps the ultimate act of trust. Few of us can stand around and watch while our engine is being built. Our old engine is removed, shipped off to an engine shop, and a month or two later a freshly overhauled "black box" arrives to be installed in our aircraft. (In the case of a factory rebuilt engine, the black box arrives first, and the core is returned later.) If we fly the typical 100 hours a year, we'll be entrusting our lives and those of our families and friends to the workmanship in that black box, sight unseen, for the next 15 or 20 years.

Choosing an overhaul shop to trust with our engine is a much more profound decision than choosing a mechanic to trust with our aircraft. If we choose a mechanic and subsequently have second thoughts about our choice, we can usually change horses any time we like. But when we choose an overhaul shop, we know we'll be stuck with the consequences of our decision for decades, so we darn well better get it right. So how do we choose an overhaul shop? Price? Warranty? Size? Reputation?

The way things once were

When I first became an aircraft owner in the late 1960s, the answer to this question was relatively simple: If you wanted the best engine money could buy, you'd buy a factory rebuilt engine from Continental or Lycoming. These so-called "factory remans" were very pricey, but you could be confident that the quality of a factory engine was top-notch, and the factory warranty was clearly superior to anything you could get from an independent field overhaul shop. Even if you ran into a problem after your factory engine was technically out of warranty, the factories would generally take care of it anyway on their nickel, just to keep you happy and ensure you'd go for another factory engine when yours reached TBO.

> I've never found any credible evidence that boutique engines are more powerful, longer lasting, or perceptibly smoother than an ordinary overhaul by a competent shop.

In those days, field overhauls were considered second-class engines that appealed to economy-minded aircraft owners who were unable or unwilling to pay the premium price for a factory engine, and who were satisfied with an engine that had fewer new parts and a lesser warranty. As a world-class skinflint, I cast my lot with this group in 1990 when I had both engines on my Cessna 310 overhauled.

Then came the implosion

When the general aviation market imploded in the 1980s, Continental and Lycoming went through a period of denial, waiting for their OEM market to come back to life. But by 1990, both companies recognized that this just wasn't going to happen. They realized that they were now basically in the overhaul business, and that their competition was the field overhaul shops.

Both engine manufacturers responded to this realization by slashing the prices of their factory rebuilt engines while simultaneously hiking the prices of many engine parts, thereby putting field overhaul shops at a huge disadvantage—and driving many of them out of business, including such then-well-known names as Mid-States (Tulsa), Schneck (Rockford), T.W. Smith (Cincinnati) and Western Skyways (Portland).

The overhaul shops that managed to survive this debacle could no longer compete with factory engines solely on price. They repositioned themselves by marketing their overhauls as being better than factory engines, albeit no cheaper (and sometimes more expensive). Some high-profile shops like Victor and Mattituck started offering high-priced "boutique" engines featuring powder-coated crankcases, chrome-plated rocker covers, and porting, polishing and balancing that bordered on anal-retentive.

I've never found any credible evidence that boutique engines are more powerful, longer lasting, or perceptibly smoother than an ordinary overhaul by a competent shop. Nevertheless, the boutique shops found plenty of customers for their high-priced engines—presumably owners who cared more about snob appeal and bragging rights than about getting a good engine at a reasonable price.

Meanwhile, back at the factory...

The loss of their OEM market and the reduced prices of factory rebuilt engines meant that Continental and Lycoming suffered a severe decline in profitability during this period. Something had to be done to rectify this situation. One way both companies addressed the problem was to take steps to reduce their warranty exposure. No longer could the buyer of a factory engine expect to receive the benefit of the doubt on a warranty claim. Many stories emerged in those bad old days about Continental and Lycoming dodging out of making good on claims for engines still under warranty on the basis of some technicality.

For example, Continental's "Top Care" cylinder warranty, much touted in their advertising, had a provision in the fine print that required the cylinders to be borescoped annually. Few mechanics borescoped cylinders back then unless there was some other indication of a problem (like weak compression), and very few piston GA mechanics even owned a borescope in those days. So if your Continental factory engine flunked a compression check 13 months into your 24-month Top Care warranty and you made a warranty claim against Continental, but your shop didn't borescope your cylinders during the last annual inspection, you were out of luck!

The way things are now

In the past three decades, the engine overhaul business has been turned upside down from the way things used to be. For years, factory engines were among the cheapest ways to go, but quality was spotty and obtaining warranty coverage from the factory was often like trying to get blood from a stone.

Factory engine quality is better now, but factory warranties are not exactly magnanimous. Continental Motors covers their rebuilt engines for 18 months and their new engines for 24 months, with most engine accessories covered for just 12 months. Lycoming covers both new and rebuilt engines for 24 months and factory overhauled engines for 12 months. Lycoming also extends these warranty periods by 12 months for high-utilization engines that fly at least 40 hours per month (which benefits a relative handful of working airplanes but doesn't do much for owner-flown airplanes). Happily, both manufacturers now seem to be more lenient and less nit-picky about dealing with warranty claims than they used to be.

It's interesting to contrast these factory aircraft engine warranties with today's new car powertrain warranties. Autos from Hyundai, Kia and Mitsubishi now come with a 10-year/100,000-mile powertrain warranty. A GA airplane that cruises at 150 knots flies roughly 15,000 statute miles per year, so a 24-month warranty covers you for about 30,000 miles.

For a time, Continental offered a 60-month warranty on their more expensive "Platinum-series" engines but discontinued that program because it was eating their lunch. Superior Air Parts also tried offering a 60-month warranty on field overhauled engines that used Superior cylinders and other parts and were done by a small group of preferred engine shops, but they terminated that program for the same reason. I think the industry learned its lesson, and isn't likely to be offering piston aircraft engine warranties longer than 24 months any time soon.

Warranties on field overhauled engines vary all over the map, most between 6 and 24 months. Warranty policies also very a lot. There are a handful of engine shops that have a reputation for no-questions-asked, customer-is-always-right warranty coverage—Powermaster Engines, RAM Aircraft, and Zephyr Aircraft Engines are in this category. Other shops—including some big-name boutique ones—have the opposite reputation.

If you're looking for a generous warranty policy, you need to be prepared to pay a premium price. The way I look at it is that part of the premium price you pay is an insurance premium for a policy that covers you for unplanned engine problems. If you elect not to pay the premium price, you also have to be prepared to self-insure.

Is the warranty worth the cost?

For most owners, paying less for an overhaul and self-insuring may not be an unreasonable way to go. Virtually all engine warranties cover "infant mortality" problems that occur during the first six to 12 months. After that, it has been my experience that most owner-flown engines that develop premature problems do so at around 500 or 600 hours. For the average owner who flies 100 hours a year or so, these problems won't arise until the engine is five years old, by which time the engine will be long out of warranty no matter who built it. To my way of thinking, if a premium warranty is likely to be useless, then there's no point in paying a premium price to get one.

So who should you trust to overhaul your engine when the time comes? Unless you've got excess disposable income burning a hole in your pocket, my advice is to avoid the high-priced boutique engines and look for a shop that offers a good-quality overhaul at a fair price, has plenty of satisfied customers, and is likely to be around for a while.

Factory engine or field overhaul?

If your old engine was a "lemon" or has a known problem that will result in a costly upcharge at overhaul time (such as a light or cracked crankcase or a non-VAR crankshaft), consider exchanging it for a factory rebuilt engine. A factory engine is also worth considering if you can't tolerate being down for 60 to 90 days, because that's realistically what you're looking at if you choose a field overhaul. (That's why big commercial operators so often favor factory exchange engines.)

On the other hand, if your engine has been good to you over the years and you hate to part with it, or if you have a strong preference for the parts that will be used such as specific cylinders, a tricked-out camshaft, or some other special feature not offered by the factory, have it field overhauled by a trustworthy shop with a decent reputation.

Whether you choose a factory engine or a field overhaul, don't expect to get any meaningful warranty coverage after the first year or so. Once the infant mortality period is over, you're pretty much on your own. Your best insurance against engine problems is frequent flying, meticulous maintenance, regular oil filter and borescope inspections, and close attention to CHTs. And don't forget to install that engine monitor (if you didn't already have one). Think of all this as your insurance policy. If used correctly, it will give you fair warning of incipient problems before they become serious and allow your expensive piston powerplant to live long and prosper. It's certainly worked well for me over my 50 years as a piston aircraft owner.

Epilogue
Some Things Have Changed, Some Haven't

The creation of this book took me on a trip down memory lane. It involved reviewing hundreds of my articles about piston aircraft engines, deciding which ones were still relevant, and bringing them up to date. In the process, I couldn't help but reflect on how much has changed in the 50 years since I bought my first airplane… and how much has stayed the same.

For the most part, the engines themselves haven't changed much. The Continental TSIO-550-K that powers the new-production Cirrus SR22T is not much different from the Continental TSIO-520-BBs that power my 1979 Cessna Turbo 310, and those are both close second cousins to the Continental O-470-R that powered my first airplane, a 1968 Cessna 182 Skylane. The Lycoming IO-360-M1A that powers the new-production Diamond DA40 XLS is nearly indistinguishable from the Lycoming IO-360-A1B6 in the 1971 Cessna 177RG Cardinal owned by my editor Colleen Keller.

The Continental O-470, TSIO-520 and TSIO-550 type certificates were issued in 1952, 1956, and 1980 respectively. The Lycoming IO-360 was certified in 1960 as a fuel-injected variant of the carbureted O-360 certified in 1955. It's hard not to wonder why these Jurassic engines are still being manufactured and installed in new-production aircraft, instead of basking proudly in some aviation museum next to the Curtis OX-5, Le Rhône rotary, Warner-Scarab, and Wright Whirlwind.

The Lycoming IO-360 engine was certified in 1960.

The reason, of course, is that the cost and effort involved in certifying a new aircraft engine design is so daunting that only companies with seriously deep pockets can consider it. Even among those deep-pocket companies, the track record for innovation is dismal. Does anyone remember the Continental 6-285 "Tiara" engine that was certified in 1969? It was a novel, highly innovative, clean-sheet piston engine design that cost Continental millions to develop and certify, but it turned out to be a disastrous commercial flop that wound up powering a handful of Piper Pawnee agplanes and then quietly faded into obscurity. In 1986, the company spent hundreds of thousands to certify a liquid-cooled TSIOL-520 "Voyager" engine that was also a resounding commercial failure. In 1997, Lycoming certified an IO-580 engine for the restart Cessna 206 Skywagon, but Cessna had so much trouble with it that they decided to go with the tried-and-true IO-540 engine (certified in 1960) instead, and the IO-580 was relegated to the scrap heap.

Retarded ignition

What amazes me even more is that our engines still use fixed-timed magneto ignition systems. This technology was introduced in automotive engines in 1897 by Robert Bosch (this was before cars had an electrical system) and gained popularity in aviation during the 1920s and 1930s (before airplanes had them). The four big Continental/Bendix S-1200 mags on my Cessna 310 are nearly identical to the ones produced by the Bendix/Scintilla factory in 1935.

The first electronic ignition system (EIS) was developed by Delco-Remy in 1948 and was offered by Pontiac in 1963. Fully solid-state EIS with microprocessor-controlled variable timing became the norm in automobiles during the 1990s. Aviation EIS from E-MAG and Light Speed Engineering have been de rigueur on kitplanes and LSAs for the past decade. Yet almost all of today's certified new-production all-electric GA airplanes with their fully redundant electrical systems are still being delivered with old-fashioned fixed-timed "tractor mags" on their engines. Why?

This WWII-era design Bendix magneto is still the mainstay of certified new-production aircraft.

Electroair began working with the FAA in early 2009 to obtain STC approval of its EIS. It took the company more than two years of work before the FAA approved the EIS on four-cylinder Continental and Lycoming engines in mid-2011, and nearly three more years to gain approval for six-cylinder engines in early 2014. Even after this five-year effort, the FAA would only approve replacing one magneto with the Electroair EIS; the other "tractor mag" must be retained even though it's demonstrably far less reliable. Lots of dual Electroair installations are flying in homebuilts with great success, but for certified aircraft the FAA remains a significant obstacle to innovation.

The promise of full-authority digital engine control (FADEC) systems for piston aircraft engines—the subject of heavy investment by both Continental and Lycoming in the late 1990s—has gone almost nowhere. Lycoming's EPIC system, developed in partnership with Unison, never made it into production. Continental acquired the FADEC system developed by Aerosance and succeeded in getting it certified for both Continental and Lycoming engines, but only a tiny handful have been delivered (most of them on Liberty XL2 trainers) and by any measure the program has been an abject commercial failure. (Do you detect a pattern here?)

Outside the box

A notable exception involves an obscure Austrian subsidiary of a Canadian company mostly known for its two-stroke snowmobile, motorbike and ATV engines that quietly redefined the small (under 150hp) four-stroke piston aircraft engine. Few noticed until 2004 when the FAA approved the LSA rule, and sexy factory-built Special Light Sport Aircraft started showing up at Oshkosh, Lakeland and Sebring. Almost all were powered by Rotax 912-series engines.

The Rotax 912 story began in 1989, when the first 80 hp engines were shipped to customers in Europe for use on ultralights and motorgliders. Rotax already dominated those markets with its two-stroke engines derived from their snowmobile engines. But almost nobody outside the ultralight community took notice, especially since the 912 had a ridiculously short 600-hour TBO. Yet Rotax pressed forward, and in 1994 obtained FAA certification of the 80 hp engine. A 115 hp turbocharged version (the 914), a 100 hp ultralight version (the 912ULS), and a certified 100 hp version (912S) followed in quick succession. TBO was increased to 1,000 hours, then 1,200 hours, then 1,500 hours. These engines started selling like hotcakes abroad, but never gained much traction in the U.S. until the FAA adopted the LSA rule in 2004 and factory-built S-LSAs became all the rage.

Designers of S-LSAs overwhelmingly chose the Rotax 912 because it was substantially lighter, more compact and more efficient than traditional 100 hp engines like

Fuel-injected FADEC-controlled Rotax 912iS

the Continental O-200 and the Lycoming O-235. The Rotax design is lightyears ahead of these competitors. It uses much smaller cylinders turning much higher revs, something made possible by an integral reduction gearbox. The cylinder heads are liquid cooled, the ignition system is electronic, and the engine is designed to run on unleaded mogas instead of 100LL. What's not to like?

By 2009, the TBO of both the 912 and 914 engines was increased to 2,000 hours, and I'm told they usually look pristine inside when torn down at that point. In 2012, the fuel-injected FADEC-controlled 912iS started shipping. By 2014, Rotax had shipped 50,000 of these engines. As I write this, the latest version of this engine is the 135 hp turbocharged 915iS. (I sure wish Rotax would come up with a 300 hp engine!)

Squeezebangs

There are other new engine designs emerging. To me, the most exciting are compression ignition (diesel) engines designed to run on Jet A.

Cessna made a huge splash in 2013 when it announced a new version of the venerable Cessna 182—dubbed the Skylane JT-A—with a French-built SMA turbodiesel engine. Both Cessna and SMA made major investments in the development and certification of this aircraft, but just as first customer deliveries were scheduled to begin Cessna suddenly pulled the plug on the Skylane JT-A. Cessna never fully explained the reason behind its decision, although we do know that one of their flight test aircraft suffered an in-flight turbocharger failure and made a successful off-airport landing. Whatever the reason for the program cancellation, I found it to be a huge disappointment; the Skylane JT-A would have been a big step forward in putting diesel-powered GA airplanes on the map.

In Germany, Thielert Aircraft Engines GmbH developed a "Centurion" line of aircraft engines that borrowed heavily from automotive diesels from Mercedes-Benz. Despite being certified, the engines proved to have serious issues with reduction gearbox longevity. Then Thielert went bankrupt, and China's AVIC acquired the rights to these engines. They were resurrected via AVIC's German subsidiary Technify Motors GmbH, who apparently resolved the gearbox issues. The engines are now being marketed through Continental Motors Group (also owned by AVIC).

The Continental/Technify 135 hp CD-135 and 155 hp CD-155 four-cylinder turbodiesel engines are STC'd for the Cessna 172 Skyhawk, the Diamond DA40 Star and DA42 Twin Star, and the Piper PA28-161 Warrior and PA-28-181 Archer. In 2014, Piper started shipping the new-production Archer DX powered by a CD-155 diesel, and in 2018 the company announced it would produce a twin-engine Piper Seminole DX powered by a pair of 170 hp CD-170 diesels.

In Austria, Diamond Aircraft Industries was an early adopter of the Thielert diesels in its DA42 Twin Star. When Thielert went belly up, Diamond suspended production of the DA42 for about a year, and owners of these now-unsupported engines were in a world of hurt. Diamond restarted DA42 production in 2009 using a 168 hp AE 300 diesel engine built by its own Austro Engine subsidiary in Austria.

Meanwhile, two small Wisconsin startup firms—DeltaHawk Engines, Inc. and Engineered Propulsion Systems, Inc.—are working hard to achieve FAA certification of two very different clean-sheet diesel aircraft engines. The DeltaHawk DH180 is a lightweight compact 180 hp two-stroke turbodiesel, while the EPS Graflight V-8 is a 350 hp brute aimed at high-performance singles and twins.

Continental CD-155 turbodiesel.

As I write this, it's still too early to tell whether any of these aviation diesels will reach critical mass sufficient to be commercially successful. I've got my fingers and toes crossed for the early adopters of these engines, because I see diesels as the best hope for the future of piston GA.

Leaner isn't meaner

Although most of us are still flying behind engines designed in the 1950s with ignition systems designed in the 1930s, the way we operate these powerplants has changed, and for the better. I vividly recall when in 1967 I received my first checkout in an airplane with a controllable-pitch prop—it was a Cessna 182—my CFI explained to me that there were two things that must be avoided at all costs because they would destroy the engine: (1) running "oversquare" by allowing manifold pressure to exceed RPM divided by 100, and (2) touching the red mixture control knob below 5,000 feet MSL. That was the conventional wisdom back then, and I believed it and even passed it on to my students after I became a CFI myself.

With the benefit of 20-20 hindsight, we now know that this was terrible advice. It turns out that the most efficient way to operate our piston aircraft engines is at the highest manifold pressure and lowest RPM that the engine manufacturer allows—typically two to three inches "oversquare" for most normally-aspirated engines, and 10+ inches "oversquare" for most turbocharged engines. I cruise my Cessna Turbo 310 at 32 inches of manifold pressure and 2,200 RPM and have achieved truly exceptional engine and cylinder longevity operating that way.

In the late 1990s, I was a very early adopter of lean-of-peak EGT (LOP) operation and the position-tuned GAMIjectors® that made it possible to operate my Continental engines that way without unacceptable roughness. At the time, almost everyone in the GA community—engine manufacturers, overhaul shops, CFIs and pilots—believed that LOP operation was terribly abusive to cylinders (especially exhaust valves) and that anyone who operated a piston aircraft engine that way was cruisin' for a bruisin'.

There's been a revolution in how we're taught to use the red knob

My friend and mentor George Braly, a brilliant aeronautical engineer and iconoclast, invented position-tuned fuel nozzles and served as the archbishop of the Church of Lean-of-Peak. George created a world-class computerized engine test stand in Ada, Oklahoma, torture-tested engines while gathering reams of data, and started preaching the superiority of LOP and backing it up with hard evidence. I was an unabashed apostle, but most people in the industry considered him to be a dangerous crackpot.

It has taken 20 years, but George's sermons have at last become accepted as the new conventional wisdom. Manufacturers like Beech and Cirrus now call for LOP operation in their POHs. Continental Motors now offers its own position-tuned fuel nozzles to facilitate LOP operation, and its latest FADEC engines automatically go LOP during cruise power settings. That this transformation has taken two decades is an indication of just how change-averse the GA industry is.

Keeping them healthy

While George has been busy preaching the gospel of LOP, I've been on my own crusade to bring the benefits of reliability-centered maintenance (RCM) to the piston GA community. Like George, I've faced an uphill battle. One of the fundamental tenets of RCM is that invasive maintenance (like cylinder replacement and engine overhaul) should be performed only when demonstrably necessary, not on a fixed-interval timetable. The data clearly shows that the risk of piston aircraft engine failure is greatest when the engine is young, not when it's old, and that we should worry more about pediatrics than geriatrics.

One obvious corollary is that the traditional practice of overhauling engines at a fixed TBO—regardless of condition—is counterproductive. We now know that doing so often forces perfectly healthy engines to shift from a low-risk geriatric state into a high-risk pediatric one fraught with infant-mortality perils. FAA regulations permit non-commercial operators to ignore the manufacturer's recommended TBO and operate engines strictly "on-condition," overhauling them only when they start exhibiting signs of illness. However, many A&P mechanics are still very uncomfortable with this approach, having been trained that manufacturer's maintenance guidance should always be followed to the letter. Sadly, manufacturer's guidance is decades behind the times in this area, just as it was with LOP operations.

Another corollary of RCM is that mechanics should be much more hesitant to remove cylinders from piston aircraft engines than they now are. History clearly shows that cylinder removal in the field is risky business in which even the subtlest gaffe can lead to catastrophe. We're talking major-league MIFs. The obvious takeaway is that cylinders shouldn't be removed unless absolutely necessary, and "top overhauls" where multiple cylinders are removed simultaneously—something that greatly increases the risk—should be avoided like the plague.

Until about 15 years ago, mechanics relied almost solely on WWII-era technology—the ubiquitous differential compression test—to determine when cylinders were unairworthy and required replacement. That test is so crude and unreliable that many

Borescopes have revolutionized how we assess engine health.

tens of thousands of perfectly sound cylinders were removed unnecessarily based on findings of substandard compression. Nowadays, however, articulating high-resolution electronic borescopes—some costing less than $200—give us a much more accurate and discriminating way to assess cylinder condition. The challenge now is to train more GA mechanics how to capture and interpret borescope images—something that presently isn't covered in most A&P school curricula—and how many common cylinder problems like burned valves and stuck rings can often be remedied without cylinder removal.

Better education of mechanics is the key to dealing with these issue, but that's not easy to accomplish. When I first became an A&P—after having been a pilot and CFI for decades—I was shocked to learn that there's absolutely no FAA requirement for mechanics to receive any sort of recurrent training the way pilots must do, and in fact most A&Ps receive nothing but informal on-the-job training once they earn their certificate. The result is that many mechanics—especially the most senior ones—have knowledge that's often quite stale and out-of-date. I've made no secret of my conviction that the FAA should mandate regular recurrent training for mechanics but so far, the agency has shown no inclination to do so.

Meantime, it's up to aircraft owners like you to be unapologetic activists when it comes to the care, feeding, and especially maintenance of their piston aircraft engines. The fact that you've made it through this book proves that you're a dyed-in-the-wool information junkie dedicated to learning all you can about this subject. My compliments!

Be mindful that you're the first line of defense when it comes to keeping tabs on the health of your engine, detecting incipient problems early before they become safety-critical, and documenting symptoms meticulously to provide the information necessary for you and your mechanic to diagnose those problems accurately and correct them in the most minimally invasive way possible. Do all that and your engine will live long and prosper.

If you need help, my company Savvy Aviation, Inc. specializes in assisting aircraft owners like you with everything from 24/7 breakdown assistance and engine monitor data analysis to troubleshooting help, shop recommendations, second opinions, and full-service concierge maintenance management. We're on the web at www.SavvyAviation.com. Don't be a stranger!

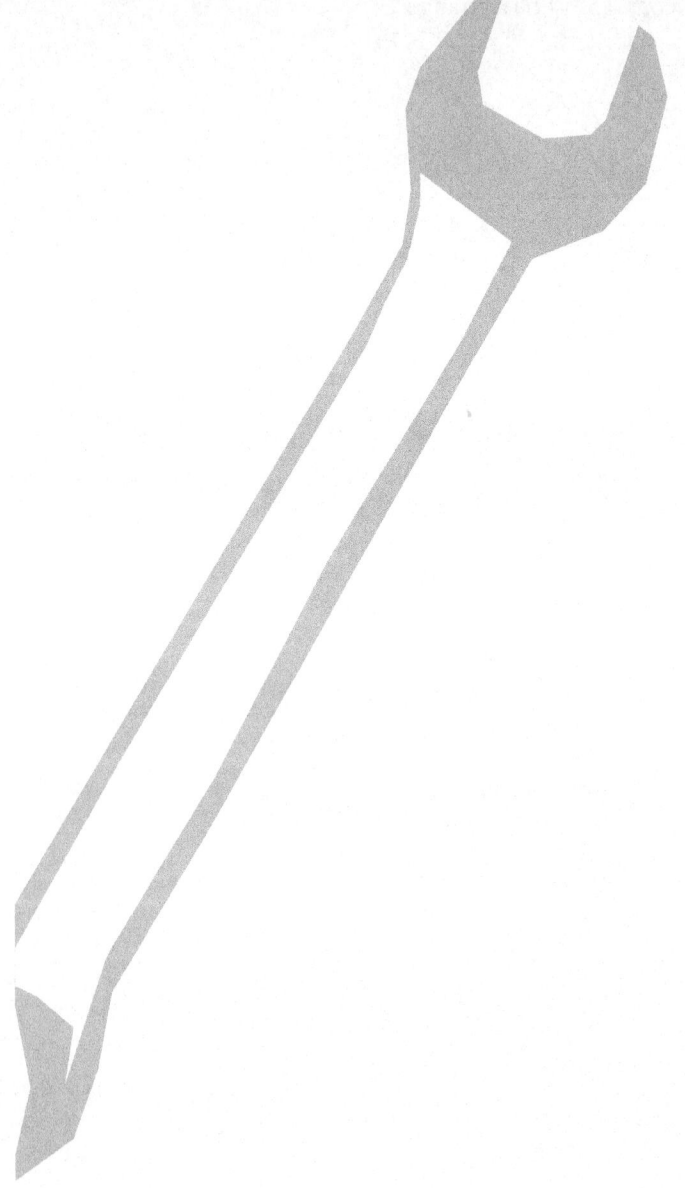

Index

A

A&P mechanics, 471
absolute (EGT), 18
absolute pressure controller (APC), 200
abutment (head-to-barrel junction), 82-83
accelerator pump (carburetor), 164
accessories, 90, 93
accessory
  case, 93
  losses, 12, 15
ACF-50, 233
acknowledgements, xi
actionable intelligence, 301, 303, 308
additives (aftermarket oil), 145-147
adhesion (plating), 45
adjustment procedure (Continental fuel injection), 180-183
advanced
  ignition timing, 84, 157
  leaning, 219-226
Advanced Pilot Seminars, 209, 221
Advisory Circular
  AC 20-105B, 442
  AC43-12a, 379
AeroKroil, 233, 365
Aeroshell
  15W-50, 112, 141-145, 230-231, 250, 292
  oil, 257
  W100 Plus, 141-145, 231, 292
  W100, 141-145, 230, 250, 260-261, 292
Aero Test Inc., 182
aftermarket additives (oil), 145-147
air-fuel
  charge, 4-5, 7-8, 25, 32
  mixture, 329, 339
  ratio, 161, 337
  ratio stoichiometric, 161
air-oil separator, 354
  M-20, 354
  Walker, 354
aircraft
  engine, 4
  insurance, 440
Aircraft Spark Plug Service, 154
Aircraft Spruce, 113-114
air leak (baffling), 245
AirVenture, 259
airworthiness
  certificate, 434
  limitations, 435, 439
  requirement, 434
Airworthiness Limitations Section (ALS), 436-438
Airworthiness Directive, 400, 416, 436-437
  AD 87-21-02 fuel filler ports, 192
  AD 97-26-17 airmelt crankshaft, 415
  AD 2004-10-14 prop strike, 132
  AD 2014-05-29 Millennium cylinders, 419

AD 2016-16-12 ECi Titan cylinders, 420
cylinders, 78-80
ECi Titan cylinders, 78-80
Superior Millennium cylinders, 78-80
Air Power Inc., 56
alarm (digital engine monitor), 82
Alcor, 18
TCP, 387
alert, 303-304
alternate air door, 288
Alternative Means of Compliance (AMOC), 419
altitude-compensating
aneroid, 173-174, 204
Continental fuel injection, 173-174
fuel pump, 183
pump, 173
pump aneroid adjustment, 183
altitude-compensating pump (adjustments), 174
aluminum, 289
aluminum particles (oil filter), 285
aluminum alloy, 90
tensile strength, 81-82
Amazon.com, ix
analysis
big data, 295-309
digital engine monitor, 68-69, 72
engine monitor data, 295-309
oil, 55, 71-72, 107, 112-113, 125, 281-293
anecdotal evidence, 45
arc-over (magneto), 158
arm (rocker), 42, 96-97
arrangements (cylinder), 89
art of troubleshooting, 313-326
ashless dispersants (oil), 141-145
ASL CamGuard, 145-147, 231, 257-262, 374, 378
assessing
cylinder condition, 271-279, 472
cylinder health, 271-279
engine condition, 265-270
engine health, 265-270
asymmetry (exhaust valve), 67
Atkinson (Walter), 209, 221
auto-lean
capability, 173
carburetor, 164
automatic wastegate control, 198
automotive engine, 4
AvBlend, 145-147, 257
avgas, 192-195

100LL, 192-195, 467
color, 194
density, 194
Aviation Laboratories, 285-286
Avidyne
Entegra EMAX, 29-30
MFD, 321

B

babbitt, 104-105, 291
Babbitt (Isaac), 104
bad mag check, 156-157
baffle, 49, 57
cooling, 245
inspection, 235
inter-cylinder, 246-248
rigid metal, 246
seals, 49, 235, 245
ball bearing, 99
banks of cylinders, 89
bar-graph display, 19
barrel (cylinder), 38
barrel-style lifter, 114
base flange (cylinder), 39, 91
basic leaning, 209-217
bathtub curve, 443
bearing, 89, 99
ball, 99
crush, 105-106
displaced, 106
failure modes, 106-107
main, 94-95
needle, 99
plain, 99-101, 107
roller, 99
rolling element, 99
saddles, 105-106
shells, 100
sleeve, 100
spun, 106
supports, 90, 94-95, 105-106
trimetal, 104-105
bed mount, 92-93
Beechcraft
Bonanza, 216, 413-414, 416
Debonair, 405-408
Mustketeer, 217
bellmouthing, 63

bending moments (crankshaft), 90
Bendix
   magneto, 152, 465
   S-1200 magneto, 152, 465
Bendix-RSA fuel injection, 166-167, 169
Bendix-Stromberg pressure carburetor, 165
Bendix/Scintilla, 465
bent pushrod, 63
Benz (Karl), 3, 162
Bernoulli's principle, 164
best-economy mixture, 243
best-power mixture, 242-244
big-end (connecting rod), 95
big data, 295-309
big mixture pull, 226, 343
black smoke, 367
Blackstone Laboratories, 289-292, 377
blade creep, 235
Blodget (Bob), 295
blow-by, 13, 83, 375
   excessive, 352
   maximum allowed, 353
blowdown, 8
blowout, 362-363
bolts
   connecting rod, 96
   flange, 90
   through, 90
boost pump, 33
bootstrapping, 202-204, 360-364, 369
borescope, x, 34, 53, 277, 353, 472
   inspection, 65-69, 72, 84, 110, 112-113, 273-276, 328, 396, 453
Bosch (Robert), 465
bottom (end), x, 55, 87, 89, 112
bottom dead center (BDC), 7
boundary
   layer, 33
   lubrication, 102, 139-140, 142
Braly (George), 190-191, 209, 221, 470-471
brass particles (oil filter), 286
break-even point (cylinder replacement), 56-57
breakaway torque, 426
breakdown assistance, 473
Briggs & Stratton, 269
bronze particles (oil filter), 286
Bruce's Custom Covers, 254
brutally lean mixture, 64
BTU, 10-11

burned exhaust valve, 51, 53, 61, 63-64, 472
bushings, 100
butterfly (throttle), 162
butylated triphenyl phosphate (bTPP), 141-142, 292
byproducts (unburned combustion), 63

C

cad-plating, 424-425
cam
   corrosion, 109-113, 118
   distress, 117-119
   follower, 96-97
   lobe cracks, 120-121
   lobe pitting, 120-121
   lobes, 229
   lobe sharp pick test, 120-121
   lobe spalling, 396
   spalling, 55, 109-113, 118
CamGuard, 145-147, 257-262
cam inspection (Continental SID05-1), 118-121
camshaft, 89-90, 93, 96-97, 99
   gear, 93, 96-97
   inspection, 55
   lobe, 96-97
cap (rotator), 42
carb heat, 219
carbon
   dioxide, 13
   monoxide, 14
carbon particles (oil filter), 285
carburetor, 162-164
   accelerator pump, 164
   auto-lean, 164
   Bendix-Stromberg, 165
   evaporative, 162
   float-type, 163
   float chamber, 162
   fuel discharge nozzle, 164
   heat, 216
   idle circuit, 164
   Marvel-Schebler, 162-164
   mixture control, 164
   pressure, 165
   proportioning, 162, 164
   sidedraft, 163
   throttle body, 162
   updraft, 163
   venturi, 164

carburizing, 118-119
  cam lobe, 96-97
case
  accessory, 93
  split the, 89
case-hardened, 39
centrifugal loads, 90, 93
Cermicrome, 43-44
Cessna
  172 Skyhawk, 236, 468
  177 Cardinal, 463
  182 Skylane, 220, 235-236, 245, 296-297, 338, 394, 399, 463, 468-469
  206 Stationair, 287
  210 Centurion, 300
  310 (twin), 220, 227, 258, 273, 290, 313, 341, 373, 465
  340 (pressurized twin), 236, 306, 308-309
  414 Chancellor, 236
  P210 Pressurized Centurion, 235
  T210 Turbo Centurion, 349-350
  TTx, 213, 215, 222
  Turbo 310, 463, 470
Champion Aerospace, 152-153, 381
changing the oil, 373-379
channel chrome cylinders, 40
cheeks (crankshaft), 94-95
chemical
  energy, 5, 9, 37
  losses, 13
choke (cylinder barrel), 39
choosing an overhaul shop, 457-462
chrome particles (oil filter), 286
chrome plated (cylinder barrels), 40
chromium, 289
CHT, 22, 24, 28-29, 31, 33-34, 48-49, 68-69, 82-83, 85, 236, 240-241, 243-244, 248, 298
  alarm, 248
  cold, 217
  control, 235
  excessive, 217, 220, 304, 320, 325
  factory gauge, 240
  indication error, 321-322
  instrumentation, 85
  management, 239-248
  maximum, 222, 299, 305
  redline, 234
  spread, 300, 303
Cirrus

SR20, 29-30
SR22, 51, 213, 215, 222, 274, 300-301, 303, 306-307, 320
SR22T, 192, 463
cleaning fuel nozzles, 189-191
clearance, 232, 250
climb
  initial, 240
  ROP, 223
clogged fuel nozzles, 185-192, 332, 342
coating (manganese phosphate), 45
coefficient of expansion, 82-83
cohort, 296, 301, 303, 305, 308
cold-start, 228, 231-233, 237
color code (cylinder base flange), 40
combustion, x, 37
  byproducts, 214, 375
  event, 339
  pressure, 48, 50
compression
  check, 189, 364, 472
  ratio, 4-5, 7-8, 14-16, 22-23, 85
  stroke, 4, 96-97
  test, 328
compression ignition (diesel), 161
compression ring (plasma-coated), 45
compression test (differential), 50-53
compressor, 197
concentricity (valve-to-seat), 45, 62-63
condensation, 230
condition monitoring, x, 65, 68-69, 125, 281, 283, 450-451, 453, 455
  cylinder, 271-279
  engine, 265-270
connecting rod, 27, 89, 94-95
  big-end, 95
  bolts, 96
  small end, 95
consumption (oil), 48, 147-148
contact
  cleaner, 189
  metal-to-metal, 46-47, 101
contact area (valve seat), 63
Continental, 29-30, 45, 54, 56, 61, 63-64, 68-69, 80-81, 152, 268, 299, 412, 415-416, 425-426, 437, 439, 458-459, 466
  bottom-induction injected engines, 333
  continuous flow fuel injection system, 180
  crossflow, 333

fuel injection, 166-167, 169-184
fuel nozzles, 245
IO-240, 170, 173, 176
IO-360, 173, 176
IO-470, 170, 173-174, 176
IO-520, 170, 173-174, 176, 250
IO-550, 170, 173-174, 176, 183
magneto wiring, 343
Maintenance Manual M-0, 50, 181, 231, 233, 447
NIC3 cylinders, 40, 44
O-200, 466
O-470, 463
O-470-series, 338
recommended maximum CHT, 215
red line CHT, 215
Standard Practice Maintenance Manual M-0, 274, 400-401
starter drive adapter, 453-454
Top Care Cylinder warranty, 459
top induction system, 177
TSIO-360, 236
TSIO-520, 212, 227, 236, 275, 447, 463
TSIO-550, 463
Continental fuel injection
  adjustment procedure, 180-183
  altitude-compensating, 173-174
  fuel control unit, 174-176
  fuel nozzles, 178-179
  fuel pump, 170-174
  manifold valve, 176-177
Continental fuel injection system (setup), 180
Continental SB03-3 (compression test and borescope inspection), 273-274, 279
Continental SB96-11B (prop strike), 131, 133
Continental SB M84-15 (old compression test), 272
Continental SB M89-9 (crankcase pressure), 352-353
Continental SB M91-8 (overhaul periods), 441
Continental SID05-1 (cam inspection), 118-121
Continental SID97-3 (fuel injection adjustment), 181
Continental SIL98-9 (time between overhauls), 441
Continental Motors, 408, 470
  formerly TCM, 265
continued-time cylinder, 54
controllable-pitch propeller, 469
cooling
  drag, 246
  efficiency, 300
  fins, 38, 41

  oil, 138
copper (oil analysis), 291
copper particles (oil filter), 286
corrosion, x, 48, 84-85, 109-113, 118, 219, 228-230, 255, 258-259, 354, 378, 414, 450
  cam, 109-113, 118
  lifter, 109-113, 118
  protection, 262
Corrosion-X, 233
counterweights (crankshaft), 94-95
cover (rocker box), 42
cowl
  flap, 49
  flaps, 246
crack (growth rate), 77
crack initiation (head-to-barrel junction), 83
cracks (cam), 120-121
crank angle, 27
crankcase, x, 55, 89-90, 94-95
  bearing saddles, 94-95
  bearing supports, 94-95
  breather, 355
  crack, 77, 455
  fretting, 285
  halves, 90-91
  pressure, 352
  pressurization, 353-354
  split the, 89
crankcase breather (positioning), 353
crankpins, 99
  crankshaft, 94-95
cranks (crankshaft), 94-95
crankshaft, 25-27, 89-90, 93-95, 99
  airmelt, 415
  bending moments, 90
  cheeks, 94-95
  counterweights, 94-95
  crankpins, 94-95
  cranks, 94-95
  failure, 441
  fracture, 440
  gear, 93, 96-97
  journals, 94-95
  non-VAR, 461
  throws, 94-95
  vacuum arc remelt VAR, 415-416
critical altitude check, 361, 368
crocus cloth, 121-122
crosshatch, 278

cruise
- best power, 223
- flight, 242
- LOP, 223
- phase, 226
- power setting, 239, 471
- ROP, 223
- speed, 303

cruise-climb, 201
crush (bearing), 105-106
cycle-to-cycle variation (CCV), 338, 340
cylinder, x, 3, 37, 90, 112
- accelerated wear, 268
- airworthiness, 274
- arrangements, 5, 89
- banks of, 89
- barrel, 38-39
- barrel choke, 39
- barrel hardness, 45-46
- barrel honing, 45
- barrel wear, 353
- base flange, 39, 91
- base O-rings, 93
- change, 427
- choke, 252
- chrome plated, 230
- condition assessment, 271-279
- condition monitoring, 271-279
- continued-time, 54
- corrosion pitting, 353
- crosshatch, 353
- deck contamination, 408
- decks, 91
- glazed, 267
- head, 5, 38, 41
- hold-down studs, 91, 93
- honing, 40
- installation, 424
- longevity, 38, 43, 46, 49
- nickel plated, 230
- non-firing, 363
- pads, 91
- reconditioned, 56, 85
- removal, 272, 411, 471-472
- replacement, x, 471
- replacing, 53-54
- scuffing, 232, 252
- skirt, 39, 91
- volumetric efficiency, 241
- walls, 229

warranty, 44
cylinder(break-in), 267
cylinder compression (zero), 363-364, 369
cylinder head temperature (CHT), 22, 24, 28-29, 31, 33-34, 48-49, 68-69, 82-83, 85, 239-248
cylinder pressure (peak), 26-29
cylinder work (risks of), 245, 419-424, 426-428

D

Daimler (Gottlieb), 3
dampeners (harmonic), 95
Deakin (John), 209, 221
deciding when to overhaul, 449-456
decks (cylinder), 91
defect (manufacturing), 45
Delco-Remy, 465
DeltaHawk, 468
DeltaHawk Engines Inc., 468
deposit build-up (valve stem), 63
de Rochas (Alphonse Beau), 3
descent phase, 226
dessicant plugs, 231
destructive detonation, 25, 31
detonation, 14-15, 24-25, 28-29, 31-33, 37, 84, 86, 195
- destructive, 25, 31
- margin, 10, 197

diagnosis, 188, 319-320, 327, 472
- differential, 313-326

Diamond
- DA40 Star, 213, 215, 222, 463, 468
- DA42 Twin Star, 468

Diamond Aircraft Industries, 468
diesel engine, 4-5, 8, 15, 161, 468-469
DIFF, 22-23
differential diagnosis (DDx), x, 313-326, 370
differential compression test, 50-53, 266-267, 271-272, 274-275, 471
digital engine monitor, x, 24, 34, 49, 82, 85-86, 123, 154-157, 217, 240, 242, 277, 308
- analysis, 65, 68-69, 72
- data, 276, 295
- data analysis, 68-69
- normalize mode, 154-155

dipstick level, 148-149
dirt in oil, 234, 237
displaced bearing, 106
displacement (engine), 10
dissimilar metals, 250

distress (exhaust valve), 72
distribution (mixture), 15
distributor valve, 170, 176
disuse, 112
dry-start, 228, 231-233, 237
dry particle analysis, 286
Durden (Rick), 373
dynafocal mount, 92-93
dynamic seal, 272-273
dynamometer, 52

E
E-MAG, 465
EAA AirVenture, 219, 258, 373
ECi, 43, 45, 81, 420
    Titan Airworthiness Directive, 78-80
    Titan cylinders, 40, 44
economics (jug), 53
educated guesswork, 318
efficiency, x, 5, 8-9
    fuel, 12
    Otto-cycle, 11, 14
    thermal, 10-11, 15
    volumetric, 10-11, 15
EGT, 17-24, 28-29, 31, 68-69
    absolute, 18
    drop, 187
    excessive, 214, 342
    indicated, 19
    oscillation, 68-69
    peak, 212, 226, 331-332
    probe temperature, 20-21
    relative, 18-19
    rise, 154-157
    spread, 17
    trace, 307
EI
    UBG-16, 19
    Ultimate Scanner, 19
Electroair, 465-466
electronic ignition, 15-16, 465
Electronics International, 19
elevated nickel, 71-72
Ellison TBI, 166
embeddability, 103-105
emergency toolkit, 343-344
end gas, 340
energy, 9
    chemical, 5, 9
    mechanical, 5
engine
    aftermarket modifications, 241
    airworthy, 265-267, 270
    boutique, 458-459
    break-in, 320
    carbureted, 212, 216, 219, 245, 337
    catastrophic failure, 420, 428, 443
    cold start, 249, 251
    condition monitoring, 265-270
    Continental 470-series, 416
    Continental 520-series, 416, 419, 422
    Continental 550-series, 416, 419, 422
    Continental IO-360, 421
    Continental IO-470, 413-414
    Continental IO-550, 405
    Continental O-470, 399
    Continental TSIO-520, 349
    cover, 254
    factory exchange, 461
    factory rebuilt, 458-459, 461
    factory remanufactured, 458, 460
    failure, 446, 471
    failures, 445
    fuel-injected, 241, 337
    ground runs, 230
    health assessment, 265-270
    inactivity, 259-261
    instrumentation, 240
    longevity, 227-237
    Lycoming IO-360, 377
    monitor, x, 325, 328, 347
    monitor data analysis, 68-69, 295-309, 453, 473
    mount, 92-93
    normally-aspirated, 204, 470
    oil, 137-149
    preheat, 249
    preservation, 231
    rebuilt, 56
    repair, 400
    roughness, 330
    rough running, 188, 327, 335-336, 470
    stoppage, 440
    teardown, 449, 455
    temperature control, 239
    transient roughness, 187
    turbocharged, 241, 470
engine-failure accident data (NTSB), 444
Engineered Propulsion Systems Inc., 468

engine monitor (data sample rate), 329
engine mount
   bed, 92-93
   dynafocal, 92-93
error (maintenance), 405-412
evaporative carburetor, 162
evidence (anecdotal), 45
evidence-based maintenance, 413-417
excessive oil consumption, 349-355
exhaust, 4
   leak, 270, 362, 369
   stains, 362
   stroke, 4
   system, 57
   system inspection, 362
   valve, 7-8, 59-60, 96-97, 219, 240, 272, 274, 470
   valve distress, 72
   valve failure, 59, 62-63, 446
exhaust gas temperature (EGT), 17-24, 28-29, 31, 68-69
exhaust valve
   burned, 51, 53, 214
   failing, 276-277
   failure, 306-307
   leaking, 399
   sodium-filled, 41, 299
exhaust valve failure
   premature, 62-63
   preventing, 65, 68-69
expansion (coefficient of), 82-83
explosion, 25
extension (life), 56
extreme-pressure additives (oil), 141-145
Exxon Elite 20W-50, 141-145, 230-231, 250
   oil, 257

F

FAA, 79-80, 465-466, 472
   National Policy Notice N8900.410, 441
fact-based maintenance decisions, 413-417
factory
   CHT probe, 214
   rebuilt engine, 457
FADEC, 167, 466-467, 471
failure
   exhaust valve, 59
   modes, 446
   patterns, 442-443

failure modes (bearing), 106-107
failures (maintenance-induced), 393-397
FAR
   43 Appendix A, 379
   43.13, 435
   43.16, 435
fastener
   preload, 422
   torque, 421
fatigue
   failure, 77, 85
   life, 77, 81, 84-85
   limit, 76-77
   metal, 75-77
   strength, 103
fault isolation, 324
Feingold (Gordon), 223
Fenton (Howard), 145-147
ferrous metal, 76, 84
FEVA, 306, 308-309
field overhaul, 56, 266
film (oil), 46-48, 50
filter (oil), 125
fine-wire spark plug, 154
fins (cooling), 38, 41
flange
   bolts, 90
   propeller mounting, 93
flight-test profiles, 327-334
float-type carburetor, 163
float chamber (carburetor), 162
floatplanes (salt water), 230
Florida (corrosion), 85
flow divider, 170, 176
flower pot mag timing tool, 325-326
FOD, 368-369
four-stroke, x, 3
friction, 38, 103, 137-140, 423-424
friction band (head-to-barrel junction), 82-83
Fuchs (Roger D.), 421-422, 424-426
fuel
   atomization, 178, 180, 338
   contamination, 187, 192-195
   control unit, 170, 174-176, 179
   cutoff, 180
   efficiency, 12, 303
   filler restrictor rings, 195
   flow, 23, 31, 33-34
   injection, 166-167, 169, 203, 336

leak, 270
metering, 161-168
metering unit, 176
nozzle air bleed, 179
nozzles, 24, 32, 170, 177-178, 180, 189-190, 331, 470
pressure regulator, 205
pump, 336
pump dry bay drain tube, 170
reserve, 214
screen, 174-175
spider, 176
system, x
tester, 194
vaporization, 178
fuel control unit (Continental fuel injection), 174-176
fuel discharge nozzle (carburetor), 164
fuel flow
   full-power, 241-242
   max takeoff, 242
fuel injection
   Bendix-RSA, 166-167, 169
   Continental, 166-167, 169-184
   schematic, 206
fuel injection system adjustment
   procedure, 182
   suggested interval, 181
fuel nozzle air bleed
   plugged, 179
   turbocharged engine, 179
fuel nozzles
   cleaning, 189-191
   clogged, 185-192
   Continental fuel injection, 178-179
   non-tuned, 333
   partially clogged, 188
   position-tuned, 337-338
   tuned, 245
   turbocharged, 204
fuel pump
   adjustable orifice, 172
   adjustable pressure relief valve, 172
   altitude-compensating, 170
   bypass check valve, 171
   Continental fuel injection, 170-174
   cutaway, 171
   engine driven, 170
   output adjustment, 172
   vapor ejector, 171

fuel system (adjustment), 241
full-authority digital engine control (FADEC), 167
full-flow oil filter, 126
full-power
   adjustments, 181
   fuel flow, 205, 248
full-rated power, 267, 270
full rich, 175

G

galleries (oil), 102-103
galling, 103
GAMI, 190-191, 222
   lean test, 331
   spread, 22-23, 332-333, 337-338, 340
GAMIjectors, 181, 189, 220, 245, 470
gap (spark plug), 158
Garmin G-1000, 296
Gasket Maker, 408-412
GATS jar, 194
gear, 89
   camshaft, 93, 96-97
   crankshaft, 93, 96-97
   trains, 93
glow plug, 32
GPS-coupled fuel totalizer, 214
Graphic Engine Monitor (GEM), 19
groundspeed, 296-298
growth rate (crack), 77
guesswork, x, 318
guide (valve), 60
gyroscopic loads, 90, 93

H

half harness (ignition), 346
halves (crankcase), 90-91
hangaring, 230
hardness (cylinder barrel), 45-46
harmonic dampeners, 95
Harrison Engine Service, 40
Hartzell, 432-434
head (cylinder), 38
head-to-barrel
   junction, 38, 82-83
   separation, 73-80, 84
headwind, 295-298
health assessment, x

cylinder, 271-279
engine, 265-270
heat
   sink, 61-62
   treating, 45
Heli-Coil inserts, 41, 389
high-altitude misfire, 157-158
high-MAP run, 334
high oil consumption, 349-355
hold-down studs, 91, 93
holed piston, 33
honing (cylinder barrel), 40, 45
horizontally opposed, 5-6, 37-38, 89
horsepower-to-weight ratio, 89
hot
   section, 37
   spot, 32, 67, 72
HotBand preheat system (Reiff), 253
hot spot (exhaust valve), 62
how-to's, 435
HP-per-cube ratio Mooney 252, 236
Huerta (Michael), 79
Hundere (Al), 18, 24
hydraulic
   lifter, 96-97
   tappets, 96-97
hydraulics, 16, 138
hydrodynamic lubrication, 46-47, 101-102, 139-140

I

ICP (internal combustion pressure), 222
idle
   cutoff, 175
   mixture adjustment, 182
   RPM adjustment, 181
   RPM unmetered fuel pressure, 182
   speed adjustment screw, 182
idle circuit (carburetor), 164
ignition
   compression, 161
   diesel, 161
   electronic, 15-16
   harness, 345-347
   magneto, 15
   spark, 161
   stress test, 329-331
   system, x, 8, 151-159
   temperamental, 341

timing, 15, 24, 31, 34, 49, 157, 299
troubleshooting, 341-347
ignition harness
   half harness, 346
   Slick, 346
ignition timing
   advanced, 84
   split, 331
in-flight
   LOP mag check, 329
   mag check, 155-156
indicated EGT, 19
induction
   air filter, 191, 234
   leak, 24, 234, 331-332, 334, 359-362, 364, 369
   leak test, 329, 333
   system, 57
   tuned, 15
infant mortality, 77, 443, 461-462, 471
injection
   fuel, 166-167
   intake port, 166-167
   throttle body, 165
injector line (length), 177
Insight GEM 610, 19
insoluble, 291
inspect and repair as necessary (IRAN), 133, 325, 402-403
inspection
   borescope, 66-67, 72, 110, 112-113
   camshaft, 55
   lifter, 112-113
   oil filter, 55, 107, 110, 126
   teardown after prop strike, 129-134
instructions for continued airworthiness (ICA), 435, 438
instrumentation, 15
insurance (aircraft), 440
intake, 4
   stroke, 4
   valve, 7-8, 15, 96-97
intake valve
   oil seal, 351
   oily stem, 351
interference fit, 38, 82-83
intermittent problems, 323
internal
   combustion, 3
   cylinder pressure, 212

internal combustion pressure (ICP), 8
intercooler, 241
interval
   maintenance, 431-438
   oil change, 149
investment-cast, 44
IRAN (inspect and repair as necessary), 133, 325, 402-403
iron, 259-262
   oil analysis, 291

J

J.P. Instruments (JPI), 19
Jet A, 15, 192-195, 467
   color, 194
   contamination, 192-195
   density, 194
   effect on piston engine, 195
   nozzle, 193
   odor, 194
   paper towel test, 193-194
journals (crankshaft), 94-95
JPI, 19, 123
   EDM 700, 19
   EDM 760, 189, 192
jug economics, 53
junction (head-to-barrel), 82-83

K

keepers (valve spring), 41
Keller (Colleen), xi, 279, 463
Kennon Products Inc., 254
kit (pickle), 113-114
knock, 28
Kollin (Ed), 257, 259, 374
KS Avionics, 19

L

labor cost (cylinder replacement), 57
language (watch your), 399-403
leak
   exhaust, 314
   induction, 24
   induction system, 314
lean-find mode, 226
lean-of-peak (LOP), 11, 15-17, 23, 49, 64, 84, 155-156, 211-212, 244-245
leaning, x, 16, 209-226
   advanced, 219-226
   basic, 209-217
lean range, 332
LED, 19
Lenkite, 145-147
Lenox Autoscope, 273, 277-278
life extension, 56
lifter, 96-97
   barrel-style, 114
   corrosion, 109-113, 118
   faces, 229
   inspection, 112-113
   mushroom-style, 114
   spalling, 55, 109-113, 117-119
light sport aircraft (LSA), 465
Light Speed Engineering, 465
limits (service), 54
Liston (Joseph), 13
litigation, 411
loads
   centrifugal, 90, 93
   gyroscopic, 90, 93
lobe (camshaft), 96-97
longevity, x
   cylinder, 38, 43, 46, 49
   engine, 227-237
LOP, 17, 23, 211-212, 244-245
   lean-of-peak, 11, 15-16, 49, 64, 84, 155-156, 212, 220, 470-471
   operation, 216, 219, 340, 343-344, 470
lopsided (exhaust valve), 67
losses
   accessory, 12, 15
   chemical, 13
   mechanical, 12, 15
   mixture, 12, 15
   thermal, 13
low-MAP run, 334
Lowry (John T. Ph.D.), 11-12
LPS-2, 233
LSA (light sport aircraft), 465
lubricant, 425, 427
lubrication, 100, 137-149
   boundary, 102, 139-140
   hydrodynamic, 46-47, 101-102, 139-140
   pressure, 140
   splash, 140

system, x
upper-cylinder, 46-47
Lycoming, 45, 56, 61, 63, 68-69, 81, 268, 299, 333, 338, 416, 425-426, 437, 439, 458-459, 466
    carbureted, 245
    CHT, 298
    IO-360, 463-464
    IO-540, 465
    O-235, 466
    O-360, 217, 278-279, 464
    recommended maximum CHT, 215
    red line CHT, 215
Lycoming SB 388C (valve wobble test), 123, 351
Lycoming SB 533C (prop strike), 132
Lycoming SI 1009 (overhaul periods), 441
Lycoming SI 1191A (compression test), 279
Lycoming SI 1425A (rope trick), 123
Lycoming SI 1492D (metal in oil filter), 284-285
Lycoming SL 180B (pickling), 231, 233

M

magneto, 15, 24, 31, 34, 93, 151-159, 336
    advanced timing, 325
    Bendix, 152
    ignition, 465
    internal inspection, 345
    maintenance, 152-153
    pressurized, 158-159
    Slick, 152-153
    timing, 325, 330
mag check, 154-155, 329
    bad, 156-157
    in-flight, 155-156, 188, 342, 344, 346-347
    LOP, 329, 346
    pre-flight, 346
mag drop, 154-155
    excessive, 157
main bearing, 94-95
maintenance, x
    error, 405-412
    evidence-based, 413-417
    interval, 431-438
    magneto, 152-153
    manual, 435-437
    spark plug, 381-391
    terminology, 399-403
maintenance-induced failure (MIF), 393-397, 471
maintenance decisions (fact-based), 413-417

major overhaul, xi, 54-56, 111
    deciding when, 449-456
making
    engines last, 227-237
    metal, 123-127
manganese phosphate coating, 45
Manifesto, ix, xi, 55, 393, 397
manifold (valve), 170, 176-177, 179-180
manifold valve
    air vent, 178
    Continental fuel injection, 176-177
    cutoff threshold, 177
manifold pressure, 314, 360-361, 470
    at idle, 59
    erratic fluctuations, 364, 369
    gauge, 358
    in-flight loss, 363
    redline, 201
manually-controlled wastegate, 198
manufacturer's
    guidance, 435
    recommended TBO, 471
manufacturing defect, 45
margin (detonation), 10
Marvel-Schebler carburetor, 162-164
Marvel Mystery Oil, 145-147, 257
massive-electrode spark plug, 153
Maybach (Wilhelm), 3, 162
mechanical
    energy, 5
    losses, 12, 15
MEK, 189
metal
    fatigue, 75-77
    making, 123-127
metal-to-metal contact, 46-47, 101
metallurgy, 81
metered fuel, 179
metered fuel pressure adjustment (full power), 182
metering (fuel), 161-168
methane, 14
methods techniques and practices, 435
micro-welds, 139-140
Microlon, 145-147, 257
microscopic particles, 125-126
mid-time engine, 115
MIF (maintenance-induced failure), 393-397, 471
MIL-C-6529 Type II (preservative oil), 231
Millennium cylinders (Superior Air Parts), 40

minimum brake specific fuel consumption (BSFC) LOP (lean-of-peak), 211
minimum lower heat value (of 100LL), 10
minteral oil, 137
misfire (high-altitude), 157-158
misfueled, 192-195
mixture, 10
   best economy, 211, 243-244
   best power, 243-244
   brutally lean, 64
   control, 470
   control valve, 174-176, 179
   cruise lean, 211
   distribution, 15, 216
   distribution test, 329, 331
   full-rich, 225
   losses, 12, 15
   maldistribution, 336-337
   management, 224-225
   recommended lean, 243-244
   too rich, 63
   ultra-lean, 188
   very rich, 243-244
mixture control (carburetor), 164
mixture distribution
   carbureted engines, 338
   uneven, 245, 333
Model 20 ATM-C Porta Test Unit, 181-182
monitoring (condition), 65, 68-69
monograde oil, 49, 137, 141-145
Mooney M20, 235, 303-304
Moore (Gordon), 277
Moore's Law, 277
morning sickness, 123, 351
Moseley (Bob), 122, 239-240, 265-266, 270
motion
   reciprocating, 94-95
   rotary, 94-95
Mouse Milk, 365
multi-point electric heater, 253
multi-probe digital engine monitor, 240, 248
multigrade oil, 137, 141-145, 230
mushroom-style lifter, 114

N

needle bearing, 99
new, 401, 403
   limits, 402

Ni-resist alloy, 41
NIC3 cylinders (Continental), 40, 44
nickel
   elevated, 71-72
   oil analysis, 291
nickel-carbide plated (cylinder barrel), 40, 44
nitralloy valve guide, 64
nitrided, 39, 95
non-AD oil, 141-145
non-ferrous metal, 76, 81
normalize mode, 330
   digital engine monitor, 154-155
normally-aspirated, 5, 201
nozzle air bleeds, 204
NTSB, 79, 132, 407-408
   engine-failure accident data, 444
Nu-Chrome, 40, 43-44

O

O-rings (cylinder base), 93
Oasis Scientific, 278
octane, 8, 14
oil, 90, 112-113, 137-149
   additive, 145-147
   additive package, 376
   analysis, x, 55, 65, 68-69, 71-72, 107, 112-113, 125, 234, 240, 258-259, 261, 281-293, 377, 395-396, 451-453
   anti-corrosion additives, 231
   ashless dispersants, 141-145
   change, 373-379
   change interval, 149, 373, 375-377, 452
   choosing type, 377
   consumption, x, 48, 147-148, 415, 451
   contamination, 375
   control ring tension, 45
   cooling, 138
   corrosion protection, 230
   dirty, 228, 233
   excessive consumption, 266
   extreme-pressure additives, 141-145
   film, 46-48, 50
   filter, 125, 234, 281
   filter inspection, 55, 107, 110, 126, 396, 451-453, 455
   galleries, 102-103
   insolubles, 377
   leak, 455

leaks, 270, 355
level, 148-149
mineral, 137
monograde, 49, 137, 141-145, 250, 378
multigrade, 137, 141-145, 250, 378
non-AD, 141-145
obtaining samples, 293
on the belly, 266, 350, 352, 354-355
petroleum-based, 141-145
polymers, 143
preservation, 231
pump, 281
semi-synthetic, 15
spectrographic analysis, 282, 286-289
synthetic, 15, 137, 141-145
system, 281
viscosity, 47-48
viscosity index improvers, 141-145
oil consumption
   excessive, 349-355
   high, 349-355
   maximum acceptable, 268
   permissible, 350
   troubleshooting, 350
   ultra-low, 268
oil filter
   contents evaluation, 284-285
   cutter, 283-284
   full-flow, 126
   inspection, 234, 282-284, 287
oil leak
   turbocontroller, 314
   wastegate actuator, 314
oil pressure
   relief valve, 323
   transient fluctuations, 323
oil screen
   pressure, 281-282
   suction, 281-283
oil temperature (excessive), 304
oil total base number (TBN), 376
on-condition, xi, 471
   maintenance, 442, 445, 450, 456
   overhaul, 55
operational issues, 84
Operations Specifications, 440
oscillation (rhythmic EGT), 68-69
Oshkosh, 466
Otto
   cycle, 3, 5-6, 25
   Nikolaus A., 3
Otto-cycle efficiency, 11, 14
overhaul, xi, 402-403, 470
   deciding when, 449-456
   engine, 399-400, 457, 471
   factory, 461
   field, 56, 424, 458, 461
   fixed-interval, 445
   major, 54-56, 111, 228, 319, 417
   manual, 402
   on-condition, xi
   top, 53, 57, 319, 349, 352-353, 414-415, 419, 422, 428, 471
overhauled, 401
overhaul shop (choosing), 457-462
overkill, 319
overlap interval (valve), 8, 96-97
oversquare, 470
O'Brien (Bill), 402

P

P-V diagram, 7
pads (cylinder), 91
PAO (polyalphaolefin), 141-145
paper towel test (for Jet A contamination), 193-194
particles
   aluminum, 285
   brass, 286
   bronze, 286
   carbon, 285
   chrome, 286
   copper, 286
   steel, 285
parts manufacturer approval (PMA), 43
Part 91, 439-440
Part 121/135, 439-440
peak
   cylinder pressure, 26-29, 38
   pressure, 422
petroleum-based oil, 141-145
Phillips
   oil, 257
   X/C 20W-50, 141-145, 250
phosphorus (elevated), 292
pickle kit, 113-114
pickling (preservation), 113-114

Pierce (Burt), 162
piggyback probe, 215
pilot's operating handbook (POH), 31, 154-155, 210-211
Piper
   Archer, 468
   Dakota, 235-236
   Super Cub, 221
   Warrior, 468
piston, 3, 5, 94-95
   clearance, 251
   crown, 33
   pin, 99
   pin plug scuffing, 285
piston pin bushing (Lycoming), 45
pitting, 109-113, 118
   cam, 120-121
plain bearing, 99-101, 107
plasma-coated compression ring, 45
plating
   adhesion, 45
   cadmium, 425
PMA (parts manufacturer approval), 43
pneumatics, 16
POH (pilot's operating handbook), 31, 154-155, 210-211, 236, 242-243
polyalphaolefin (PAO), 141-145
polymers (oil), 143
poppet valve, 199-201, 316
position-tuned fuel nozzles, 470
positive fuel cutoff, 178
post-maintenance test flight, 345, 397
power
   setting, 50, 229, 235, 237
   stroke, 4, 96-97
powerplant management, x
pre-ignition, 24-25, 32-33, 37, 84, 86
prebuy examination, 112-114
preheat, 252
preheating, x, 228, 232, 249-255
   continuous, 255
preload, 423-428
preservation by pickling, 113-114
preservative oil (MIL-C-6529 Type II), 231
pressure
   carburetor, 165
   check, 361
   combustion, 48, 50
   controller, 198-199
   cooling, 246
   lubrication, 140
   regulator, 206
   screen, 125-126
pressure ratio controller (PRC), 200
pressurized magneto, 158-159
preventing failure (exhaust valve), 65, 68-69
preventive maintenance, x, 345, 379, 397
probe temperature (EGT), 20-21
product liability, 411
profiles (flight-test), 327-334
propeller, 37
   cover, 254
   hand turning, 230
   mounting flange, 93
   overspeed, 455
propeller strike, 455
   AD 2004-10-14, 132
   Continental SB96-11B, 131, 133
   definition, 131
   Lycoming SB 533C, 132
   teardown inspection, 129-134
proportioning carburetor, 162, 164
PTFE (Teflon), 145-147
pump (boost), 33
pushing TBO, 55
pushrod, 96-97
   bent, 63

R

racing engine, 4
radial, 5
RAM Aircraft, 44
rebuilt, 401-403
   engine, 56
reciprocating
   loads, 90
   motion, 94-95
recommended lean mixture, 243-244
reconditioned cylinders, 56, 85
Red
   Box, 216, 219, 221-224
   Fin, 219, 223-226
Reiff HotBand preheat system, 232, 254
Reiff Preheat Systems, 253
relative EGT, 18-19
reliability-centered maintenance, 219, 397, 442, 456, 471

repair, 402-403
repaired, 401-402
replacing cylinders, x, 53-54
report card, 301, 303, 305
residual fraction effect, 340
Resnikoff Conundrum, 445
restrictor port, 193
retarded ignition timing, 157
rhythmic EGT oscillation, 68-69
rich-of-peak (ROP), 17, 50, 84, 86, 211
ring
   land, 33
   plasma-coated compression, 45
   reversal area, 141
   trapezoidal, 48
risks of cylinder work, 245, 419-424, 426-428
rocker
   arm, 42, 96-97
   box cover, 42
   shaft, 42, 99
rod bolts, 96
roller bearing, 99
rolling element bearing, 99
ROP, 17, 50, 84, 86
   rich-of-peak, 211
rotary motion, 94-95
rotation (spark plug), 153
rotator cap, 42
Rotax, 466-467
   912, 151
Rotec TBI, 166
rough running, x, 59, 187-188
   troubleshooting, 188-189, 335-340
RPM, 138, 470
running torque, 426
runout engine, 115
rust, 109-113, 118, 228-229, 231, 237, 259, 378, 396

S

S-LSA (special light sport aircraft), 466
S-N curve, 76, 84
saddles (bearing), 105-106
sandcast, 44
SavvyAnalysis, 295, 301, 303, 305-306, 321, 325, 330
Savvy Aviation Inc., ix, 432, 452, 473
scanning electron microscopy, x, 453
Schebler (George), 162
screen
   pressure, 125-126
   suction, 125-127
sealant, 408
seal band (head-to-barrel junction), 82-83
seat (valve), 61
Second OilPinion (Howard Fenton), 145-147, 285-286
semi-synthetic oil, 15
separation (head-to-barrel), 73-80, 84
service
   bulletin, 435, 438, 440
   limits, 54, 402
setting (power), 50
Shadin digital fuel flow, 185, 188
shaft (rocker), 42
sharp pick test (cam lobe), 120-121
shells (bearing), 100
shop (overhaul), 457-462
shotgunning, 318, 335, 353, 358, 365, 370
shrink band (head-to-barrel junction), 82-83
sidedraft carburetor, 163
silicon, 287
   contamination, 234
   excessive, 288
   oil analysis, 291
Simkinson (Bill), 19
since major overhaul (SMOH), 45, 55
single-mag operation, 330
sink (heat), 61-62
skirt (cylinder), 39, 91
sleeve bearing, 100
Slick
   50, 145-147, 257
   magneto, 152-153
Slick Electro, 153
slope controller, 200
small end (connecting rod), 95
SMOH (since major overhaul), 45, 55
sodium-filled exhaust (valve), 61
sodium-filled exhaust valve (Lycoming), 41
spalling
   cam, 109-113, 118
   cam and lifter, 55
   lifter, 109-113, 117-119
spark, 5
   ignition, x, 161
spark plug, 8, 24, 32-34, 381
   anti-seize, 389
   bomb tester, 388

cleaning, 387
common types, 384
contamination, 385
copper gasket, 388
damage by dropping, 391
fine-wire, 154, 340, 381, 384, 387
fouling, 343, 385, 387
gap, 158
gapping, 391
gasket probe, 214
heat rating, 383
inspection, 84
installation torque, 390
iridium, 384
lead-fouling, 217
longevity, 388
maintenance, 381-391
massive-electrode, 153, 340, 381, 384
non-firing, 336, 342-343
numbering, 381-382
preventive maintenance, 386
projected-core BY type, 384
reach, 383
rotation, 153
thread chaser, 389
special light sport aircraft (S-LSA), 466
spectrographic oil analysis program (SOAP), 125
Spiralock rod nuts, 96
splash lubrication, 140
split the case, 89
springs (valve), 41-42, 96-97
spun bearing, 106, 251, 428
squawks, 328
staking valve, 273
startup wear, 101
static seal, 272-273
steel particles (oil filter), 285
Steen Skybolt, 279
stem (valve), 96-97
stoichiometric air-fuel ratio, 161
stretch bolts, 96
strike (valve), 64
stroke, 94-95
   compression, 96-97
   Otto cycle, 4
   power, 96-97
stuck
   rings, 472
   valve, 63-64

studs (cylinder hold-down), 91, 93
suck-squeeze-bang-blow, 4
suction screen, 125-127
sulfur, 13
   dioxide, 14
sulfuric acid, 14
Superior Millennium
   Airworthiness Directive, 78-80
   cylinders, 40, 44
   investment-cast cylinders, 419
Superior Air Parts, 43, 45, 54, 56, 81, 265
supports (bearing), 90, 105-106
swallowed valve, 62-64, 446
symmetry (exhaust valve), 66
symptoms, 327
synthetic oil, 15, 137, 141-145

T

tailwind, 295-298
takeoff, 240
   fuel flow, 183
Tanis
   preheat system, 232, 254
   TAS100, 253
Tanis Aircraft Services, 250, 253
tappet, 96-97
TBO, xi, 43, 53-56, 107, 112, 133-134, 227, 237, 239, 268, 413, 417, 431-447, 449-450, 455, 466-467, 471
   going beyond, 55, 445
   on-condition maintenance, 451
   time between overhauls, 58
TCDS (type certificate data sheet), 433-434
TDC (top dead center), 25-27, 31, 48
teardown inspection after prop strike, 129-134
Teflon (PTFE), 145-147
temperature
   control, x
   management, 234, 239-248
Tempest Plus, 381
tensile strength (aluminum alloy), 81-82
tension (oil control ring), 45
terminology (maintenance), 399-403
test pilot, 328
tetraethyl lead, 217
   TEL, 8, 14, 375
therapy, 319
thermal

efficiency, 10-11, 15
  losses, 13
thermometer, 303
The Aviation Consumer, 378
Thielert, 468
throttle
  butterfly, 162
  control valve, 174-176, 179
throttle body (carburetor), 162
throttle body injection
  Ellison, 166
  Rotec, 166
  TBI, 165
through-bolt, 90-91, 93, 407, 423, 425-426, 428
through-hardened, 39, 44
throws (crankshaft), 94-95
time between replacement (TBR), 436
time between overhauls, 43, 53-56
  TBO, 58, 107, 112, 133-134, 431-447
timing
  advanced, 157
  ignition, 15, 24, 31, 34, 49, 157
  retarded, 157
tin (oil analysis), 291
TIT
  excessive, 342, 344
  oscillation, 308
  redline, 235
  turbine inlet temperature, 24, 236
Titan cylinders (ECi), 40, 44
too-rich mixture, 63
top
  end, x, 35
  overhaul, 53, 57
top dead center (TDC), 7, 25-27, 31, 48
torque, 423-424, 427
  breakaway, 426
  dry, 423
  plates, 427
  procedures, 425
  running, 426
  sequence, 427
  wet, 424-425
  wrench, 423, 427
transient anomalies, 187
trapezoidal ring, 48
trend
  analysis, 305
  monitoring, 288, 292

trim (electric), 323
trimetal bearing, 104-105
troubleshooting, x, 24, 313-315, 318, 320, 322-324, 326, 357, 366, 369-370, 415, 473
  art of, 313-326
  diagnostic, 319
  ignition issues, 341-347
  rough running, 188-189, 335-340
  turbocharging issues, 357-370
  turbosystem, 357-370
true airspeed, 296-298
Tubbs (Jimmy), 80
tuned induction, 15
turbine, 197-198
  blade scrape, 368-369
  engines, 443, 445
  wheel stretch, 368
turbine inlet temperature (TIT), 24
turboboosting, 197
turbocharged, 5
turbocharger, 198, 200-201, 314, 360
  failure modes, 366-368
  oil leak, 368
  oil seal, 351
turbocharging, x, 15, 24, 48, 197-206, 357
  troubleshooting, 357-370
turbocontroller, 200, 314-316, 359-360, 362-363, 366
  troubleshooting, 365
turbonormalizing, 197
turbosystem, 197-206, 314, 357
  troubleshooting, 357-370
type certificate (TC), 434
type certificate data sheet (TCDS), 433-434

U

Ulrich (Nathan Ph.D.), 444
ultrasonic cleaner, 190
unburned combustion byproducts, 63
Unison Industries, 153, 466
unleaded avgas, 8, 15
unmetered fuel, 174
updraft carburetor, 163
upper-cylinder lubrication, 46-47
upper deck
  air, 191
  pressure, 190, 199, 201, 359, 366

V

VA-400 (ViVidia Ablescope), 277-279
valve, 3, 41, 472
   burned, 51, 61, 63-64
   duration, 96-97
   exhaust, 7-8, 96-97
   guide, 38, 60, 240
   guide wear, 61, 214
   intake, 7-8, 15, 96-97
   lift, 96-97
   overlap interval, 8, 96-97
   seat, 38, 61
   seat contact area, 63
   spring keepers, 41
   springs, 41-42, 96-97
   staking, 273
   stem, 62, 96-97
   stems, 228
   strike, 64
   stuck, 63-64
   timing, 96-97
   wobble test, 123
valve-to-seat concentricity, 45, 62-63
valve guide (nitralloy), 64
valve stem (sodium-filled), 61
Van's RV, 299
variable absolute pressure controller (VAPC), 200
venturi (carburetor), 164
very rich mixtures, 244
vibration characteristics, 89
viscosity (oil), 47-48, 291
viscosity index improvers (oil), 141-145
ViVidia Ablescope VA-400, 277-279
volumetric efficiency, 10-11, 15

W

warranty, 458-459, 461
   Continental, 30-31, 460
   cylinder, 44
   Lycoming, 460
   Superior Air Parts, 460
wastegate, 198, 201, 315, 359-360, 363, 366
   actuator, 200, 369
   butterfly, 199
   problems, 364
   sticky, 364-365, 369
   stuck, 314
watch your language, 399-403

water
   byproduct of combustion, 13
   corrosive effect, 13
wear, 38, 137-140
   startup, 101
   valve guide, 61
wet torque, 425
what's changed and what hasn't, 463-473
when-to's, 435
when to overhaul (deciding), 449-456
white metal, 104-105
wobble test (valve), 123
won't idle smoothly, 183
Wrather (Chris), 295
www.factorycylinders.com, 56
www.factoryengines.com, 56

Y

Youngquist (John), 19

Z

zero-time logbook, 402

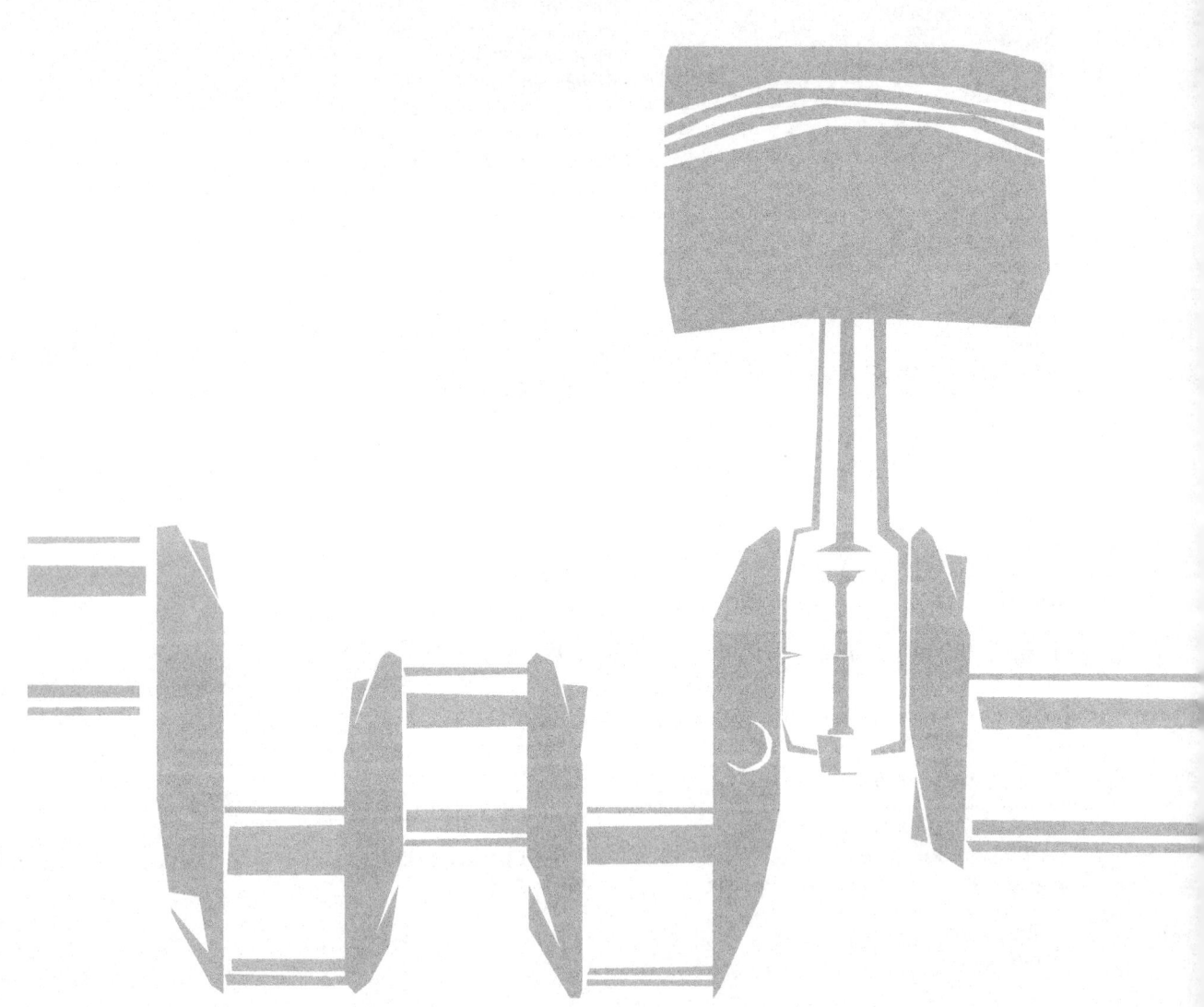

Printed in Great Britain
by Amazon